Multivariate Analysis: TECHNIQUES FOR EDUCATIONAL AND PSYCHOLOGICAL RESEARCH

Multivariate Analysis:
TECHNIQUES FOR EDUCATIONAL AND PSYCHOLOGICAL RESEARCH

Maurice M. Tatsuoka
UNIVERSITY OF ILLINOIS

John Wiley & Sons, Inc. New York . London . Sydney . Toronto

Library of Congress Catalogue Card Number: 78-151729

ISBN 0-471-84590-6

Printed in the United States of America.

10 9 8 7 6 5 4 3 2

To the memory of
Phillip Justin Rulon

Preface

This book developed from a series of mimeographed class notes distributed in advanced-statistics and multivariate-analysis courses, which I taught in the Department of Educational Psychology and the Department of Psychology, respectively, at the University of Illinois at Urbana-Champaign. I am indebted to the many students in these courses for having pointed out ways in which my presentation could be improved, and for detecting and correcting numerical errors in some of the examples.

However, I am most grateful to the late Phillip J. Rulon of the Graduate School of Education, Harvard University, under whose stimulating tutelage I was first introduced to multivariate analysis. His untimely death in June 1968 has been a source of great dismay and sadness for me. I had planned to dedicate this book to Professor Rulon as a token of my deep gratitude to him, and to show him that his wayward former student had returned to multivariate statistics after teaching science and mathematics at the undergraduate level for several years. Now, alas, I can only dedicate it to his memory.

I am also grateful to many other former teachers of mine, among whom Professors John B. Carroll, Frederick Mosteller, and David V. Tiedeman have given me constant encouragement and moral support by expressing interest in my work and by offering suggestions for topics to be included. My colleagues at the University of Illinois Psychology Department, Ledyard R Tucker and Arthur Whimbey, each read parts of my manuscript and offered valuable suggestions.

I sincerely thank Edward H. Wolf, Rutgers University, and George P. H. Styan, McGill University, who reviewed the manuscript thoroughly and made many excellent suggestions. I followed most of these suggestions and made revisions. I alone am responsible for any remaining defects and inaccuracies.

Urbana, Illinois

Maurice M. Tatsuoka

Acknowledgments

I am indebted to the Literary Executor of the late Sir Ronald A. Fisher, F.R.S., to Dr. Frank Yates, F.R.S., and to Oliver and Boyd, Edinburgh, for permission to reprint Tables E·2 and E·4 from their book, *Statistical Tables for Biological, Agricultural and Medical Research*. Table E·3 was abridged from a table originally prepared by Catherine M. Thompson, and has been reproduced with the permission of E. S. Pearson of the *Biometrika* Trustees.

M.M.T.

Contents

Multivariate Analysis: TECHNIQUES FOR EDUCATIONAL AND PSYCHOLOGICAL RESEARCH

Chapter One

Introduction

Multivariate statistical analysis, or *multivariate analysis* for short, is that branch of statistics which is devoted to the study of multivariate (or multidimensional) distributions and samples from those distributions. This, at least, is how the mathematical statistician would characterize this discipline. For applied statisticians and researchers who use statistics as a tool, however, this characterization—although technically correct—would leave much to be desired in communicative value; it may even sound tautologous.

1.1 SCOPE OF THIS BOOK AND PREREQUISITES FOR ITS STUDY

In applied contexts, particularly in educational and psychological research, multivariate analysis is concerned with a group (or several groups) of individuals, each of whom possesses values or scores on two or more variables such as tests or other measures. We are interested in studying the interrelations among these variables, in looking for possible group differences in terms of these variables, and in drawing inferences relevant to these variables concerning the populations from which the sample groups were chosen.

The above, admittedly rather loose, description of the subject matter of this book is a very broad one. It would include, for example, multiple regression analysis, or even bivariate regression, which involves one criterion (or dependent variable) and one predictor (or independent variable). But these topics are almost always covered in two-semester sequences of graduate courses in educational and psychological statistics. We therefore omit such a discussion in this book except for a brief review in the context of introducing some elements of matrix algebra in Chapter Two.

The exclusion of multiple regression analysis serves both to delineate a little further the scope of multivariate analysis covered in this book and to indicate the statistical background expected of the reader. With regard to the first point, it is more accurate to say that, with one exception, we deal only with situations in which there is a multiplicity of *dependent* variables. (The single exception is the analysis of covariance with several covariates but only one dependent variable, which is included for reasons described below.) Thus, factor analysis—which is a multivariate technique *par excellence*, but in which no variables are clearly designated as dependent variables—is also excluded, although obviously not for reasons of elementariness. Instead, the treatment of factor analysis in any but the most superficial manner requires a whole book by itself. Needless to say, many excellent treatises on this subject are available.

As for the extent of statistical background required, a two-semester graduate (or advanced undergraduate) course in applied statistics, using such texts as Glass and Stanley (1970), McNemar (1969), or Hays (1963) is presumed. Thus the reader is expected to be familiar with statistical inference and hypothesis testing involving one dependent variable. Included in these rubrics are such matters as interval estimation, univariate significance tests using the normal, chi-square, t- and F-distributions, and the analysis of variance for factorial experiments. It is also assumed that the reader has a fair mastery of correlation and regression analyses in the bivariate case, and at least a passing knowledge of multiple regression—although, as stated above, a review of the latter is given for the sake of illustrating the use of matrix algebra.

The question of how much mathematical background is necessary for tackling any textbook in applied statistics is always a difficult one to answer accurately. Many authors like to assure their prospective readers that "a working knowledge of high school algebra" is all that is required. This is usually true in the sense that nothing beyond high school algebra is involved in the actual development presented. Nevertheless, it often turns out, unhappily, that the student who has not had at least one or two college mathematics courses simply lacks (unless he is exceptionally bright) the mathematical "sophistication" or "way of thinking" to follow the text adequately. The same condition applies, perhaps with greater force than usual, to this book.

Although extensive use is made of matrix algebra, its elements are discussed in considerable detail in Chapter Two, and its slightly more advanced aspects are treated in Chapter Five. Once the basic rules of matrix algebra are learned, there is actually little beyond the skills of high school algebra that is needed to achieve, with practice, a working competence in matrix operations. However, to fully understand the geometric implications of those aspects of matrix algebra treated in Chapter Five, some familiarity with analytic geometry is necessary.

It is true that some differential calculus is used on a few occasions, but this is solely for the benefit of those who are familiar with the calculus. Those who are not may take the results on faith, and the chain of reasoning will not be broken, since differentiation is involved only in one step in each instance of its use. Thus, it is probably fair to say that one or two college mathematics courses and an ability to follow mathematical thinking (or the willingness to acquire this ability) are the mathematical prerequisites for studying this book.

1.2 EXPOSITORY APPROACH AND ORGANIZATION OF THIS BOOK

Pointing out the analogy between a given multivariate technique and the corresponding univariate method is one of the principal didactic strategies used throughout this book, especially in Chapters Three, Four, and Seven. The reader should soon come to appreciate the utility of matrix notation in highlighting this analogy, and he will thus be amply rewarded for the trouble he may have to take in learning the elements of matrix algebra in Chapter Two. Besides the drawing of analogies, other plausible arguments are of course presented. But rigorous mathematical proofs and derivations are held at a minimum except in the two chapters specifically devoted to supplying the necessary background in matrix algebra. The reader who is interested in pursuing more of the mathematics of multivariate analysis should consult such books as those by Anderson (1958) and Rao (1965). A recent book by Morrison (1967) is less mathematical than these, but somewhat more advanced than the present book.

It has already been mentioned that Chapter Two presents an introduction to the elements of matrix algebra and a review of multiple regression analysis. Readers who are familiar with matrix algebra and the matrix algebraic solution of the multiple regression problem (which involves writing the set of normal equations in matrix form and solving it by means of matrix inversion) may skim or omit this chapter.

Chapter Three is devoted to the analysis of covariance, first with one covariate and then with more than one. This, as was mentioned earlier, is the only multivariate technique discussed in this book that does not involve a multiplicity of dependent variables. It is taken up at this point for two reasons, despite the fact that the calculations are more complicated than those involved in the significance tests treated in the following chapter. First, since it involves only one dependent variable, it is conceptually simpler even if computationally more difficult than the multivariate significance tests. Many students will have studied covariance analysis (with one covariate at least) in their second statistics course—and even those who have not should find it relatively easy to follow the basic ideas of this technique because of its close relationship with analysis of variance and regression analysis. Second, this technique provides a striking example of the close analogy between a multivariate technique and its univariate counterpart that is rendered explicit by the use of matrix algebra. To these might be added a third reason, that it would probably be beneficial (in the long run) for the student to undertake fairly involved calculations with matrices soon after he has been introduced to them. However, those who do not agree with these reasons (especially the third, which makes a virtue of a certain amount of tedious drill) may omit this chapter, or return to it later, without any great loss of continuity. This chapter is somewhat isolated from the subsequent ones in that it deals with a technique involving only one dependent variable. The only reference made to this chapter in the sequel is in connection with the definition and computation of the within-groups sums-of-squares-and-cross-products matrix.

In Chapter Four we discuss the multivariate counterparts of the familiar t-test for the significance of the difference between two group means (Hotelling's T^2 test) and of the F-test for the significance of overall differences among several group means (simplest case of Wilks' Λ-ratio test). As a preliminary to understanding the multivariate significance tests, the first half of the chapter is devoted to a study of the multivariate normal distribution and the problem of establishing confidence regions (corresponding to univariate confidence intervals) for the population centroid when the population variance-covariance matrix Σ is known. Once this is thoroughly grasped, it is a short step to understand the rationale of the T^2 test, even though the proof that this statistic follows an F-distribution is beyond our scope. The transition from the univariate test when σ is known to that when σ is unknown is very closely paralleled by the transition, in the multivariate situation, from the case when Σ is known to that when Σ is unknown. And this parallelism is clearly shown by the use of matrix notation. The jump to the Λ-test for more than two groups is over a greater gap that cannot be bridged by elementary means. We therefore only show that a simple function of Λ is a natural multivariate extension of the ratio of between-groups to within-groups mean squares used in the univariate situation and point out that, when the number of groups is two, the Λ-ratio reduces to a simple function of Hotelling's T^2.

A further discourse on matrix algebra is the subject matter of Chapter Five. Here, the reader is introduced to the problem of determining the eigenvalues and eigenvectors of a matrix in a manner not usually found in books on matrix algebra per se. This is because the problem is developed in the statistical context of determining a linear combination of a given set of variables that has a larger variance than any other linear combination; it is also tied in with the geometric problem of axis rotation. Thus, this chapter actually presents the essential mathematics of principal components analysis, and may hence be studied in conjunction with a course in factor analysis, although this subject itself is not treated in this book. The first half of Section 5.6, in particular, is devoted to showing how principal components analysis fits into the broader context of factor analysis, and the points made are illustrated by a numerical example in Appendix D. Some of the proofs in this chapter may be a little difficult to follow for readers with limited mathematical background. However, the results are stated in the form of numbered theorems and properties, so that students may skim or omit the proofs if desired, at least on a first reading, and yet absorb the content of the theorems that are utilized in the sequel.

In Chapter Six, the mathematically related techniques of discriminant analysis and canonical correlation analysis are discussed. It should be emphasized that their relationship is mathematical rather than substantive, for their purposes are quite distinct. In discriminant analysis we seek linear combinations of a set of variables that best differentiate among several groups, while in canonical analysis we determine a linear combination of each of two sets of variables such that correlate most highly with each other. The relationship between the two techniques lies in the fact that the problem of discriminant analysis can be formally recast into a canonical correlation problem (although no computational benefit is gained by such recasting). It is with the hope that understanding either one of the two techniques will facilitate an understanding of the other that they are treated in the same chapter.

Chapter Seven is a short one on multivariate analysis of variance (MANOVA) for factorial experiments. The brevity of this chapter does not imply that this technique is any simpler than those discussed in the other chapters, but only that the basic statistical principles involved have been sufficiently developed by this time, so that a brief exposition suffices. A detailed numerical example supplements the brief description, since the actual calculations are probably the most difficult aspect of MANOVA, once the rationale of Wilks' Λ-criterion is thoroughly understood—as we hope it will be after its repeated applications in Chapters Four and Six.

The last chapter deals with a rather special application of multivariate analysis, namely, to classification problems. It is addressed mainly to those readers who are concerned with educational or vocational guidance, although the versatile researcher will no doubt find its contents applicable, with some modification, to other fields of endeavor, such as clinical diagnosis and personality research.

In concluding our overview of the contents and organization of this book, an explanation must be given for the conspicuous absence of computer programs herein. One reason for this is that there already exists an excellent book by Cooley and Lohnes (1971), which gives detailed programming guides as well as actual FORTRAN programs for most of the multivariate techniques useful in behavioral-science research. Another reason stems from the author's belief that, despite the rapidly increasing availability of electronic computers today, much is to be gained by the student's going through the calculations "by hand" (that is, with only a desk calculator to assist him) for at least a few miniature examples. The student who has undergone this sort of learning experience will be more likely to develop a thorough understanding of the major steps involved in a sequence of computations than one who, from the outset, leaves "all the dirty work" to the computer. He will consequently be in a better position later, when confronted with a real-life problem, to make judicious and efficient use of available computer programs for constituent parts of its solution in the event that a tailor-made program for the entire solution is not readily available. He will be less likely, for instance, to be stumped by the lack of a complete discriminant-analysis program when programs for matrix multiplication and inversion, and for eigenvalue problems are readily available. (The latter are much more widely available than the former.) The point may be brought home by the true story (related to the author by a colleague of his) of a graduate student whose thesis work was held up for several days while he constructed a program for computing point-biserial correlation coefficients, a large number of which he needed to calculate. All the while, a program for the usual product-moment correlation coefficient was available at the computer center to which he had access.

Chapter Two

Mathematical Preliminaries: Some Matrix Algebra

The reader probably already has a general idea of what a *matrix* is. He has heard of a correlation matrix, a matrix of sociometric choices, and the like. He will have associated the term "matrix" with a rectangular (most often square) arrangement of numbers in rows and columns. Perhaps, however, he has thought of a matrix as being no more than a tabular display of certain types of data, and has not regarded it as a mathematical entity amenable to manipulations or operations (like addition and multiplication) analogous to those he is familiar with in the realm of numbers themselves.

As a starting point, the notion of a matrix as a convenient rectangular display of numerical data, is convenient and useful; it will remind us of the concrete meaning of the abstract entity with which we shall deal more and more as we proceed in this book. To be able to understand and apply the several multivariate statistical techniques described in the sequel, however, we must go beyond the mere tabular-display notion, and develop some facility with the algebra of matrices. Although this chapter and Chapter Five should provide the reader with sufficient understanding of matrix algebra for the purposes of this book, those who wish to pursue more advanced phases of multivariate analysis will find it profitable to consult such texts as Horst (1963) and Searle (1966).

2.1 DEFINITION AND BASIC TERMINOLOGY

A *matrix* is a rectangular arrangement of numbers in several *rows* (horizontal arrays) and several *columns* (vertical arrays). In applicational contexts, each row and each column has a specific referent in the field of interest—such as a particular test or experimental variable in educational or psychological measurement.

A familiar example is a *correlation matrix*, which is a square matrix (that is, one in which the number of rows and the number of columns are equal) displaying the coefficients of correlation between pairs of tests. For instance, if three tests, say an algebra test (A), an English test (E), and a physics test (P), are given to a group of college applicants, the coefficients of correlation between pairs of these tests, for this group, may be displayed in a square matrix with three rows and three columns:

$$\begin{bmatrix} 1.00 & .32 & .47 \\ .32 & 1.00 & .35 \\ .47 & .35 & 1.00 \end{bmatrix}$$

This is called a 3×3 matrix (read "three by three"); the first numeral refers to the number of rows and the second, to the number of columns. In this example the first row refers to Test A, the second row to Test E, and the third row to Test P; the three columns also refer to these three tests in the same order.

The numerical entries are called the *elements* of the matrix. A particular element is specified by the number of the row and the number of the column in which it stands. Thus, the ".35" found in the second row and third column of the above correlation matrix is called its (2, 3)-element; again, the first numeral refers to the row, and the second, to the column.

Often it is necessary to denote a matrix and its elements without specifying actual numerical values for the latter. It is then customary to denote the matrix by an uppercase letter in boldface type, like **A**, **B**, or **R** (often used for correlation matrices); its elements are then denoted by the corresponding lowercase letter with double subscripts indicating the row and column numbers. Thus, a 4×3 matrix **A** may be denoted as

$$\mathbf{A} = \begin{bmatrix} a_{11} & a_{12} & a_{13} \\ a_{21} & a_{22} & a_{23} \\ a_{31} & a_{32} & a_{33} \\ a_{41} & a_{42} & a_{43} \end{bmatrix}$$

Sometimes this denotation is abbreviated to

$$\mathbf{A} = (a_{ij}) \qquad [i = 1, 2, 3, 4; j = 1, 2, 3]$$

and the statement in brackets, indicating the range of i and j, may be omitted if these are clear from the context.

In many contexts it becomes necessary simultaneously to consider a given matrix and an associated matrix obtained by writing the elements of each *row* of the former in the corresponding *column* of the latter. The matrix thus generated is called the *transpose* of the first matrix, and is denoted by suffixing a prime (′) to the symbol denoting the first matrix. Thus, for example, if

$$\mathbf{B} = \begin{bmatrix} 3 & -1 & 2 \\ -5 & 0 & 4 \end{bmatrix}$$

its transpose is

$$\mathbf{B}' = \begin{bmatrix} 3 & -5 \\ -1 & 0 \\ 2 & 4 \end{bmatrix}$$

(The elements, 3, −1, 2, of the first row of **B** constitute, in this order, the elements of the first *column* of **B**′; and similarly with −5, 0, 4.)

From the definition of a transpose, it is evident that the transpose of an $m \times n$ matrix is an $n \times m$ matrix. It is also easy to see that the transpose *of the transpose* of a given matrix is identical to the given matrix itself. Thus, $(\mathbf{B}')'$ would be formed by writing 3 and −5 in its first column, −1 and 0 in its second, and 2 and 4 in its third column, so that we have

$$(\mathbf{B}')' = \begin{bmatrix} 3 & -1 & 2 \\ -5 & 0 & 4 \end{bmatrix}$$

which is none other than the original matrix **B** itself. We state this simple, but often useful, fact as an equation for future reference: For any matrix **A**,

$$(\mathbf{A}')' = \mathbf{A} \tag{2.1}$$

Implicit in the above use of the equality sign relating two indicated matrices is the definition of equality of two matrices: they are said to be *equal* if and only if they are identical—that is, if each element of the first matrix is equal to the corresponding element of the second matrix. Thus, when we assert that

$$\mathbf{C} = \mathbf{B}$$

where **B** is the matrix displayed in the paragraph before last, we are saying that

$$c_{11} = 3 \qquad c_{12} = -1 \qquad c_{13} = 2$$

$$c_{21} = -5 \qquad c_{22} = 0 \qquad c_{23} = 4$$

A single matrix equation involving $m \times n$ matrices, therefore, is equivalent to mn ordinary algebraic equations. Symbolically, if **A** and **B** are $m \times n$ matrices,

$$\mathbf{A} = \mathbf{B} \Leftrightarrow a_{ij} = b_{ij} \qquad [i = 1, 2, \ldots, m; j = 1, 2, \ldots, n] \tag{2.2}$$

(The double arrow "$\Leftrightarrow$" is read, "if and only if.") The power of matrix algebra lies in allowing compact expressions of sets of equations and thereby facilitating their further manipulation.

A correlation matrix, like the one displayed at the outset of this section, exemplifies a square matrix with the additional special feature that it is equal to its own transpose. Such a matrix is called a *symmetric matrix*, and is characterized by the fact that, for each i and j, the (i, j)-element and (j, i)-element are equal. (In a correlation matrix, this is so because the two elements r_{ij} and r_{ji} both represent the coefficient of correlation between the ith and jth variables.) Most of the matrices we encounter in multivariate statistical analysis will be square matrices, and many of these will be symmetric matrices—for reasons that will soon become evident.

Another kind of matrix that will be frequently encountered in the sequel is the opposite extreme from a square matrix; namely, a matrix with only one row or one column, like

$$[3, -1, 2, 4] \qquad \text{or} \qquad \begin{bmatrix} -2 \\ 1 \\ 5 \end{bmatrix}$$

It is quite proper to refer to such matrices as $1 \times n$ or $n \times 1$ matrices, but they are also called *vectors*. More specifically, a $1 \times n$ matrix is called an n-dimensional row vector and an $n \times 1$ matrix is called an n-dimensional column vector. (The reason for this geometrical nomenclature will be explained later.)

It is customary to denote a *column* vector by a lowercase letter in boldface type, and a *row* vector by a letter followed by a prime (since it is the transpose of a column vector). Thus, an n-dimensional row vector with numerically unspecified elements will be indicated as, for example,

$$\mathbf{u}' = [u_1, u_2, u_3, \ldots, u_n]$$

while an n-dimensional column vector will be written as

$$\mathbf{v} = \begin{bmatrix} v_1 \\ v_2 \\ \vdots \\ v_n \end{bmatrix}$$

2.2 ADDITION OF MATRICES; MULTIPLICATION OF A MATRIX BY A SCALAR

Suppose that each of five sixth-grade pupils is given an arithmetic test which yields two subtest scores: one for computational skills and one for ability to solve word problems. The ten scores thus obtained may be arranged in a 5 × 2 matrix, called a *score matrix*, thus:

$$\mathbf{X} = \begin{bmatrix} 30 & 15 \\ 25 & 10 \\ 28 & 12 \\ 32 & 14 \\ 22 & 13 \end{bmatrix}$$

where each row shows the scores of a particular pupil, the first column lists the computational-skills scores, and the second column lists the word-problem scores.

Next, suppose that, after a period of instruction, these same five pupils are given an alternate form of the arithmetic test. The second set of scores may be presented in another score matrix:

$$\mathbf{Y} = \begin{bmatrix} 34 & 17 \\ 27 & 12 \\ 30 & 16 \\ 35 & 17 \\ 26 & 15 \end{bmatrix}$$

where each row refers to the same pupil as did the corresponding row in the first score matrix **X**.

We may, for some reason, be interested in displaying the total score for each part of the two forms of the test, earned by each pupil. We would then report the matrix

$$\mathbf{T} = \begin{bmatrix} 64 & 32 \\ 52 & 22 \\ 58 & 28 \\ 67 & 31 \\ 48 & 28 \end{bmatrix}$$

which is an element-by-element sum of the two matrices **X** and **Y** ($64 = 30 + 34$, $32 = 15 + 17$, $52 = 25 + 27$, and so forth). This procedure for obtaining **T** from **X** and **Y** illustrates the definition of *matrix addition*: each element of the sum matrix is the sum of the corresponding elements of the two addend matrices. Symbolically, we may write,

$$\mathbf{C} = \mathbf{A} + \mathbf{B} \Leftrightarrow c_{ij} = a_{ij} + b_{ij} \qquad \text{for each } (i, j)\text{-pair} \tag{2.3}$$

From this definition, it should be clear that two matrices can be added if and only if they have the same number of rows *and* the same number of columns (briefly, if they are of the same *order*). Thus, the expression **A** + **B** would be meaningless if, for example, **A** is a 3 × 4 matrix and **B** is a 3 × 3 matrix (or anything other than a 3 × 4 matrix).

Instead of the total score earned by each pupil on each part of the two forms of our arithmetic test, we may want to report the average score on each part for each pupil. Each average would, of course, be obtained by dividing the relevant element of **T** by 2—or, equivalently, by multiplying each element of **T** by 1/2. The resulting "score" matrix (in which each "score" is really the average of two scores) is:

$$\mathbf{A} = \begin{bmatrix} 32.0 & 16.0 \\ 26.0 & 11.0 \\ 29.0 & 24.0 \\ 33.5 & 15.5 \\ 24.0 & 14.0 \end{bmatrix}$$

In matrix notation, the relation between **A** and **T** is expressed as

$$\mathbf{A} = (1/2)\mathbf{T}$$

The rule for multiplying any matrix by a number (which, to emphasize its distinction from matrices and vectors, is called a *scalar*) is exemplified in the above equation. The rule is simply to multiply each and every element of the matrix by that scalar. A general, symbolic statement of this is, therefore, as follows:

$$\mathbf{B} = k\mathbf{A} \Leftrightarrow b_{ij} = ka_{ij} \qquad \text{for each } (i, j)\text{-pair} \tag{2.4}$$

Finally, we may want to take the difference between corresponding scores on Form Y (the form administered after the instructional period) and Form X (the first-administered form) of our arithmetic test as a measure of the gain occurring between the pre- and postinstruction performances. An element-by-element subtraction of matrix **X** from matrix **Y** would give us the desired gain-score matrix, **G**:

$$\mathbf{G} = \mathbf{Y} - \mathbf{X} = \begin{bmatrix} 4 & 2 \\ 2 & 2 \\ 2 & 4 \\ 3 & 3 \\ 4 & 2 \end{bmatrix}$$

The rule for subtracting one matrix from another, just exemplified, is really a consequence of the addition rule (2.3) and a special case of the scalar-multiplication rule (2.4) in which k is given the value -1. This is because the difference, $a - b$, between any two numbers, a and b, is equivalent to the sum

$$a + (-1) \cdot b$$

When this fact is applied, element by element, in forming the difference, **A** − **B**, between two matrices, we are naturally led to the rule,

$$\mathbf{C} = \mathbf{A} - \mathbf{B} \Leftrightarrow c_{ij} = a_{ij} - b_{ij} \qquad \text{for each } (i, j)\text{-pair} \tag{2.3a}$$

EXERCISES

Given three matrices,

$$\mathbf{A} = \begin{bmatrix} 6 & 0 & -9 \\ 3 & -5 & 1 \end{bmatrix} \quad \mathbf{B} = \begin{bmatrix} -1 & 0 & 3/2 \\ 2 & 1 & -1/2 \end{bmatrix} \quad \text{and} \quad \mathbf{C} = \begin{bmatrix} 0 & 5/2 \\ 0 & 1/6 \\ 0 & -1/3 \end{bmatrix}$$

answer the following questions.

1. Which of the following matrix expressions are meaningful?

   (a) $\mathbf{A} + \mathbf{B}$ (b) $\mathbf{B} + \mathbf{C}$ (c) $\mathbf{C}' + \mathbf{A}$ (d) $\mathbf{B}' - \mathbf{C}$ (e) $\mathbf{A} - \mathbf{C}$

   Write each of the meaningful expressions as a single matrix, and tell why the others are not meaningful.
2. Determine the matrices

$$\mathbf{D} = (1/3)\mathbf{A}$$

   and

$$\mathbf{E} = 2\mathbf{B}$$

3. Verify that the matrix $\mathbf{D} + \mathbf{E}$ can be written as a scalar multiple of $\mathbf{C}'$.
4. By computing each side separately, verify the following equation:

$$(\mathbf{A} + \mathbf{B})' = \mathbf{A}' + \mathbf{B}'$$

2.3 MULTIPLICATION OF TWO MATRICES

So far, our operations with matrices have been straightforward "natural" extensions of the corresponding operations with numbers. The addition and subtraction of two matrices, and the multiplication of a matrix by a scalar, were each defined by an element-by-element operation using the corresponding operations on numbers, and nothing else.

In matrix multiplication we encounter, for the first time, an operation that goes beyond a "commonsense" extension of the corresponding operation with numbers. First, there is a rather peculiar restriction on the forms of pairs of matrices that can be multiplied to yield products. Moreover, this restriction is not that the two matrices must be of the same order (which is a rather "natural" restriction), as in the case of matrix addition. Instead, the condition to be satisfied by two matrices **A** and **B**, in order that their product **AB** can be formed, is that *the number of columns in* **A** is equal to *the number of rows in* **B**. Thus, for instance, if **A** is a 2×3 matrix, then **B** must be a $3 \times n$ matrix for the product **AB** to be defined (where n may be any natural number). An illustration will clarify both the procedure of matrix multiplication and the reason for the restriction.

Let

$$\mathbf{A} = \begin{bmatrix} 2 & -3 & 1 \\ -1 & 4 & 0 \end{bmatrix}$$

and

$$\mathbf{B} = \begin{bmatrix} 3 & 1 \\ 4 & 2 \\ 5 & -3 \end{bmatrix}$$

Then, since **A** has three columns and **B** has three *rows*, the product **AB** = **C** (say) can be formed. The procedure is as follows:

1. Each element of the first *row* of **A** is multiplied by the corresponding element of the first *column* of **B**, and the resulting products are added. This sum is c_{11}, the (1, 1)-element of the product matrix **C**. Thus, for the present example,

$$c_{11} = (2)(3) + (-3)(4) + (1)(5) = 6 - 12 + 5 = -1$$

2. The elements of the first row of **A** are multiplied by the respectively corresponding elements of the second column of **B** and the products added, to obtain c_{12}:

$$c_{12} = (2)(1) + (-3)(2) + (1)(-3) = 2 - 6 - 3 = -7$$

3. Similar operations with the second-row elements of **A** and the first-column elements of **B** yield c_{21}:

$$c_{21} = (-1)(3) + (4)(4) + (0)(5) = -3 + 16 + 0 = 13$$

4. Similarly,

$$c_{22} = (-1)(1) + (4)(2) + (0)(-3) = -1 + 8 + 0 = 7$$

Arranging in a matrix the four elements obtained above, we have the product matrix

$$\mathbf{AB} = \mathbf{C} = \begin{bmatrix} -1 & -7 \\ 13 & 7 \end{bmatrix}$$

These operations for computing the elements of a product matrix may be summarized in the general statement,

$$\mathbf{C} = \mathbf{AB} \Leftrightarrow c_{ij} = \sum_{k=1}^{s} a_{ik}b_{kj} \qquad [i = 1, 2, \ldots, m; j = 1, 2, \ldots, n] \quad (2.5)$$

where **A** is an $m \times s$ matrix, and **B** is an $s \times n$ matrix.

Careful examination of the operations involved in matrix multiplication, illustrated above, should reveal several things:

1. The reason why the number of *columns* in the first factor matrix (**A** in our example) must agree with the number of *rows* of the second factor

(**B** in our example) is that the elements of each row of the former are "paired off" with elements of each *column*, in turn, of the latter.

2. The product matrix (**C**, above) has as many rows as does the first factor, and as many columns as does the second factor. Thus, if a $p \times q$ matrix is multiplied by a $q \times r$ matrix, the product is a $p \times r$ matrix.

3. The (i, j)-element of the product matrix is the result of "pairing" the ith row of the first factor with the jth column of the second factor.

4. The order of multiplication is important. That is, unlike the multiplication of two numbers (where $ab = ba$), **AB** and **BA** are usually not equal to each other; in fact, even when one of these products exists, the other may be undefined. (Consider, for example, the case when **A** is a 3×4 matrix and **B** is 5×3. Which order of multiplication is possible in this case?)

The property of matrix multiplication noted above constitutes an important departure of matrix algebra from the algebra of numbers. Mathematicians describe this difference by saying that, although the multiplication of numbers is *commutative* ($ab = ba$), the multiplication of matrices is *noncommutative* ($\mathbf{AB} \neq \mathbf{BA}$, in general). In contrast, it should be noted that the commutative law holds for matrix addition, as should be clear from definition (2.3). That is, for any two matrices of the same order, $\mathbf{A} + \mathbf{B} = \mathbf{B} + \mathbf{A}$, just as $a + b = b + a$ for any two numbers.

In view of the noncommutativity of matrix multiplication, it would be ambiguous to speak simply of "multiplying **A** and **B**." Reference must be made to the order in which the factors are taken. Thus, if the product **AB** is intended, we say that "**A** is *postmultiplied* by **B**" or that "**B** is *premultiplied* by **A**;" the letters "**A**" and "**B**" are reversed in both of these descriptions when the product **BA** is implied.

An Example of Statistical Import. The matrix operations previously expounded may be applied to an example foreshadowing the important role of matrices in multivariate statistical analysis. Consider, once again, the 5×2 score matrix displayed in the beginning of Section 2.2. That is,

$$\mathbf{X} = \begin{bmatrix} 30 & 15 \\ 25 & 10 \\ 28 & 12 \\ 32 & 14 \\ 22 & 13 \end{bmatrix}$$

whose elements are the scores earned by five pupils on two subtests (computation and word problems) of an arithmetic test.

Now, the means of the five scores on the two subtests are 27.4 and 12.8, respectively. Let us define a 5×2 *mean-score matrix*

$$\bar{\mathbf{X}} = \begin{bmatrix} 27.4 & 12.8 \\ 27.4 & 12.8 \\ 27.4 & 12.8 \\ 27.4 & 12.8 \\ 27.4 & 12.8 \end{bmatrix}$$

The elements in the first column are all equal to the mean of the computations subtest, and those in the second column are equal to the mean of the word-problems subtest. By Eq. 2.3a, defining the difference between two matrices, then, the *deviation-score matrix* may be written as

$$\boldsymbol{x} = \mathbf{X} - \bar{\mathbf{X}} = \begin{bmatrix} 2.6 & 2.2 \\ -2.4 & -2.8 \\ 0.6 & -0.8 \\ 4.6 & 1.2 \\ -5.4 & 0.2 \end{bmatrix}$$

Next, let us form the product matrix $\boldsymbol{x}'\boldsymbol{x}$. Following the rules for matrix multiplication given earlier, we find,

$$\boldsymbol{x}'\boldsymbol{x} = \begin{bmatrix} 63.2 & 16.4 \\ 16.4 & 14.8 \end{bmatrix}$$

The important point is that the elements of this product matrix are familiar statistical quantities, as may be readily verified by retracing how each element was computed. For instance, the (1, 1)-element was obtained as

$$(2.6)^2 + (-2.4)^2 + (0.6)^2 + (4.6)^2 + (-5.4)^2 = 63.2$$

Thus, this element is precisely the "sum-of-squares" (that is, the sum of squared deviations from the mean) for the first subtest, which, following customary statistical notation, may be denoted by $\sum x_1^2$. (A more complete notation would require a second subscript, indicating the pupil number, after the "1", and we would write $\sum_{i=1}^{5} x_{1i}^2$, but it is conventional to abbreviate this as $\sum x_1^2$ when the context specifies the number of scores involved.) Similarly, the (2, 2)-element gives the value of $\sum x_2^2$, the sum-of-squares for the second subtest. The two off-diagonal elements (which are equal in value) give the sum-of-cross-products $\sum x_1 x_2$ (and, equivalently $\sum x_2 x_1$) between the two subtests.

Thus, the elements of $\boldsymbol{x}'\boldsymbol{x}$ are such that, when divided by the number of degrees of freedom (n.d.f. = 4, in this example), they yield the sample variances (along the diagonal) and the sample covariances (off-diagonal) for the tests involved. Symbolically, we may write, for the general case of p tests,

$$x'x = \begin{bmatrix} \sum x_1^2 & \sum x_1x_2 & \cdots & \sum x_1x_p \\ \sum x_2x_1 & \sum x_2^2 & \cdots & \sum x_2x_p \\ \vdots & & \ddots & \vdots \\ \sum x_px_1 & \sum x_px_2 & \cdots & \sum x_p^2 \end{bmatrix} \qquad (2.6)$$

Such a matrix will be referred to as a *sums-of-squares-and-cross-products matrix*, or an *SSCP matrix*, for short. It is evident that any SSCP matrix is symmetrical, just as is a correlation matrix (with which the former is closely related, as shown in Exercise 2 at the end of this chapter).

Another point worth mentioning in connection with an SSCP matrix is that it can be computed directly from the raw-score matrix and the mean-score matrix without first constructing the deviation-score matrix. This is accomplished by using the following identity, which may be easily verified by recalling a corresponding identity for the sum-of-squares of a single test.

$$x'x = \mathbf{X}'\mathbf{X} - \bar{\mathbf{X}}'\bar{\mathbf{X}} \qquad (2.7)$$

The corresponding equation for the univariate case is

$$\sum x^2 = \sum X^2 - N(\bar{X})^2$$

where N is the sample size, that is, the number of scores at hand. The analogy becomes closer when this equation is rewritten as

$$\sum x^2 = \sum X^2 - \sum \bar{X}^2$$

No summation sign appears in the matrix counterpart Eq. 2.7 because the summation is already "built-in" in the matrix products.

A special case of matrix multiplication is the multiplication of a row vector and a column vector of the same dimensionality. Multiplications in both orders are possible, but the results are quite different. Thus, for instance, if

$$\mathbf{u}' = [1, -2, 3]$$

and

$$\mathbf{v} = \begin{bmatrix} 3 \\ 1 \\ 2 \end{bmatrix}$$

then

$$\mathbf{u}'\mathbf{v} = [1, -2, 3] \begin{bmatrix} 3 \\ 1 \\ -2 \end{bmatrix} = (1)(3) + (-2)(1) + (3)(-2) = -5$$

which is a scalar. For this reason, $\mathbf{u}'\mathbf{v}$ is called the *scalar product* of $\mathbf{u}$ and $\mathbf{v}$. It is also called the *inner product* or *dot product* of these two vectors.

On the other hand, multiplication in the reverse order yields what is called the *matrix product*:

$$\mathbf{vu}' = \begin{bmatrix} 3 \\ 1 \\ -2 \end{bmatrix} [1, -2, 3] = \begin{bmatrix} 3 & -6 & 9 \\ 1 & -2 & 3 \\ -2 & 4 & -6 \end{bmatrix}$$

a 3×3 matrix. (Note that a "degenerate case" of the rule for matrix multiplication is applied here: each "row" of $\mathbf{v}$ contains just one element, as does each "column" of $\mathbf{u}'$.)

Although we emphasized above a difference between the multiplication of matrices and the multiplication of numbers, there are also important points of similarity. One is the property of *associativity*. The reader has long been familiar with the fact that, given any three numbers a, b, and c, their triple product abc may be formed equivalently as either $(ab)c$ or $a(bc)$. (For example, $(2 \times 3) \times 4 = 6 \times 4 = 24$; and $2 \times (3 \times 4) = 2 \times 12 = 24$.) This same property of associativity holds also for matrix multiplication. That is, if $\mathbf{A}$, $\mathbf{B}$, and $\mathbf{C}$ are three matrices such that the product $\mathbf{AB}$ can be formed, and this product matrix, in turn, can be postmultiplied by $\mathbf{C}$ to yield a triple product $(\mathbf{AB})\mathbf{C}$, then $\mathbf{A}(\mathbf{BC})$ (that is, postmultiplying $\mathbf{A}$ by the product $\mathbf{BC}$) is also a legitimate triple product, and the two triple products are identical. We state this property in the equation,

$$(\mathbf{AB})\mathbf{C} = \mathbf{A}(\mathbf{BC}) \tag{2.8}$$

Because of this property, either triple product may, without ambiguity, be denoted as $\mathbf{ABC}$, omitting the parentheses. As an exercise, the reader should verify Eq. 2.8 by computing both $(\mathbf{AB})\mathbf{C}$ and $\mathbf{A}(\mathbf{BC})$ and noting that they are equal, using the following three matrices:

$$\mathbf{A} = \begin{bmatrix} 3 & 1 \\ 2 & -1 \\ -1 & 2 \end{bmatrix} \qquad \mathbf{B} = \begin{bmatrix} 5 & 2 \\ 3 & 1 \end{bmatrix} \qquad \mathbf{C} = \begin{bmatrix} 1 & 2 & 3 & 4 \\ 1 & -1 & 0 & 2 \end{bmatrix}$$

A type of triple product that occurs very frequently in the formulas of multivariate analysis is that having the general form $\mathbf{u}'\mathbf{Au}$, where $\mathbf{u}$ is an n-dimensional column vector, $\mathbf{u}'$ is its transpose, and $\mathbf{A}$ is an $n \times n$ square matrix. Since $\mathbf{u}'$ consists of only one row and $\mathbf{u}$ of only one column, the indicated product $\mathbf{u}'\mathbf{Au}$ denotes a 1×1 "matrix," that is, a scalar. For example, if

$$\mathbf{u}' = [3 \quad -1 \quad 2]$$

and

$$\mathbf{A} = \begin{bmatrix} 5 & 2 & -1 \\ -3 & 4 & 2 \\ 1 & 2 & 3 \end{bmatrix}$$

then

$$\mathbf{u'Au} = [3 \quad -1 \quad 2]\begin{bmatrix} 5 & 2 & -1 \\ -3 & 4 & 2 \\ 1 & 2 & 3 \end{bmatrix}\begin{bmatrix} 3 \\ -1 \\ 2 \end{bmatrix}$$

$$= [20 \quad 6 \quad 1]\begin{bmatrix} 3 \\ -1 \\ 2 \end{bmatrix} = 56$$

It is interesting to note that, when the elements of **u** and **A** are indicated by letters, the result of carrying out the indicated product **u′Au** can be written as a rather simple algebraic expression involving the u_i and a_{ij}:

$$\mathbf{u'Au} = a_{11}u_1^{\ 2} + a_{22}u_2^{\ 2} + a_{33}u_3^{\ 2} + (a_{12} + a_{21})u_1u_2$$
$$+ (a_{13} + a_{31})u_1u_3 + (a_{23} + a_{32})u_2u_3 \qquad (2.9)$$

An expression of this form is called a *quadratic form* (of the u_i). Observe that each term of this expression consists either of the square of an element of **u** or of the product of two different elements of **u**, each multiplied by a coefficient that comes from **A** according to the following rules:

1. The coefficient of $u_i^{\ 2}$ is a_{ii}, for each i. (An element such as a_{ii}—where the row and column indices are equal—of a square matrix is called a *diagonal element.* Such elements stand on the diagonal line extending from the upper left corner to the lower right corner of the matrix.)

2. The coefficient of u_iu_j (where $i \neq j$) is $(a_{ij} + a_{ji})$, for each (i, j)-pair. (It may be mnemonically better to imagine the term $(a_{ij} + a_{ji})u_iu_j$ decomposed into $a_{ij}u_iu_j + a_{ji}u_ju_i$.)

These rules provide us with an alternative method for evaluating triple products of the stated form, without going through the two-stage matrix multiplication indicated. It is good computational policy to calculate quadratic forms in both ways, as an arithmetic check. As an exercise, verify the result **u′Au** = 56 for the example given earlier, using the alternative computational scheme.

Another property common to the multiplication of matrices and the multiplication of numbers is that of *distributivity* of multiplication over addition. This principle, quite familiar to the reader in the realm of numbers, asserts that $a(b + c) = ab + ac$. We state this property for matrices in the equations

$$\mathbf{A(B + C) = AB + AC} \qquad (2.10a)$$

$$\mathbf{(B + C)A = BA + CA} \qquad (2.10b)$$

Two equations are needed to make it clear that distributivity holds both when the indicated sum (**B** + **C**) occurs as the second factor and when it comes first.

One other principle of matrix algebra connected with products is the rule concerning the transpose of the product of two or more matrices. This rule may be stated, for the case of two matrices, as follows:

$$(\mathbf{AB})' = \mathbf{B}'\mathbf{A}' \tag{2.11}$$

That is, the transpose of the product of two matrices is equal to the product of their respective transposes, multiplied in the *reverse order*. Verification of this rule is left to the reader. To remember that $(\mathbf{AB})'$ is *not* equal to $\mathbf{A}'\mathbf{B}'$, as one might unwarily think it is, we need only recall the following. If, for instance, $\mathbf{A}$ is 3×4 and $\mathbf{B}$ is 4×5, then $\mathbf{AB}$ is 3×5; so $(\mathbf{AB})'$ is 5×3. $\mathbf{B}'\mathbf{A}'$ is $(5 \times 4) \times (4 \times 3) = 5 \times 3$, while $\mathbf{A}'\mathbf{B}'$ is not even defined.

EXERCISES

Do the following problems with reference to the three matrices,

$$\mathbf{A} = \begin{bmatrix} 3 & 1 \\ 2 & 0 \\ -1 & 2 \end{bmatrix} \quad \mathbf{B} = \begin{bmatrix} 1 & -2 & 1 \\ -1 & 2 & 1 \end{bmatrix} \quad \mathbf{C} = \begin{bmatrix} 2 & 3 & -1 \\ 1 & -2 & 1 \end{bmatrix}$$

and whatever other matrices that may be given in the respective problems.

1. Compute $\mathbf{A}(\mathbf{B} + \mathbf{C})$ and $\mathbf{AB} + \mathbf{AC}$, thereby verifying Eq. 2.10a.
2. Verify Eq. 2.10b by computing $(\mathbf{B} + \mathbf{C})\mathbf{A}$ and $\mathbf{BA} + \mathbf{CA}$.
3. Compute $(\mathbf{AB})\mathbf{C}'$ and $\mathbf{A}(\mathbf{BC}')$. What law do the results verify?
4. Verify that $((\mathbf{AB})\mathbf{C}')' = \mathbf{CB}'\mathbf{A}'$. Which two equations does this result verify?
5. Let, further,

   $$\mathbf{u}' = [1 \quad 2 \quad -1]$$

   Compute the quadratic form $\mathbf{u}'(\mathbf{AB})\mathbf{u}$ in two ways.
6. Let $\mathbf{D} = \begin{bmatrix} 2 & 0 \\ 0 & 3 \end{bmatrix}$

   Compute $\mathbf{DC}$.
   How is this product related to $\mathbf{C}$? (That is, state a simple rule by which you could immediately write out $\mathbf{DC}$, given $\mathbf{C}$.)
7. Compute $\mathbf{AD}$, where $\mathbf{D}$ is as given in question 6. What do you notice about this product in relation to $\mathbf{A}$?
8. Let

   $$\mathbf{E} = \begin{bmatrix} 5 & 0 & 0 \\ 0 & 5 & 0 \\ 0 & 0 & 5 \end{bmatrix}$$

   Compute $\mathbf{EA}$.
   How does this result compare with $\mathbf{A}$?

2.4 THE IDENTITY MATRIX AND MATRIX INVERSION

The last three exercises in the preceding section involved square matrices that were peculiar in that their elements were all 0's except along the diagonal from the upper left corner to the lower right corner (which is called that *main diagonal* of the matrix). Such a matrix is known as a *diagonal matrix*, and the reader has observed the effects of premultiplying and postmultiplying a given matrix by a diagonal matrix. Namely, premultiplying any $n \times m$ matrix **A** by an $n \times n$ diagonal matrix

$$\mathbf{D} = \begin{bmatrix} d_1 & 0 & 0 & \cdots & 0 \\ 0 & d_2 & 0 & \cdots & 0 \\ \vdots & & & \ddots & \vdots \\ 0 & 0 & 0 & \cdots & d_n \end{bmatrix}$$

has the following effect: each element in the ith *row* of **A** gets multiplied by d_i the (i, i)-element of **D**. Symbolically,

$$\mathbf{B} = \mathbf{DA} \Leftrightarrow b_{ij} = d_i a_{ij} \qquad (i = 1, 2, \ldots, n; j = 1, 2, \ldots, m) \qquad (2.12a)$$

Similarly, if **A** is an $m \times n$ matrix, then

$$\mathbf{B} = \mathbf{AD} \Leftrightarrow b_{ij} = d_j a_{ij} \qquad (2.12b)$$

That is, when **A** is postmultiplied by a diagonal matrix **D**, the elements in each *column* of **A** get multiplied by the corresponding diagonal element of **D**.

The last exercise involved a diagonal matrix **E** with the further special property that all its diagonal elements are equal. It follows from Eqs. 2.12a and 2.12b that either pre- or postmultiplying by a matrix like **E** gives rise to a matrix each of whose elements is equal to a constant multiple of the corresponding element of the original matrix, the multiple being the common value of the diagonal elements of **E** (which was 5 in Exercise 8). But, referring to Eq. 2.4, we see that this is the same result as would be obtained when a matrix is multiplied by a scalar. For this reason, a diagonal matrix with all its diagonal elements equal is called a *scalar matrix*.

A special kind of scalar matrix is one whose diagonal elements are all equal to unity. It should be clear from the foregoing discussion that the effect of pre- or postmultiplying any matrix by this special scalar matrix is to leave the given matrix unchanged. That is, the effect is the same as that of multiplying by 1 in the realm of numbers. ($x \cdot 1 = x$, for any number x.) A scalar matrix of the form

$$\begin{bmatrix} 1 & 0 & \cdots & 0 \\ 0 & 1 & \cdots & 0 \\ \vdots & \vdots & \ddots & \\ 0 & 0 & \cdots & 1 \end{bmatrix}$$

is, therefore, called an *identity matrix*, and is denoted by the letter **I**, with a subscript to indicate its order if necessary. Thus, if **A** is any $n \times n$ matrix,

$$\mathbf{I}_n\mathbf{A} = \mathbf{A}\mathbf{I}_n = \mathbf{A} \tag{2.13}$$

(If **A** is not a square matrix but an $n \times m$, then $\mathbf{I}_n\mathbf{A} = \mathbf{A}$ and $\mathbf{A}\mathbf{I}_m = \mathbf{A}$.)

Having established the matrix analogue of the number 1, a natural question to ask next would be whether there exists an analogue to the reciprocal of a number. That is, given a matrix **A**, can we find another matrix **X** such that $\mathbf{AX} = \mathbf{I}$ (just as, for any number $a \neq 0$, there exists a reciprocal $x = 1/a$ such that $ax = 1$)? It turns out that, subject to certain restrictions (which will become clear later), the answer is affirmative. But the determination of such an **X** for a nonsquare matrix **A** requires more advanced techniques than we now have at our command. Its discussion will therefore be deferred to Chapter Five, and for the present we confine our attention to square matrices. For example, if

$$\mathbf{A} = \begin{bmatrix} 2 & 4 & 3 \\ -3 & 2 & 1 \\ -1 & 3 & 2 \end{bmatrix}$$

the reader may readily verify, by forming the product **AX**, that

$$\mathbf{X} = \begin{bmatrix} 1 & 1 & -2 \\ 5 & 7 & -11 \\ -7 & -10 & 16 \end{bmatrix}$$

is its "reciprocal" matrix, which is usually known as its *inverse matrix*. In analogy with the exponential notation for the reciprocal of a number ($1/a = a^{-1}$), the inverse of a given matrix **A**, when it exists, is denoted as $\mathbf{A}^{-1}$. It is true, only for square matrices, that either pre- or postmultiplying by the inverse yields the identity. That is, $\mathbf{A}^{-1}\mathbf{A} = \mathbf{A}\mathbf{A}^{-1} = \mathbf{I}$.

To compute the inverse of a given square matrix from first principles, one needs a working familiarity with the concept of a *determinant*. A brief summary of the rules for computing a determinant are given in Appendix A, which should suffice as a working guide. (However, the reader who has never before been introduced to determinants may want to consult a college algebra text such as Richardson (1966), for a more detailed presentation.) Here we shall assume a knowledge of determinants and illustrate the basic computational method for finding the inverse of a 3×3 matrix. The method will be shown to reduce almost to triviality for a 2×2 matrix. For practical computation of the inverse of a matrix of order greater than 3×3, however, a systematized routine, such as that described in Appendix B, becomes a near-necessity. Other computational routines may be found in Horst's *Matrix Algebra for Social Scientists*.

EXAMPLE 2.1. Find the inverse of

$$\mathbf{A} = \begin{bmatrix} 1 & -1 & 2 \\ 3 & 4 & -2 \\ -2 & 1 & 3 \end{bmatrix}$$

Step 1. Evaluate the determinant $|\mathbf{A}|$ of the matrix **A**.

As stated in Appendix A, this may be done by expanding the determinant with respect to any one of its rows or columns. We illustrate by expanding along the first row. The required *cofactors* are as follows:

$$A_{11} = \begin{vmatrix} 4 & -2 \\ 1 & 3 \end{vmatrix} = (4)(3) - (-2)(1) = 14$$

$$A_{12} = -\begin{vmatrix} 3 & -2 \\ -2 & 3 \end{vmatrix} = -[(3)(3) - (-2)(-2)] = -5$$

$$A_{13} = \begin{vmatrix} 3 & 4 \\ -2 & 1 \end{vmatrix} = (3)(1) - (4)(-2) = 11$$

The value of the determinant $|\mathbf{A}|$ is then given by

$$\begin{aligned} |\mathbf{A}| &= a_{11}A_{11} + a_{12}A_{12} + a_{13}A_{13} \\ &= (1)(14) + (-1)(-5) + (2)(11) \\ &= 14 + 5 + 22 = 41 \end{aligned}$$

If the value of the determinant of the given matrix (41 in our example) happens to be 0, no further steps are taken; *an inverse matrix simply does not exist* in this case—for reasons which will become clear in Step 3.

Step 2. Compute the *adjoint* or *adjugate* **adj(A)** of the given matrix.

For this purpose we need the cofactor of each and every element of **A**. Since we have already (Step 1) computed the cofactors of the elements of some row or column (row 1 in this illustration), we need only to compute those of elements in the remaining rows or columns. For the present example, these are:

$$A_{21} = -\begin{vmatrix} -1 & 2 \\ 1 & 3 \end{vmatrix} = 5 \qquad A_{22} = \begin{vmatrix} 1 & 2 \\ -2 & 3 \end{vmatrix} = 7$$

$$A_{23} = -\begin{vmatrix} 1 & -1 \\ -2 & 1 \end{vmatrix} = 1 \qquad A_{31} = \begin{vmatrix} -1 & 2 \\ 4 & -2 \end{vmatrix} = -6$$

$$A_{32} = -\begin{vmatrix} 1 & 2 \\ 3 & -2 \end{vmatrix} = 8 \qquad A_{33} = \begin{vmatrix} 1 & -1 \\ 3 & 4 \end{vmatrix} = 7$$

(As an arithmetic check, it should be verified that expansions along rows or columns other than that used in Step 1 yield the same value for $|\mathbf{A}|$. Thus, in our example,

$$(3)(5) + (4)(7) + (-2)(1) = 41$$

and

$$(-2)(-6) + (1)(8) + (3)(7) = 41)$$

We now collect the values of all the cofactors of **A** into a matrix, in this manner: the cofactors of the elements of the first *row* of **A** are written in the first *column* of this new matrix; those of the second *row* of **A** constitute the second *column* of the new matrix; and so on. The resulting matrix is called the *adjoint* or *adjugate* of **A**, and is denoted by **adj**(**A**). That is,

$$\mathbf{adj}(\mathbf{A}) = \begin{bmatrix} A_{11} & A_{21} & A_{31} \\ A_{12} & A_{22} & A_{32} \\ A_{13} & A_{23} & A_{33} \end{bmatrix}$$

(To repeat, A_{ij} is the *j*th *row*, *i*th *column* element of **adj**(**A**), the two subscripts referring to row and column numbers in reverse order from the usual notation.) For our example,

$$\mathbf{adj}(\mathbf{A}) = \begin{bmatrix} 14 & 5 & -6 \\ -5 & 7 & 8 \\ 11 & 1 & 7 \end{bmatrix}$$

is the adjoint of **A**.

Step 3. Divide each element of **adj**(**A**) by $|\mathbf{A}|$ (the determinant of **A**, obtained in Step 1); the resulting matrix is $\mathbf{A}^{-1}$, the desired inverse of **A**.

It should now be clear why $\mathbf{A}^{-1}$ fails to exist when $|\mathbf{A}| = 0$; division by 0 is forbidden. A matrix whose determinant is zero is called a *singular* matrix; only nonsingular matrices have inverses. For our example,

$$\mathbf{A}^{-1} = \mathbf{adj}(\mathbf{A})/|\mathbf{A}| = (1/|\mathbf{A}|)\mathbf{adj}(\mathbf{A})$$

$$= (1/41)\begin{bmatrix} 14 & 5 & -6 \\ -5 & 7 & 8 \\ 11 & 1 & 7 \end{bmatrix}$$

$$= \begin{bmatrix} 14/41 & 5/41 & -6/41 \\ -5/41 & 7/41 & 8/41 \\ 11/41 & 1/41 & 7/41 \end{bmatrix} \qquad (2.15)$$

The reader should verify that when the original matrix **A** is either pre- or postmultiplied by the matrix just obtained, the identity matrix

$$\mathbf{I}_3 = \begin{bmatrix} 1 & 0 & 0 \\ 0 & 1 & 0 \\ 0 & 0 & 1 \end{bmatrix}$$

is obtained.

Let us now apply the computational steps, outlined above, to a 2 × 2 matrix in general notation; that is,

$$\mathbf{A} = \begin{bmatrix} a_{11} & a_{12} \\ a_{21} & a_{22} \end{bmatrix}$$

Its determinant is

$$|\mathbf{A}| = a_{11}a_{22} - a_{12}a_{21}$$

The cofactors of the four elements of **A** are simply

$$\begin{cases} A_{11} = a_{22} & A_{12} = -a_{21} \\ A_{21} = -a_{12} & A_{22} = a_{11} \end{cases}$$

Hence, the adjoint of **A** is

$$\mathbf{adj(A)} = \begin{bmatrix} a_{22} & -a_{12} \\ -a_{21} & a_{11} \end{bmatrix}$$

(Comparing this with **A** itself, we see that the rule for finding the adjoint of a 2 × 2 matrix may be stated as follows: "Switch around the diagonal elements, and change the signs of the off-diagonal elements.") Then, provided $a_{11}a_{22} - a_{12}a_{21} \neq 0$, the inverse of **A** is given by

$$\mathbf{A}^{-1} = 1/(a_{11}a_{22} - a_{12}a_{21}) \begin{bmatrix} a_{22} & -a_{12} \\ -a_{21} & a_{11} \end{bmatrix} \tag{2.16}$$

if $a_{11}a_{22} - a_{12}a_{21} = 0$, **A** is singular and has no inverse.

EXERCISES

Find the inverse of each of the nonsingular matrices given in problems 1 and 2 below.

1. (a) $\begin{bmatrix} 5 & 3 \\ 6 & 4 \end{bmatrix}$ (b) $\begin{bmatrix} 1 & -2 \\ -2 & 3 \end{bmatrix}$ (c) $\begin{bmatrix} -4 & 2 \\ -6 & 3 \end{bmatrix}$ (d) $\begin{bmatrix} 3 & 0 \\ 0 & -5 \end{bmatrix}$

2. (a) $\begin{bmatrix} 2 & 2 & -4 \\ 10 & 14 & -22 \\ -14 & -20 & 32 \end{bmatrix}$ (b) $\begin{bmatrix} 1 & 1 & 1 \\ 2 & -1 & -2 \\ 1 & -2 & -1 \end{bmatrix}$ (c) $\begin{bmatrix} 1 & -3 & 2 \\ 3 & 1 & 0 \\ 2 & -1 & 1 \end{bmatrix}$

3. Problem 1d, above, exemplifies the simple rule for finding the inverse of a diagonal matrix. State the rule and prove it for diagonal matrices of arbitrary order.
4. Comparing the solution of problem 2a, above, with the original matrix given in the example on p. 22, the reader should note two properties of matrix inversion:

$$(\mathbf{A}^{-1})^{-1} = \mathbf{A} \tag{2.17}$$

$$(c\mathbf{A})^{-1} = (1/c)\mathbf{A}^{-1} \tag{2.18}$$

   where c is any scalar not equal to 0.
   Prove these properties algebraically.
5. Let

$$\mathbf{A} = \begin{bmatrix} 5 & 2 \\ 8 & 4 \end{bmatrix}$$

   and

$$\mathbf{v}' = \begin{bmatrix} -2 & 6 \end{bmatrix}$$

   Find the value of the quadratic form $\mathbf{v}'\mathbf{A}^{-1}\mathbf{v}$.
6. If **A** and **B** are both nonsingular square matrices of the same order, prove that

$$(\mathbf{AB})^{-1} = \mathbf{B}^{-1}\mathbf{A}^{-1} \tag{2.19}$$

   That is, prove that the inverse of the product of two matrices is equal to the product of their respective inverses, in *reverse order*.

2.5 MULTIPLE REGRESSION: REVIEW AND MATRIX FORMULATION

It is assumed that the reader is familiar with the concepts of multiple regression and multiple correlation, at least to the extent that a brief and rather incomplete review will suffice to revive a working knowledge of them. Given measurements on a set $X_1, X_2, \ldots, X_p$ of predictor variables and on one criterion variable Y for a group of N individuals, the problem of multiple regression is to construct a *linear function*

$$\tilde{Y} = a + b_1X_1 + b_2X_2 + \cdots + b_pX_p \tag{2.20}$$

having the property that the *sum of squared errors*,

$$\epsilon^2 = \Sigma(Y - \tilde{Y})^2 = \Sigma(Y - a - b_1X_1 - b_2X_2 - \cdots - b_pX_p)^2 \tag{2.21}$$

is as small as possible for the data at hand. More specifically, the problem is to determine the values of $a, b_1, b_2, \ldots, b_p$ so as to minimize the quantity ϵ^2. This is done by taking the partial derivative of the last expression for ϵ^2 with respect to each of the variables $a, b_1, b_2, \ldots, b_p$, setting each derivative equal to zero, and solving the resulting set of linear equations for $a, \ldots, b_p$.

We illustrate the procedure below for the case of $p = 2$ (that is, when we have two predictor variables), and indicate the generalization to any number of predictors.

The equations obtained by setting the respective partial derivatives of ϵ^2 equal to zero are as follows:

$$\begin{cases} Na + (\Sigma X_1)b_1 + (\Sigma X_2)b_2 = \Sigma Y \\ (\Sigma X_1)a + (\Sigma X_1{}^2)b_1 + (\Sigma X_1 X_2)b_2 = \Sigma X_1 Y \\ (\Sigma X_2)a + (\Sigma X_2 X_1)b_1 + (\Sigma X_2{}^2)b_2 = \Sigma X_2 Y \end{cases} \tag{2.22}$$

These are called the *normal equations* (in raw-score form). The reader who is not familiar with differential calculus will find the following mnemonic device convenient for remembering how to write out a set of normal equations or even for just seeing the pattern involved in them.

Imagine writing, for each of the N individuals in our sample, an equation just like Eq. 2.20, except that the $\tilde{Y}$ is replaced by Y. Using the subscript i to indicate the serial number for individuals, we would have N equations of the form,

$$a + X_{1i}b_1 + X_{2i}b_2 = Y_i \tag{2.23}$$

Adding the respective members of these N equations, we get

$$Na + \left(\sum_{i=1}^{N} X_{1i}\right) b_1 + \left(\sum_{i=1}^{N} X_{2i}\right) b_2 = \sum_{i=1}^{N} Y_i$$

which is the first one of the set of normal equations (2.20). The second and subsequent equations are obtained in the following manner. For the second, multiply each of the N equations (2.23) by the appropriate X_{1i}. We would have:

$$\begin{aligned} X_{11}a + (X_{11}{}^2)b_1 + (X_{11}X_{21})b_2 &= X_{11}Y_1 \\ X_{12}a + (X_{12}{}^2)b_1 + (X_{12}X_{22})b_2 &= X_{12}Y_2 \\ &\vdots \\ X_{1N}a + (X_{1N}{}^2)b + (X_{1N}X_{2N})b_2 &= X_{1N}Y_N \end{aligned}$$

On adding the respective members of these N equations, we get the second normal equation. For the third, each of the N equations (2.23) is multiplied by the appropriate X_{2i} before addition, and so on (if there are more than two predictor variables).

The set of normal equations (2.22) may be solved by matrix methods just as they stand. But it is somewhat simpler first to eliminate a by expressing it in terms of the bs from the first equation, and by substituting this expression in the subsequent equations. Thus, by dividing both sides of the first equation by N and then transposing all terms except a to the right, we have

$$a = \overline{Y} - \overline{X}_1 b_1 - \overline{X}_2 b_2 \tag{2.24}$$

Substituting this expression in the second normal equation, with the coefficient of a rewritten as $N\bar{X}_1$, we obtain, after rearranging terms,

$$(\Sigma X_1^2 - N\bar{X}_1^2)b_1 + (\Sigma X_1 X_2 - N\bar{X}_1\bar{X}_2)b_2 = (\Sigma X_1 Y - N\bar{X}_1\bar{Y})$$

But each of the expressions in parentheses, the reader will recognize, may be written in terms of deviation scores, thus:

$$(\Sigma x_1^2)b_1 + (\Sigma x_1 x_2)b_2 = \Sigma x_1 y$$

Similarly, the third normal equation, with a eliminated and deviation-score notation introduced, becomes

$$(\Sigma x_2 x_1)b_1 + (\Sigma x_2^2)b_2 = \Sigma x_2 y$$

The two (or, more generally, p) equations thus derived from the original three (or $p + 1$) normal equations are called the *normal equations in deviation-score form*.

Now, these equations may be written in matrix notation, as follows.

$$\begin{bmatrix} \Sigma x_1^2 & \Sigma x_1 x_2 \\ \Sigma x_2 x_1 & \Sigma x_2^2 \end{bmatrix} \begin{bmatrix} b_1 \\ b_2 \end{bmatrix} = \begin{bmatrix} \Sigma x_1 y \\ \Sigma x_2 y \end{bmatrix}$$

as may be readily verified by carrying out the matrix multiplication indicated on the left-hand side. The extension to the general case of p predictor variables is immediate. The matrix of coefficients of the bs will then be a $p \times p$ SSCP matrix (see p. 17), and the normal equations are represented by the single matrix equation

$$\begin{bmatrix} \Sigma x_1^2 & \Sigma x_1 x_2 & \cdots & \Sigma x_1 x_p \\ \Sigma x_2 x_1 & \Sigma x_2^2 & \cdots & \Sigma x_2 x_p \\ \vdots & & \ddots & \vdots \\ \Sigma x_p x_1 & \Sigma x_p x_2 & \cdots & \Sigma x_p^2 \end{bmatrix} \begin{bmatrix} b_1 \\ b_2 \\ \vdots \\ b_p \end{bmatrix} = \begin{bmatrix} \Sigma x_1 y \\ \Sigma x_2 y \\ \vdots \\ \Sigma x_p y \end{bmatrix} \tag{2.25}$$

After this equation has been solved for the vector $\mathbf{b} = [b_1, b_2, \ldots, b_p]'$, we determine a by substituting back in the generalization of Eq. 2.24.

$$a = \bar{Y} - \bar{X}_1 b_1 - \bar{X}_2 b_2 - \cdots - \bar{X}_p b_p \tag{2.26}$$

Thus, Eqs. 2.25 and 2.26, taken together, will provide a complete solution of the multiple regression problem.

Before we proceed to the solution of Eq. 2.25, it may be well to give a numerical example covering the development so far.

EXAMPLE 2.2. Suppose that, in the example first introduced in Section 2.2 and subsequently elaborated on in the "example of statistical import" in Section 2.3, we further had another test Y (which might be a mathematical reasoning test given at the end of the school year). Let the scores of the five pupils on this test be as follows:

$$\mathbf{Y} = \begin{bmatrix} 34 \\ 25 \\ 30 \\ 38 \\ 26 \end{bmatrix}$$

The SSCP matrix required for writing the left-hand side of the appropriate instances of Eq. 2.25 was already computed in the example in Section 2.3. However, for the purpose of numerical illustration of the step going from the normal equations in raw-score form (see Eqs. 2.22) to those in deviation-score form (summarized, in matrix notation, in Eq. 2.25, let us start out with the former set of normal equations. Appropriate substitutions in Eqs. 2.22 yield the following set of equations (the normal equations in raw-score form):

$$\begin{cases} 5a + 137b_1 + 64b_2 = 153 \\ 137a + 3817b_1 + 1770b_2 = 4273 \\ 64a + 1770b_1 + 834b_2 = 1990 \end{cases}$$

from the first equation of this set, we obtain:

$$a = 30.6 - 27.4b_1 - 12.8b_2$$

When this expression is substituted for a in the other two equations of the set, we obtain:

$$137(30.6 - 27.4b_1 - 12.8b_2) + 3817b_1 + 1770b_2 = 4273$$

and

$$64(30.6 - 27.4b_1 - 12.8b_2) + 1770b_1 + 834b_2 = 1990$$

Upon collecting like terms, these equations reduce to:

$$\begin{cases} 63.2b_1 + 16.4b_2 = 80.8 \\ 16.4b_1 + 14.8b_2 = 31.6 \end{cases}$$

or, equivalently,

$$\begin{bmatrix} 63.2 & 16.4 \\ 16.4 & 14.8 \end{bmatrix} \begin{bmatrix} b_1 \\ b_2 \end{bmatrix} = \begin{bmatrix} 80.8 \\ 31.6 \end{bmatrix}$$

in matrix notation. Note that the coefficient matrix on the left-hand side is precisely the SSCP matrix computed for the example in Section 2.3; also, the reader should verify that the elements of the vector on the right-hand side are equal, respectively, to $\sum x_1 y$ and to $\sum x_2 y$.

Solving the Normal Equations. To discuss the matrix-algebraic solution of Eq. 2.25, it is convenient to introduce the following abbreviations for the SSCP matrix and the two column vectors involved: Denote the SSCP matrix among the predictor variables by $\mathbf{S}_{pp}$, the vector of regression weights by $\mathbf{b}$, and the vector of sums-of-cross-products between the predictors and the criterion by $\mathbf{S}_{pc}$. Equation 2.25 may then be written symbolically as

$$\mathbf{S}_{pp}\mathbf{b} = \mathbf{S}_{pc} \tag{2.27}$$

This equation is formally analogous with a simple linear equation,

$$ab = c$$

where a and c are known numbers, and b is the unknown. To solve such an equation we simply divide both sides by a (which is the same as multiplying both sides by the reciprocal, $1/a$, of a) to get

$$b = (1/a)c$$

Similarly, to solve the matrix Eq. 2.27, we premultiply both sides by $\mathbf{S}_{pp}^{-1}$ (assuming that it exists), and obtain

$$\mathbf{S}_{pp}^{-1}(\mathbf{S}_{pp}\mathbf{b}) = \mathbf{S}_{pp}^{-1}\mathbf{S}_{pc} \tag{2.28}$$

which reduces to

$$\mathbf{b} = \mathbf{S}_{pp}^{-1}\mathbf{S}_{pc} \tag{2.29}$$

because the left-hand side of Eq. 2.28 may be transformed, successively, as

$$\begin{aligned} \mathbf{S}_{pp}^{-1}(\mathbf{S}_{pp}\mathbf{b}) &= (\mathbf{S}_{pp}^{-1}\mathbf{S}_{pp})\mathbf{b} && \text{(associative law)} \\ &= \mathbf{I}_p\mathbf{b} && \text{(definition of inverse matrix)} \\ &= \mathbf{b} && \text{(property of identity matrix)} \end{aligned}$$

Thus, Eq. 2.29 gives the desired solution to the normal equations represented by (2.27)—provided that the SSCP matrix (for the predictor variables) is nonsingular. For real data, except when there is some artificial restriction among the predictor variables (such as their sum being constant for all individuals), we may safely assume that the SSCP matrix is nonsingular, and hence that $\mathbf{S}_{pp}^{-1}$ exists. Methods for handling cases when $\mathbf{S}_{pp}$ is singular will be discussed later.

Once the *b*s (the regression weights) have thus been obtained, it is a simple matter to calculate the additive constant a to complete the regression Eq. 2.20. We simply substitute the values of the *b*s in Eq. 2.26 for a.

EXAMPLE 2.3. To continue with the example given above in illustrating the process of constructing the normal equations in deviation score form, let us compute the inverse of the SSCP matrix obtained there. We find:

$$\mathbf{S}_{pp}^{-1} = \begin{bmatrix} 63.2 & 16.4 \\ 16.4 & 14.8 \end{bmatrix}^{-1} = (1/666.40)\begin{bmatrix} 14.8 & -16.4 \\ -16.4 & 63.2 \end{bmatrix}$$

$$= \begin{bmatrix} .0222 & -.0246 \\ -.0246 & .0948 \end{bmatrix}$$

For the vector **b**, we therefore obtain

$$\mathbf{b} = \begin{bmatrix} .0222 & -.0246 \\ -.0246 & .0948 \end{bmatrix}\begin{bmatrix} 80.8 \\ 31.6 \end{bmatrix} = \begin{bmatrix} 1.016 \\ 1.008 \end{bmatrix}$$

Then, substituting in the expression for a, we get

$$a = 30.6 - (27.4)(1.016) - (12.8)(1.008) = -10.141$$

(The reader should substitute the values for a, b_1, b_2, obtained above, in the original raw-score normal equations for this example to see that they are satisfied within rounding error.) Thus, the desired multiple regression equation for this example is:

$$\tilde{Y} = -10.141 + 1.016X_1 + 1.008X_2$$

The Multiple Correlation Coefficient. As the reader is no doubt aware, the multiple correlation coefficient $R_{y.12\ldots p}$ is defined as the product-moment coefficient of correlation between the criterion variable Y and the "predicted" criterion score $\tilde{Y}$ in the sample used for constructing the regression equation. Thus, in principle, we could compute the multiple correlation coefficient (often called "multiple-R" for short) by using Eq. 2.20 to find the "predicted" criterion score $\tilde{Y}_i$ for each individual in the sample, and then correlating these with the actual criterion scores Y_i. That is,

$$R_{y.12\ldots p} = r_{y\tilde{y}} \tag{2.30}$$

by definition. However, it would obviously be very tedious to compute the multiple-R in this manner, and the reader probably recalls seeing the computational formula

$$R = \sqrt{\beta_1 r_{1y} + \beta_2 r_{2y} + \cdots + \beta_p r_{py}} \tag{2.31}$$

where, for each i, r_{iy} is the coefficient of correlation between X_i and Y, and β_i is the *standardized* partial regression coefficient of Y on X_i (often called the "beta weight"). The β_i's are related to the raw-score regression coefficients b_i, computed from Eq. 2.29, by the relations

$$\beta_i = \frac{s_i}{s_y} b_i \tag{2.32}$$

where s_i is the standard deviation of X_i, and s_y that of Y. As an exercise in matrix algebra, we shall derive formula (2.31) from the definitional Eq. 2.30.

Squaring both sides of Eq. 2.30 and using the definition of the product-moment correlation coefficient, we obtain

$$R^2 = r_{y\tilde{y}}^2 = \frac{(\Sigma y\tilde{y})^2}{(\Sigma y^2)(\Sigma \tilde{y}^2)} \tag{2.33}$$

where y and $\tilde{y}$ are, respectively, the actual and predicted criterion scores in deviation-score form. To express $\tilde{y}$ in terms of observed scores, we first need to find $\bar{\tilde{Y}}$. Substitution of the right-hand side of Eq. 2.26 for a in Eq. 2.20 yields

$$\begin{aligned}\tilde{Y} &= (\bar{Y} - b_1\bar{X}_1 - b_2\bar{X}_2 - \cdots - b_p\bar{X}_p) + b_1X_1 + b_2X_2 + \cdots + b_pX_p \\ &= \bar{Y} + b_1(X_1 - \bar{X}_1) + b_2(X_2 - \bar{X}_2) + \cdots + b_p(X_p - \bar{X}_p) \\ &= \bar{Y} + b_1x_1 + b_2x_2 + \cdots + b_px_p\end{aligned}$$

Summing both sides of this equation over the entire sample and dividing by N, we find

$$\bar{\tilde{Y}} = \bar{Y}$$

Hence

$$\begin{aligned}\tilde{y} &= \tilde{Y} - \bar{\tilde{Y}} \\ &= \tilde{Y} - \bar{Y} \\ &= b_1x_1 + b_2x_2 + \cdots + b_px_p\end{aligned}$$

Therefore, the expression inside the parentheses in the numerator of the rightmost member of Eq. 2.33 becomes

$$\begin{aligned}\Sigma y\tilde{y} &= \Sigma y(b_1x_1 + b_2x_2 + \cdots + b_px_p) \\ &= b_1\Sigma x_1y + b_2\Sigma x_2y + \cdots + b_p\Sigma x_py\end{aligned}$$

or, writing the right-hand side in vector notation,

$$\Sigma y\tilde{y} = \mathbf{b}'\mathbf{S}_{pc} \tag{2.34}$$

where $\mathbf{b}$ and $\mathbf{S}_{pc}$ are as defined immediately before Eq. 2.27.

Next, in order to see how $\sum\tilde{y}^2$ can be rewritten, we examine the case when $p = 3$:

$$
\begin{aligned}
\sum\tilde{y}^2 \\
&= \sum(b_1x_1 + b_2x_2 + b_3x_3)^2 \\
&= \sum({b_1}^2{x_1}^2 + {b_2}^2{x_2}^2 + {b_3}^2{x_3}^2 + 2b_1b_2x_1x_2 + 2b_1b_3x_1x_3 + 2b_2b_3x_2x_3) \\
&= {b_1}^2\sum{x_1}^2 + {b_2}^2\sum{x_2}^2 + {b_3}^2\sum{x_3}^2 \\
&\quad + 2b_1b_2\sum x_1x_2 + 2b_1b_3\sum x_1x_3 + 2b_2b_3\sum x_2x_3
\end{aligned}
$$

Comparing the right-hand side here with that of Eq. 2.9, which gives the expanded expression for the quadratic form $\mathbf{u'Au}$, we see that

$$
\begin{aligned}
\sum\tilde{y}^2 &= [b_1, b_2, b_3]\begin{bmatrix} \sum{x_1}^2 & \sum x_1x_2 & \sum x_1x_3 \\ \sum x_2x_1 & \sum{x_2}^2 & \sum x_2x_3 \\ \sum x_3x_1 & \sum x_3x_2 & \sum{x_3}^2 \end{bmatrix}\begin{bmatrix} b_1 \\ b_2 \\ b_3 \end{bmatrix} \\
&= \mathbf{b'S}_{pp}\mathbf{b} \qquad (2.35)
\end{aligned}
$$

Substituting the right-hand members of Eqs. 2.34 and 2.35 into Eq. 2.33, we get

$$R^2 = \frac{(\mathbf{b'S}_{pc})^2}{(\sum y^2)(\mathbf{b'S}_{pp}\mathbf{b})}$$

This expression may be simplified by utilizing the fact that, by definition, $\mathbf{b}$ satisfies Eq. 2.27:

$$\mathbf{S}_{pp}\mathbf{b} = \mathbf{S}_{pc}$$

from which, on premultiplying both sides by $\mathbf{b'}$, we get

$$\mathbf{b'S}_{pp}\mathbf{b} = \mathbf{b'S}_{pc}$$

Therefore, the second factor in the denominator of the expression just written for R^2 may be replaced by its equivalent, $\mathbf{b'S}_{pc}$. Thus

$$R^2 = \frac{(\mathbf{b'S}_{pc})^2}{(\sum y^2)(\mathbf{b'S}_{pc})}$$

or

$$R^2 = (1/\sum y^2)\mathbf{b'S}_{pc} \qquad (2.36)$$

This equation provides a computational formula for the multiple-R alternative to (2.31)—one that is, in fact, more convenient than the latter when we use Eq. 2.25 and 2.26 for determining the multiple regression equation in raw-score form. A few further steps will suffice to show that Eq. 2.36 is algebraically equivalent to Eq. 2.31. We give an outline of these steps below, and leave the details to be completed by the reader as an exercise.

First, note from (2.32) that

$$b_i = \frac{s_y}{s_i}\beta_i = \frac{\sqrt{\sum y^2}}{\sqrt{\sum x_i^2}}\beta_i$$

Hence, if the vector $[\beta_1, \beta_2, \ldots, \beta_p]$ is denoted as $\boldsymbol{\beta}'$ and if we define a diagonal matrix

$$\mathbf{D} = \begin{bmatrix} \sqrt{\sum x_1^2} & 0 & 0 \\ 0 & \sqrt{\sum x_2^2} & 0 \\ 0 & 0 & \sqrt{\sum x_3^2} \end{bmatrix}$$

it may readily be shown that

$$\mathbf{b}' = (\sqrt{\textstyle\sum y^2})\boldsymbol{\beta}'\mathbf{D}^{-1} \tag{2.37}$$

Substituting this expression for $\mathbf{b}'$ in the right-hand side of Eq. 2.36 yields

$$\begin{aligned} R^2 &= (1/\textstyle\sum y^2)[(\sqrt{\sum y^2})(\boldsymbol{\beta}'\mathbf{D}^{-1})]\mathbf{S}_{pc} \\ &= (1/\sqrt{\textstyle\sum y^2})\boldsymbol{\beta}'(\mathbf{D}^{-1}\mathbf{S}_{pc}) \end{aligned}$$

It remains only to show that

$$(1/\sqrt{\textstyle\sum y^2})\mathbf{D}^{-1}\mathbf{S}_{pc} = \begin{bmatrix} r_{1y} \\ r_{2y} \\ \vdots \\ r_{py} \end{bmatrix}$$

so that the last expression for R^2 becomes

$$R^2 = \boldsymbol{\beta}' \begin{bmatrix} r_{1y} \\ r_{2y} \\ \vdots \\ r_{py} \end{bmatrix} = [\beta_1, \beta_2, \ldots, \beta_p] \begin{bmatrix} r_{1y} \\ r_{2y} \\ \vdots \\ r_{py} \end{bmatrix}$$

or

$$R^2 = \beta_1 r_{1y} + \beta_2 r_{2y} + \cdots + \beta_p r_{py}$$

Taking the square root of both sides yields Eq. 2.31.

Now let us return to Eq. 2.36 and rewrite it in a slightly different form that will be useful for subsequent reference. Taking the transpose of both sides of Eq. 2.29 gives

$$\mathbf{b}' = \mathbf{S}'_{pc}\mathbf{S}_{pp}^{-1}$$

which may be substituted in expression (2.36) to yield

$$R^2 = (1/\textstyle\sum y^2)\mathbf{S}'_{pc}\mathbf{S}_{pp}^{-1}\mathbf{S}_{pc} \tag{2.38}$$

This expression provides a second example of a statistical formula in which a quadratic form is involved. (Recall that we saw an earlier instance in Eq. 2.35 for $\sum \tilde{y}^2$.) Other examples will occur with increasing frequency as we proceed.

A slight notational change renders Eq. 2.38 into a more symmetrical form. We rewrite $\mathbf{S}'_{pc}$ as $\mathbf{S}_{cp}$, and deliberately treat $\sum y^2$ as a 1×1 "matrix" denoted by $\mathbf{S}_{cc}$, so that $1/\sum y^2 = \mathbf{S}_{cc}^{-1}$. With this notation, Eq. 2.38 becomes

$$R^2 = \mathbf{S}_{cc}^{-1}\mathbf{S}_{cp}\mathbf{S}_{pp}^{-1}\mathbf{S}_{pc} \tag{2.39}$$

[Observe that, if we mentally replace c by 1 (because there is but one criterion variable), the subscripts indicate the order (that is, the numbers of rows and columns) of each factor of this quadruple product. The fact that the second subscript of each factor equals the first subscript of the next, indicates that all the multiplications are possible. Also, we can immediately tell that the final result is a scalar, because the first and last subscripts in the sequence are both c (= 1).]

This last expression for R^2, although longer than the preceding ones, has the mnemonic property that it formally involves the inverses of two square matrices (one of which is really a scalar) and two vectors (one a row vector, and the other a column vector) that together make up an augmented SSCP matrix—augmented, because it includes an extra row and column (relating to the criterion variable) as compared to $\mathbf{S}_{pp}$. This augmented SSCP matrix is

$$\left[\begin{array}{cccc|c} \sum x_1^2 & \sum x_1x_2 & \cdots & \sum x_1x_p & \sum x_1y \\ \sum x_2x_1 & \sum x_2^2 & \cdots & \sum x_2x_p & \sum x_2y \\ \vdots & & & & \vdots \\ \sum x_px_1 & \sum x_px_2 & \cdots & \sum x_p^2 & \sum x_py \\ \hline \sum yx_1 & \sum yx_2 & \cdots & \sum yx_p & \sum y^2 \end{array}\right]$$

where the dotted lines indicated its partitioning into $\mathbf{S}_{pp}$, $\mathbf{S}_{pc}$, $\mathbf{S}_{cc}$, and $\mathbf{S}_{cp}$, in the pattern:

$$\left[\begin{array}{c|c} \mathbf{S}_{pp} & \mathbf{S}_{pc} \\ \hline \mathbf{S}_{cp} & \mathbf{S}_{cc} \end{array}\right]$$

This pattern of partitioning an SSCP matrix, embracing both predictor and criterion variables, will be useful for further reference.

EXAMPLE 2.4. To continue with the actual computation of the multiple-R, we further need $\sum y^2$, the sum-of-squares of the criterion variable Y in order to complete the calculations. Referring back to p. 29, the Y-scores for our five pupils are 34, 25, 30, 38, and 26. Hence,

$$\sum y^2 = \sum Y^2 - (\sum Y)^2/N = 119.2$$

Computing R^2 in accordance with Eq. 2.36, therefore, we obtain

$$R^2 = (1/119.2)[1.016, 1.008]\begin{bmatrix}80.8\\31.6\end{bmatrix} = (1/119.2)(113.9456)$$
$$= 0.9559$$

To illustrate the alternative computation of R^2 from Eq. 2.39, and the requisite construction and partitioning of an augmented SSCP matrix, we note that this latter matrix is, for the present example, equal to

$$\left[\begin{array}{cc|c} 63.2 & 16.4 & 80.8 \\ 16.4 & 14.8 & 31.6 \\ \hline 80.8 & 31.6 & 119.2 \end{array}\right]$$

Here the 2 × 2 submatrix on the upper left is $\mathbf{S}_{pp}$ (the SSCP matrix among the p predictors alone); the 2 × 1 matrix on the upper right is $\mathbf{S}_{pc}$ (the vector of sums-of-products between predictors and criterion); the 1 × 2 matrix on the lower left is $\mathbf{S}_{cp}$ (= $\mathbf{S}'_{pc}$, the transpose of $\mathbf{S}_{pc}$); and the 1 × 1 matrix (that is, the scalar 119.2) in the lower right corner is $\mathbf{S}_{cc}$ (= $\sum y^2$, the sum-of-squares of the criterion). Thus, substitution in Eq. 2.39 yields

$$R^2 = [119.2]^{-1}[80.8, 31.6]\begin{bmatrix}63.2 & 16.4\\16.4 & 14.8\end{bmatrix}^{-1}\begin{bmatrix}80.8\\31.6\end{bmatrix}$$

which, the reader should verify, reduces to 0.9559, as before.

EXERCISES

1. "*Square root*" *of a diagonal matrix.* In certain applications, it is necessary to determine a square matrix **X** that has the property that **XX** = **A**, where **A** is a given square matrix. Such an **X** may be called the square root of **A**, and is denoted as $\mathbf{A}^{1/2}$. Determination of $\mathbf{A}^{1/2}$ in general is extremely difficult, but it becomes a simple matter in the special case when **A** is a diagonal matrix.

   Show that, if

$$\mathbf{A} = \begin{bmatrix} a_1 & & & 0 \\ & a_2 & & \\ & & \ddots & \\ 0 & & & a_p \end{bmatrix}$$

then

$$\mathbf{A}^{1/2} = \begin{bmatrix} \sqrt{a_1} & & & 0 \\ & \sqrt{a_2} & & \\ & & \ddots & \\ 0 & & & \sqrt{a_p} \end{bmatrix}$$

2. *Relation between SSCP matrix and correlation matrix.* Let

$$\mathbf{S} = \begin{bmatrix} \sum x_1^2 & \sum x_1 x_2 & \cdots & \sum x_1 x_p \\ \sum x_2 x_1 & \sum x_2^2 & \cdots & \sum x_2 x_p \\ \vdots & & \ddots & \vdots \\ \sum x_p x_1 & \sum x_p x_2 & \cdots & \sum x_p^2 \end{bmatrix}$$

be the SSCP matrix, and

$$\mathbf{R} = \begin{bmatrix} 1 & r_{12} & \cdots & r_{1p} \\ r_{21} & 1 & \cdots & r_{2p} \\ \vdots & & \ddots & \\ r_{p1} & r_{p2} & \cdots & 1 \end{bmatrix}$$

be the correlation matrix among p variables.

If we now define a diagonal matrix

$$\mathbf{\Delta} = \begin{bmatrix} \sum x_1^2 & & & 0 \\ & \sum x_2^2 & & \\ & & \ddots & \\ 0 & & & \sum x_p^2 \end{bmatrix}$$

show that

$$\mathbf{S} = \Delta^{1/2}\mathbf{R}\,\Delta^{1/2} \tag{2.40}$$

If, furthermore, we let

$$(\Delta^{1/2})^{-1} = \Delta^{-1/2}$$

Eq. 2.40 may be solved for **R** to yield

$$\mathbf{R} = \Delta^{-1/2}\mathbf{S}\,\Delta^{-1/2} \tag{2.41}$$

3. It has been shown (Eq. 2.37) that the vector of raw-score regression coefficients **b** and that of standardized regression coefficients $\boldsymbol{\beta}$ are related by the equation

$$\begin{aligned} \mathbf{b} &= (\sqrt{\textstyle\sum y^2})\mathbf{D}^{-1}\boldsymbol{\beta} \\ &= (\sqrt{\textstyle\sum y^2})\,\Delta^{-1/2}\boldsymbol{\beta} \end{aligned}$$

in the notation of Exercise 2 above.

Use this relation and the fact that **b** satisfies the normal equation (2.27),

$$\mathbf{S}_{pp}\mathbf{b} = \mathbf{S}_{pc}$$

to prove that $\boldsymbol{\beta}$ may be determined, without first finding **b**, by solving the equation

$$\mathbf{R}_{pp}\boldsymbol{\beta} = \mathbf{R}_{pc} \tag{2.42}$$

where $\mathbf{R}_{pp}$ is the correlation matrix among the predictor variables, and $\mathbf{R}_{pc}$ is the column vector of predictor-criterion intercorrelations.

Chapter Three

Analysis of Covariance with More Than One Covariate

In this chapter we discuss a statistical technique known as the *analysis of covariance*. In essence, this technique is an extension of the analysis of variance to take into account the possible effects, on the dependent variable, of one or more uncontrolled variables, (the *covariates*). A simple illustration will clarify the general idea.

Suppose that the relative merits of two approaches to introducing the written English language to first graders are being investigated. One approach uses the traditional orthography (T.O.)—that is, the usual alphabet and the usual spelling of words (fraught with inconsistencies, as we all know)—from the very beginning. The other approach uses, in the beginning, a quasi-phonetic system known as the "initial teaching alphabet" (I.T.A.), which includes most of the 26 letters of the ordinary alphabet but is augmented by several special symbols (such as æ for the sound of "a" as in f*a*ce; ω for the "oo" in "book"; ꭃ for the "oo" in "spoon") in order more nearly to attain a one-to-one correspondence between the pronunciation and spelling of words. The pupils are weaned off from the I.T.A. to the traditional one as soon as they have achieved reasonable mastery in the former.

In experimentally testing the two approaches for their relative merits, it may be impossible (as is often the case in educational research conducted in the classroom) to assign individual pupils at random to the two treatments. Rather, we would usually have to allocate entire classes to the treatment groups. We are then confronted with the possibility that classes assigned to the two groups may differ considerably on some relevant variable, such as general intelligence. Hence, after the groups have been subjected to their respective treatments and then tested with some criterion test (a standardized reading achievement test, for instance), the interpretation of any group difference found will be problematical: the difference will have been influenced, presumably, by the difference in intelligence between the two groups as well as by the different treatments to which they were subjected. The function of analysis of covariance is to make adjustments for the effects of the uncontrolled variable(s) in comparing the group performance.

The main purpose of this chapter is to describe the technique for the case of more than one covariate, but we shall begin with a review of the case of one covariate. Readers who wish to have a more detailed exposition of the one-covariate case should consult such authors as Draper and Smith (1966), Edwards (1968), McNemar (1969), or Walker and Lev (1953).

3.1 ANALYSIS OF COVARIANCE WITH ONE COVARIATE

The question of whether the groups' differences on the covariate suffice to explain away their differences on the criterion variable, may be somewhat more technically rephrased as follows: "Does the regression of the criterion on the covariate account for all of the significant variation between groups on the criterion variable?" If the answer turns out to be in the affirmative, then there would be no grounds for asserting that the different treatments undergone by the two or more groups were responsible for the differences on the criterion. Hence, the null hypothesis—stating that the population means of the several treatment groups, *adjusted for differences on the covariate*, are equal—would not be rejected.

This indirect approach of considering "the" regression of the criterion on the covariate immediately raises the question: "*Which* regression?" With several different groups represented in our sample, there are several different regression lines conceivable. These are schematically illustrated in Fig. 3.1 for the case of two groups.

Part *a* of this figure shows a regression line of Y (the criterion variable) on X (the covariate) for each of the two groups, based on the least-squares fit for *each group separately*. Part *b* shows a regression line passing through each group centroid, but having a common "within-groups" slope b_w, computed from Eq. 3.11; this is the best estimate of the corresponding parameter under the hypothesis that the population regression coefficients for the two groups are equal. Fig. 3.1*c* shows a single regression line, obtained by disregarding the distinction between the two groups and treating all cases as forming a single composite sample.

The rationale of covariance analysis hinges on deciding which one of the three situations depicted in Fig. 3.1 is most likely to be true of the populations corresponding to the several groups being compared. If the decision is to favor situation *a*—that is, to reject the hypothesis of equal regression slopes in the populations—then no further test concerning mean differences on Y, adjusted

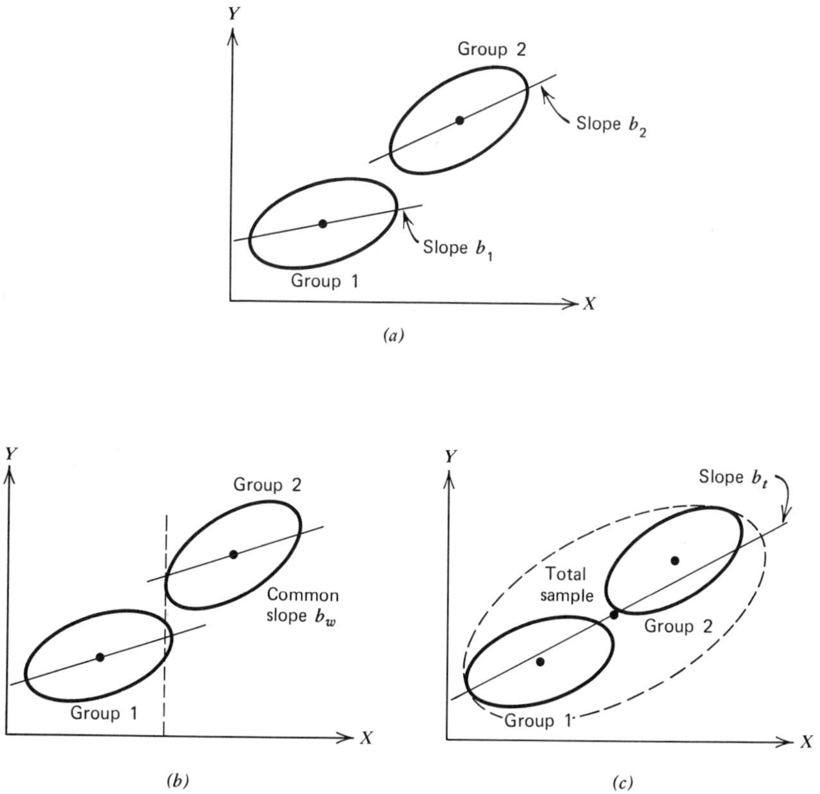

FIGURE 3.1 ***Regression lines of Y on X for two groups under three different conditions: (a) Separate regression lines, with slopes b_1 and b_2, fitted for two groups; (b) lines passing through respective group centroids, with common slope b_w; (c) single regression line, with slope b_t, fitted for total sample.***

for differences on X, is carried out.[1] If, however, no evidence is found for favoring *a* (that is, if the null hypothesis concerning slopes is tenable), then a second test is conducted to decide between *b* and *c*. If the decision based on this second test is for *b*, then we conclude that the differences on X do not

[1]In this case a test known as the Johnson-Neyman technique may be used. In brief, this method answers the question, "For what values of X (the covariate) are the *adjusted* Y-scores in the two groups significantly different, and in which direction?"—instead of arriving at a blanket conclusion of the form, "The Group 1 mean on Y is significantly greater (or smaller) than the Group 2 mean on Y when adjustments are made for X." For a description of this technique, see Johnson and Jackson (1959), pp. 432–438; or Walker and Lev (1953), pp. 398–404. A computer program for carrying out the test in the case of two covariates has been developed by Carroll and Wilson (1970).

sufficiently account for the group differences on Y. The reasoning is that, if the two regression lines have the same slope but different elevations (technically, different Y-intercepts), then even assuming that the groups had had equal X-means, we would "predict" the Y-means to be different. (See the dotted vertical line in Fig. 3.1*b*, showing what the "predicted," or "adjusted," Y-means of the two groups would have been if both groups had the same X-mean, equal to the grand mean of the total sample.) If, on the other hand, the second test finds c to be tenable, then we conclude that our data provide no grounds for believing that the differential treatment of groups would have produced different Y-means if the X-means had been equal.

Based on the rationale outlined above, the significance tests alluded to essentially consist of comparing the residual sums-of-squares of Y under the three different regression situations. To derive the requisite quantities, we introduce the following notation. Let:

Y_{ki} = criterion score earned by ith subject in kth treatment group;

X_{ki} = covariate score earned by ith subject in kth treatment group;

$Y_{k.} = \sum_{i=1}^{n_k} Y_{ki}$ = sum of Y-scores for kth group, where n_k = number of subjects in kth group;

$X_{k.} = \sum_{i=1}^{n_k} X_{ki}$ = sum of X-scores for kth group;

$Y_{..} = \sum_{k=1}^{K} Y_{k.}$ = grand total of Y-scores in entire sample of K groups combined (note that we are denoting the total number of groups by the upper case K, and letting k be the running subscript for any group);

$X_{..} = \sum_{k=1}^{K} X_{k.}$ = grand total of X-scores in entire sample of K groups combined.

The customary within-groups and total sums-of-squares of Y, needed in the analysis of variance, are then given by

$$SS_w = \sum_{k=1}^{K} \left[\sum_{i=1}^{n_k} Y_{ki}^2 - Y_{k.}^2/n_k \right]$$

and

$$SS_t = \sum_{k=1}^{K} \sum_{i=1}^{n_k} Y_{ki}^2 - Y_{..}^2/N \qquad \text{where } N = \sum_{k=1}^{K} n_k$$

respectively.

For the analysis of covariance, however, we also need the corresponding quantities for X (so the symbols SS_w and SS_t no longer suffice); we further need the sums-of-cross-products (in deviation-score form) between X and Y. Various notation systems are used for these quantities. Here we shall use a system that may at first seem unduly complicated, but which will facilitate our subsequent transfer to the case of several covariates. Besides the quantities mentioned above, we shall also need sums-of-squares and sums-of-products for each group separately (exemplified by the summand, in brackets, in the expression for SS_w, above); so we first define

$$S_{k,yy} = \sum_{i=1}^{n_k} Y_{ki}^2 - Y_{k.}^2/n_k \tag{3.1}$$

the sum-of-squares of Y for the kth group alone. Similarly, we let

$$S_{k,xx} = \sum_{i=1}^{n_k} X_{ki}^2 - X_{k.}^2/n_k \tag{3.2}$$

and

$$S_{k,xy} = S_{k,yx} = \sum_{i=1}^{n_k} X_{ki}Y_{ki} - X_{k.}Y_{k.}/n_k \tag{3.3}$$

In terms of the quantities defined in Eqs. 3.1 to 3.3, for each group separately, the corresponding pooled within-groups quantities are expressed quite simply. We use the generic symbol "W", with subscripts to specify whether the sum-of-squares of Y, the sum-of-squares of X, or the sum-of-products is being referred to. Thus,

$$W_{yy} = \sum_{k=1}^{K} S_{k,yy} \tag{3.4}$$

$$W_{xx} = \sum_{k=1}^{K} S_{k,xx} \tag{3.5}$$

$$W_{xy} = W_{yx} = \sum_{k=1}^{K} S_{k,xy} \tag{3.6}$$

For the total sums-of-squares and sum-of-products, no intermediate quantities are defined, and we merely introduce the symbols T_{yy}, T_{xx}, $T_{xy} = T_{yx}$.

That is,

$$T_{yy} = \sum_{k=1}^{K} \sum_{i=1}^{n_k} Y_{ki}^2 - Y_{..}^2/N \tag{3.7}$$

$$T_{xx} = \sum_{k=1}^{K} \sum_{i=1}^{n_k} X_{ki}^2 - X_{..}^2/N \tag{3.8}$$

and

$$T_{xy} = T_{yx} = \sum_{k=1}^{K} \sum_{i=1}^{n_k} X_{ki}Y_{ki} - X_{..}Y_{..}/N \tag{3.9}$$

The regression coefficients of Y on X that apply to different ones of the three situations a, b, and c in Fig. 3.1 may be expressed in terms of the appropriate ones of the quantities defined in Eqs. 3.1 to 3.9.

Recall that the general formula for a regression coefficient is

$$b_{yx} = \sum xy / \sum x^2$$

To compute the coefficient(s) applicable in each of the three situations, we need only use the appropriate sum-of-products and sum-of-squares of X in this formula.

Thus, in situation *a*, where each of the K groups has its own regression line with a slope specific to the group, the slope for the kth group is given by

$$b_{k,yx} = S_{k,xy}/S_{k,xx} \tag{3.10}$$

In situation *b*, where each group is fitted with a regression line passing through its centroid but having a common slope $b_{w,yx}$, this slope is given by

$$b_{w,yx} = W_{xy}/W_{xx} \tag{3.11}$$

Finally, in situation *c*, where all the groups are thrown together as a single composite group, the appropriate regression coefficient is

$$b_{t,yx} = T_{xy}/T_{xx} \tag{3.12}$$

Next, we compute the residual sum-of-squares of Y under each of the regression situations, recalling that the analysis of covariance is based on making comparisons among the different kinds of residual sums-of-squares of Y. In reference to formulas for residual sums-of-squares, the reader is probably more familiar with one involving the correlation coefficient than with one involving the regression coefficient. The customary expression in the analysis of variance (for testing the significance of lincar regression) is

$$SS_{\text{res}} = (\sum y^2)(1 - r_{xy}{}^2)$$

But, recalling that

$$r_{xy} = \sum xy / \sqrt{(\sum x^2)(\sum y^2)}$$

while

$$b_{yx} = \sum xy / \sum x^2$$

it is readily seen that

$$\begin{aligned} SS_{\text{res}} &= \sum y^2 - (\sum xy)^2 / \sum x^2 \\ &= \sum y^2 - (b_{yx}{}^2)(\sum x^2) \end{aligned} \tag{3.13}$$

Hence, the residual sum-of-squares of Y in each of the three situations *a*, *b*, and *c* may be written from Eq. 3.13 by substituting the appropriate ones of the $\sum y^2$, $\sum x^2$, and $\sum xy$ or b_{yx}.

For situation *a*, the residual sum-of-squares of Y for the kth group thus becomes

$$S_{k,yy} - (S_{k,xy})^2 / S_{k,xx}$$

By adding these quantities over all the groups, we obtain the residual sum-of-squares of Y for the entire sample, which we denote by S_1:

$$S_1 = \sum_{k=1}^{K} [S_{k,yy} - (S_{k,xy})^2/S_{k,xx}]$$

or, since $\sum_{k=1}^{K} S_{k,yy} = W_{yy}$ by definition,

$$S_1 = W_{yy} - \sum_{k=1}^{K} [(S_{k,xy})^2/S_{k,xx}] \tag{3.14}$$

For situation b, the residual sum-of-squares of Y, which we denote by S_2, is as follows:

$$S_2 = W_{yy} - (W_{xy})^2/W_{xx} \tag{3.15}$$

and for situation c we have

$$S_3 = T_{yy} - (T_{xy})^2/T_{xx} \tag{3.16}$$

For any set of data, it is necessarily the case that

$$S_1 \leq S_2 \leq S_3 \tag{3.17}$$

(The first inequality follows from the fact that S_1 is found by adding the residual sums-of-squares from regression lines constructed separately for the several groups, and is hence the aggregate of the minimum possible residual for each group.) The questions to be answered in the two tests of covariance analysis essentially amount to the following.

1. When S_1 is divided by its number of degrees of freedom (n.d.f.) ν_1, and S_2 is divided by its n.d.f. ν_2, do we get variance estimates that can be regarded as estimating the same parameter and differing only by sampling error?

2. Similarly with S_2 and S_3; that is, can S_2/ν_2 and S_3/ν_3 be regarded as differing only by sampling error? If the answer to question 1 is in the affirmative, we conclude that situations a and b are equally applicable to our data, and that, hence, the separate group regression coefficients $b_1, b_2, \ldots, b_k$ are not significantly different from one another. We can then legitimately ask the second question. If the answer here is "yes," then situations b and c are equally applicable to our data, so that group differences on the covariate are regarded as sufficient to explain away the differences on the criterion variable. If the answer is "no," on the other hand, we conclude that situation b, but not c, is applicable to our data, so that group differences on the criterion variable remain even after the covariate differences are "partialled out." The implication then is that the different treatments were probably responsible for producing the criterion-variable differences.

In seeking the answers to questions 1 and 2, above, it may seem natural to take the ratios

$$\frac{S_2/\nu_2}{S_1/\nu_1} \quad \text{and} \quad \frac{S_3/\nu_3}{S_2/\nu_2}$$

respectively, and refer these values to the appropriate F-distributions.[2] The difficulty with this idea is that the quantities S_1 and S_2, as well as S_2 and S_3, are not independently distributed. (This is readily seen in the case of S_1 and S_2 by examining Eqs. 3.14 and 3.15: S_1 and S_2 have the term W_{yy} in common.) This difficulty may be circumvented, however, by taking differences before forming the ratios.

Thus, for answering question 1, we first obtain the difference

$$\begin{aligned} S_4 &= S_2 - S_1 \\ &= \sum_{k=1}^{K} [(S_{k,xy})^2/S_{k,xx}] - (W_{xy})^2/W_{xx} \end{aligned} \tag{3.18}$$

which is independent of S_1, and S_4/σ^2, as well as S_1/σ^2, has a chi-square distribution (where σ^2 is an unknown population variance). We then take the ratio

$$F_\beta = \frac{S_4/\nu_4}{S_1/\nu_1} \tag{3.19'}$$

where $\nu_4 = \nu_2 - \nu_1$, which follows the F-distribution with ν_4 and ν_1 degrees of freedom under the null hypothesis of parallel regression lines.

Similarly, for the second test, we find

$$S_5 = S_3 - S_2 = (T_{yy} - W_{yy}) - [(T_{xy})^2/T_{xx} - (W_{xy})^2/W_{xx}] \tag{3.20}$$

and then form the ratio

$$F_\mu = \frac{S_5/\nu_5}{S_2/\nu_2} \tag{3.21'}$$

where $\nu_5 = \nu_3 - \nu_2$; this ratio follows the F-distribution with ν_5 and ν_2 degrees of freedom. It should be noted that the expression in the first pair of parentheses of Eq. 3.20 is the between-groups sum-of-squares of Y. ($SS_t - SS_w = SS_b$, in the usual analysis-of-variance notation.) S_5 is, therefore, an

[2]The reader will recall that an F-variate is formed by taking a ratio

$$\frac{u_1/n_2}{u_2/n_2}$$

where u_1 and u_2 are independently distributed as chi-squares with n_1 and n_2 degrees of freedom, respectively.

adjusted between-groups sum-of-squares. S_2, on the other hand, is an adjusted within-groups sum-of-squares, as may readily be seen from Eq. 3.15. Hence, the F-ratio (3.21′) is quite analogous to that used in simple ANOVA, which is

$$F = MS_b/MS_w$$

The foregoing completely describes the two tests of covariance analysis, except for the matter of how the various numbers of degrees of freedom are determined. A rigorous derivation of these numbers would entail too lengthy a digression, so we merely give an intuitive explanation. Consider S_3 first. There are N cases in the total sample, on the basis of which we estimate the two constants $a_{t,yx}$ and $b_{t,yx}$ of the regression equation,

$$\tilde{Y} = a_{t,yx} + b_{t,yx}X$$

for the situation shown in Fig. 3.1*c*. We thus lose two degrees of freedom, and hence the n.d.f. for S_3 is $\nu_3 = N - 2$. Similarly, for S_1 we have $n_k - 2$ degrees of freedom for the kth term of the sum defining S_1 (see equation preceding Eq. 3.14). Thus,

$$\nu_1 = \sum_{k=1}^{K} (n_k - 2) = N - 2K$$

Finally, for S_2, we note from Fig. 3.1*b* that we are estimating K separate constants $a_{k,yx}$ (one for each group) and one regression coefficient $b_{w,yx}$. Thus, $\nu_2 = N - (K + 1)$. The degrees of freedom for S_4 and S_5 are obtained by the appropriate differences, $\nu_2 - \nu_1$ and $\nu_3 - \nu_2$, respectively, corresponding to the definitional Eqs. 3.18 and 3.20 for these quantities. For ready reference we list the several n.d.f.'s in Table 3.1. Substituting the appropriate νs from this table into Eqs. 3.19′ and 3.21′, we may rewrite the formulas for the two F-ratios in their final forms as Eqs. 3.19 and 3.21, below.

TABLE 3.1 N.D.F.'s for the Various Residual Sums-of-Squares.

Sum-of-Squares	Defining Equation	N.D.F.
S_1	3.14	$\nu_1 = N - 2K$
S_2	3.15	$\nu_2 = N - K - 1$
S_3	3.16	$\nu_3 = N - 2$
S_4	3.18	$\nu_4 \; (= \nu_2 - \nu_1) = K - 1$
S_5	3.20	$\nu_5 \; (= \nu_3 - \nu_2) = K - 1$

$$F_{\beta} = \frac{S_4/(K - 1)}{S_1/(N - 2K)} \tag{3.19}$$

where S_4 and S_1 are as defined in Eqs. 3.18 and 3.14, respectively.

$$F_{\mu} = \frac{S_5/(K - 1)}{S_2/(N - K - 1)} \tag{3.21}$$

where S_5 and S_2 are as defined in Eqs. 3.20 and 3.15, respectively.

3.2 AN EXAMPLE

We now apply the results of the preceding section to a problem described in outline at the beginning of this chapter.

Two groups of first graders—one taught by the I.T.A. method, the other by the T.O.—were tested on a reading mastery test at the end of the school year, with results as follows:

	I.T.A. Group	T.O. Group
mean	20.18	16.92
s.d.	2.67	4.02
n	62	87

If a *t*-test for the significance of the difference between the two means is conducted, we would find the difference to be highly significant. However, it was also true that the I.T.A. group had a higher mean Kuhlmann-Anderson IQ than did the T.O.—both groups having been tested before their instruction in reading began. Consequently, it is desirable to use the analysis of covariance to see whether the difference in IQ is sufficient to account for the observed difference on the reading mastery test. The complete set of summary data necessary for this purpose is shown in Table 3.2.

TABLE 3.2 Data for Analysis of Covariance in the I.T.A. Versus T.O. Example—*X*, Kuhlmann-Anderson IQ (covariate) and *Y*, Reading Mastery Test (criterion).

	I.T.A. Group	T.O. Group	Total Sample
n	62	87	149
$\sum X$	6,885	9,533	16,418
$\sum X^2$	771,997	1,057,391	1,829,388
$\sum Y$	1,251	1,472	2,723
$\sum Y^2$	25,685	26,312	51,997
$\sum XY$	139,423	163,187	302,610

In computing the various sums-of-squares and cross-products needed, it is convenient to arrange these quantities in the form of matrices. Thus, we first compute the SSCP matrices $\mathbf{S}_1$ and $\mathbf{S}_2$ for the two groups separately. The results are as follows.

$$\mathbf{S}_1 = \begin{bmatrix} S_{1,yy} & S_{1,yx} \\ S_{1,xy} & S_{1,xx} \end{bmatrix} = \begin{bmatrix} 443.05 & 501.47 \\ 501.47 & 7428.85 \end{bmatrix} \tag{3.22}$$

where the elements have been computed by use of Eqs. 3.1 to 3.3. That is,

$$S_{1,yy} = 25{,}685 - (1251)^2/62 = 443.05$$

$$S_{1,xx} = 771{,}997 - (6885)^2/62 = 7428.85$$

and

$$S_{1,yx} = S_{1,xy} = 139{,}423 - (1251)(6885)/62 = 501.47$$

Similarly,

$$\mathbf{S}_2 = \begin{bmatrix} S_{2,yy} & S_{2,yx} \\ S_{2,xy} & S_{2,xx} \end{bmatrix} = \begin{bmatrix} 1406.44 & 1893.02 \\ 1893.02 & 12{,}815.26 \end{bmatrix} \tag{3.23}$$

Next the within-groups SSCP matrix $\mathbf{W}$ is computed by adding $\mathbf{S}_1$ and $\mathbf{S}_2$, in accordance with Eqs. 3.4 to 3.6.

$$\mathbf{W} = \mathbf{S}_1 + \mathbf{S}_2 = \begin{bmatrix} W_{yy} & W_{yx} \\ W_{xy} & W_{xx} \end{bmatrix} = \begin{bmatrix} 1849.49 & 2394.49 \\ 2394.49 & 20{,}244.11 \end{bmatrix} \tag{3.24}$$

Finally, the elements of $\mathbf{T}$, the SSCP matrix for the entire sample consisting of the two groups thrown together, are computed directly from Eqs. 3.7 to 3.9:

$$T_{yy} = 51{,}997 - (2{,}723)^2/149, \quad \text{etc.}$$

giving us

$$\mathbf{T} = \begin{bmatrix} T_{yy} & T_{yx} \\ T_{xy} & T_{xx} \end{bmatrix} = \begin{bmatrix} 2233.72 & 2568.30 \\ 2568.30 & 20{,}322.74 \end{bmatrix} \tag{3.25}$$

We are now ready to compute the three kinds of residual sums-of-squares of Y, defined in Eqs. 3.14 to 3.16. It should be noted that, if the various SSCP matrices are represented by the general schema

$$\begin{bmatrix} a & b \\ b & c \end{bmatrix}$$

the desired quantities S_1, S_2, and S_3 all have the general form

$$a - (b \times b)/c \qquad \text{or} \qquad a - bc^{-1}b$$

(except that S_1 is the sum of two quantities of this form). Keeping this pattern in mind will not only facilitate the calculations a great deal, but will later enable us readily to make a transition to the case of several covariates. Thus,

$$\begin{aligned} S_1 &= [443.05 - (501.47)^2/7428.85] + [1406.44 - (1893.02)^2/12{,}815.26] \\ &= 1536.00 \\ S_2 &= 1849.49 - (2394.49)^2/20{,}244.11 \\ &= 1566.26 \\ S_3 &= 2233.72 - (2568.30)^2/20{,}322.74 \\ &= 1909.15 \end{aligned}$$

Having computed these residual sums-of-squares, we first conduct the test for equality of population regression slopes, using Eqs. 3.18 and 3.19:

$$S_4 = S_2 - S_1 = 1566.26 - 1536.00 = 30.26$$

whence

$$F_\beta = \frac{30.26/(2 - 1)}{1536.00/(149 - 4)} = 2.86$$

From Table E.4 of the appendix, we find, by suitable interpolation, that the 95th centile[3] of the F distribution with 1 and 145 degrees of freedom is 3.91. Hence, the obtained value 2.86 does not lead to rejection of the null hypothesis at $\alpha = .05$. We may therefore proceed to the significance test for the adjusted Y means, using Eqs. 3.20 and 3.21:

$$S_5 = S_3 - S_2 = 1909.15 - 1566.26 = 342.89$$

whence

$$F_\mu = \frac{342.89/(2 - 1)}{1566.26/(149 - 2 - 1)} = 31.96$$

This value far exceeds the 99th centile of the F distribution with 1 and 146 degrees of freedom. We therefore reject the null hypothesis that a single regression line of Y on X adequately fits both groups, and hence conclude that the mean difference on X (IQ) between the two groups does not account for the mean difference on Y (reading mastery test). Therefore, if we are satisfied that other sources of variation were adequately controlled, we may conclude that the I.T.A. approach was more efficient in teaching the first graders to read.

[3]The term "percentile" is, unfortunately, in more common use than "centile." That the latter is the more logical word should be clear when it is recognized that we are referring to a concept analogous to "decile," "quartile," and "tertile." To speak of "percentile" is just as awkward as to say "perquartile" for "quartile." It is to be hoped that the correct term, "centile," will gradually come to supersede the incongruous "percentile."

3.3 COVARIANCE ANALYSIS WITH SEVERAL COVARIATES

Now suppose that, instead of one covariate X, we have p covariates $X_1, X_2, \ldots, X_p$. We shall denote the criterion variable by X_0 instead of Y. By an obvious extension of the notation introduced on p. 42, we let

$X_{\alpha ki}$ = the score on X_α $(\alpha = 0, 1, 2, \ldots, p)$ earned by the ith subject in the kth treatment group

$$X_{\alpha k.} = \sum_{i=1}^{n_k} X_{\alpha ki} \qquad (\alpha = 0, 1, 2, \ldots, p)$$

$$X_{\alpha ..} = \sum_{k=1}^{k} X_{\alpha k.} \qquad (\alpha = 0, 1, 2, \ldots, p)$$

The rationale of covariance analysis with p covariates exactly parallels that of the one-covariate case. We first inquire whether regression hyperplanes fitted group by group (analogous to situation a in the previous section) lead to a significantly smaller overall residual sum-of-squares of X_0 than do hyperplanes passing through the separate group centroids but having a common within-groups orientation (situation b). If the answer to this question is in the negative—that is, if the null hypothesis postulating equal $\boldsymbol{\beta}_k$'s (vectors of partial regression weights) in the populations is sustained—we then conduct a test to decide between situations b and c.[4] The latter implies that a single regression hyperplane fits all K groups treated as a single composite group, and hence that the group differences on X_0 are sufficiently accounted for by their differences on the covariates, $X_1, X_2, \ldots, X_p$. The various sums-of-squares and sums-of-products needed for conducting the two tests are given by the following equations.

$$S_{k,\alpha\beta} = \sum_{i=1}^{n_k} X_{\alpha ki}X_{\beta ki} - X_{\alpha k.}X_{\beta k.}/n_k$$
$$(k = 1, 2, \cdots, K;\ \alpha, \beta = 0, 1, 2, \cdots, p) \tag{3.26}$$

$$W_{\alpha\beta} = \sum_{k=1}^{K} S_{k,\alpha\beta} \qquad \text{(within-groups)} \tag{3.27}$$

$$T_{\alpha\beta} = \sum_{k=1}^{K}\sum_{i=1}^{n_k} X_{\alpha ki}X_{\beta ki} - X_{\alpha ..}X_{\beta ..}/N \qquad \text{where } N = \sum_{k=1}^{K} n_k \tag{3.28}$$

Note that each of these equations gives sums-of-squares when $\alpha = \beta$, and sums-of-products when $\alpha \neq \beta$. We collect these quantities into several matrices, as we did in the numerical example in the previous section (see Eqs. 3.22 to 3.25).

[4]See footnote 1, p. 41, on how to handle the problem when this null hypothesis is not tenable.

$$\mathbf{S}_k = \left[\begin{array}{c|cccc} S_{k,00} & S_{k,01} & S_{k,02} & \cdots & S_{k,0p} \\ \hline S_{k,10} & S_{k,11} & S_{k,12} & & S_{k,1p} \\ \vdots & & & \ddots & \vdots \\ S_{k,p0} & S_{k,p1} & S_{k,p2} & \cdots & S_{k,pp} \end{array}\right] \tag{3.29}$$

which, for each k, is the SSCP matrix for that particular group.

$$\mathbf{W} = \sum_{k=1}^{K} \mathbf{S}_k \tag{3.30}$$

$$= \left[\begin{array}{c|ccccc} W_{00} & W_{01} & W_{02} & W_{03} & \cdots & W_{0p} \\ \hline W_{10} & W_{11} & W_{12} & W_{13} & & W_{1p} \\ \vdots & & & & \ddots & \vdots \\ W_{p0} & W_{p1} & W_{p2} & W_{p3} & \cdots & W_{pp} \end{array}\right]$$

the sum of the K group-by-group SSCP matrices, which is the within-groups SSCP matrix.

$$\mathbf{T} = \left[\begin{array}{c|cccc} T_{00} & T_{01} & T_{02} & \cdots & T_{0p} \\ \hline T_{10} & T_{11} & T_{12} & & T_{1p} \\ \vdots & & & \ddots & \vdots \\ T_{p0} & T_{p1} & T_{p2} & \cdots & T_{pp} \end{array}\right] \tag{3.31}$$

is the total SSCP matrix.

Note that each of these SSCP matrices has been partitioned, by dotted lines, into four submatrices, just as the SSCP matrix for a multiple-regression problem was partitioned in the previous chapter (see p. 35). We label these submatrices in accordance with the following pattern:

$$\mathbf{S} = \left[\begin{array}{c|c} S_{00} & S_{0p} \\ \hline S_{p0} & S_{pp} \end{array}\right]$$

where $\mathbf{S}$ stands for any one of the SSCP matrices, $\mathbf{S}_k$, $\mathbf{W}$, or $\mathbf{T}$. The upper-left submatrix is, in each case, a scalar ($S_{k,00}$, W_{00}, or T_{00}) representing the appropriate sum-of-squares of the criterion variable. The upper-right and lower-left submatrices ($\mathbf{S}_{k,0p}$ and $\mathbf{S}_{k,p0}$; $\mathbf{W}_{0p}$ and $\mathbf{W}_{p0}$; or $\mathbf{T}_{0p}$ and $\mathbf{T}_{p0}$) are, in each case, the row vector and the column vector of the sums-of-products between the criterion variable and the p covariates. The lower-right submatrix ($\mathbf{S}_{k,pp}$, $\mathbf{W}_{pp}$, or $\mathbf{T}_{pp}$) is, in each case, the appropriate SSCP matrix among the p covariates.

The residual sum-of-squares of X_0 in the p-covariate case is computed from a natural extension of Eq. 3.13 for the one-covariate case. The expression

$\sum x^2$ is replaced by the appropriate SSCP matrix (generically denoted $\mathbf{S}_{pp}$ above) among the p covariates. The multiplication of $\sum x^2$ by $b_{yx}{}^2$, occurring in Eq. 3.13, is replaced by a premultiplication of $\mathbf{S}_{pp}$ by the row vector $\mathbf{b}'$ of partial regression weights, followed by a postmultiplication by its transpose $\mathbf{b}$. That is,

$$\begin{aligned} SS_{\text{res}} &= \sum y^2 - b_{yx}{}^2 \sum x^2 \\ &= \sum y^2 - b_{yx}(\sum x^2) b_{yx} \\ &\Rightarrow S_{00} - \mathbf{b}'\mathbf{S}_{pp}\mathbf{b} \end{aligned}$$

where $\Rightarrow$ stands for "is replaced by."

The vector $\mathbf{b}$ of partial regression weights involved in the above expression is computed in accordance with Eq. 2.29, that is,

$$\mathbf{b} = \mathbf{S}_{pp}^{-1}\mathbf{S}_{p0} \tag{2.29}$$

We thus have, for the p-covariate case,

$$\begin{aligned} SS_{\text{res}} &= S_{00} - (\mathbf{S}_{pp}^{-1}\mathbf{S}_{p0})'\mathbf{S}_{pp}(\mathbf{S}_{pp}^{-1}\mathbf{S}_{p0}) \\ &= S_{00} - \mathbf{S}_{0p}(\mathbf{S}_{pp}^{-1}\mathbf{S}_{pp})(\mathbf{S}_{pp}^{-1}\mathbf{S}_{p0}) \end{aligned}$$

where we have utilized the fact that S_{pp}^{-1} is a symmetrical matrix, and that $\mathbf{S}'_{p0} = \mathbf{S}_{0p}$, in transforming $(\mathbf{S}_{pp}^{-1}\mathbf{S}_{p0})'$. Noting further that $\mathbf{S}_{pp}^{-1}\mathbf{S}_{pp} = \mathbf{I}$, we finally get

$$SS_{\text{res}} = S_{00} - \mathbf{S}_{0p}\mathbf{S}_{pp}^{-1}\mathbf{S}_{p0} \tag{3.32}$$

as the general schema for computing each of the three kinds of residual sums-of-squares of X_0.

For situation *a*, the appropriate quantities to substitute in Eq. 3.32 are the submatrices of $\mathbf{S}_k$ for each group—the resulting group-by-group residual sums-of-squares then being added to obtain the overall residual. Using the same symbol S_1 as in the one-covariate case to denote the overall residual sum-of-squares of X_0 in this situation, we have:

$$\begin{aligned} S_1 &= \sum_{k=1}^{K} [S_{k,00} - \mathbf{S}_{k,0p}\mathbf{S}_{k,\,pp}^{-1}\mathbf{S}_{k,p0}] \\ &= W_{00} - \sum_{k=1}^{K} \mathbf{S}_{k,0p}\mathbf{S}_{k,\,pp}^{-1}\mathbf{S}_{k,p0} \end{aligned} \tag{3.33}$$

which is exactly parallel to Eq. 3.14.

For situation *b*, each occurrence of "S" in Eq. 3.32 is replaced by "**W**", and we obtain

$$S_2 = W_{00} - \mathbf{W}_{0p}\mathbf{W}_{pp}^{-1}\mathbf{W}_{p0} \tag{3.34}$$

as the appropriate residual sum-of-squares of X_0. This is exactly analogous to S_2 as defined in Eq. 3.15 for the one-covariate case.

Finally, for situation *c*, we replace each "**S**" in Eq. 3.32 by "**T**", and get

$$S_3 = T_{00} - \mathbf{T}_{0p}\mathbf{T}_{pp}^{-1}\mathbf{T}_{p0} \tag{3.35}$$

in analogy with Eq. 3.16.

Note that the basic pattern for each of the residual sums-of-squares is exactly analogous to that in the univariate case. The parallel is clearly seen by writing the general schema of the SSCP matrices and the residual sums-of-squares for the one-covariate and *p*-covariate cases side by side, as in Table 3.3.

TABLE 3.3 Schema for SSCP Matrices and Residual Sums-of-Squares of the Criterion Variable: One-Covariate and *p*-Covariate Cases Compared.

	One Covariate	*p* Covariates
SSCP Matrix	$\left[\begin{array}{c\|c} \sum y^2 & \sum yx \\ \hline \sum xy & \sum x^2 \end{array}\right]$	$\left[\begin{array}{c\|c} S_{00} & \mathbf{S}_{0p}\ [(1 \times p) \text{ vector}] \\ \hline \mathbf{S}_{p0}\ (p \times 1) \text{ vector} & \mathbf{S}_{pp}\ [(p \times p) \text{ matrix}] \end{array}\right]$
General Form of SS_{res}	$\sum y^2 - (\sum xy)^2/\sum x^2$ $= \sum y^2 - (\sum yx) \times (1/\sum x^2)(\sum xy)$	$S_{00} - \mathbf{S}_{0p}\mathbf{S}_{pp}^{-1}\mathbf{S}_{p0}$

Just as in the one-covariate case, we form the difference

$$\begin{aligned} S_4 &= S_2 - S_1 \\ &= \sum_{k=1}^{k} \mathbf{S}_{k,0p}\mathbf{S}_{k,\ pp}^{-1}\mathbf{S}_{k,p0} - \mathbf{W}_{0p}\mathbf{W}_{pp}^{-1}\mathbf{W}_{p0} \end{aligned}$$

which is distributed as a chi-square, independently of S_1. Thus, by dividing S_4 and S_1 each by its n.d.f. and forming the ratio

$$F_\beta = \frac{S_4/\nu_4}{S_1/\nu_1} \tag{3.37$'$}$$

we have a statistic, following an *F* distribution, to be used in testing the hypothesis of equality of the slopes of the *K* population regression hyperplanes.

Similarly, for the second test, we find

$$\begin{aligned} S_5 &= S_3 - S_2 \\ &= (T_{00} - W_{00}) - [\mathbf{T}_{0p}\mathbf{T}_{pp}^{-1}\mathbf{T}_{p0} - \mathbf{W}_{0p}\mathbf{W}_{pp}^{-1}\mathbf{W}_{p0}] \end{aligned} \tag{3.38}$$

and then form the ratio

$$F_\mu = \frac{S_5/\nu_5}{S_2/\nu_2} \tag{3.39'}$$

which follows an F distribution and is appropriate for testing the equality of the K population means adjusted for differences in the covariate centroids.

The appropriate n.d.f.'s to be used in Eqs. 3.37′ and 3.39′ can be determined in much the same way as was outlined on p. 47 for the one-covariate case. We have only to keep in mind that for each regression coefficient in the earlier case we now have p regression coefficients—one for each covariate. Of course, we still have just one additive constant a for each type of regression equation. Thus, for S_3 (based on the total-sample regression equation) we have estimated one additive constant and p regression coefficients, and hence the n.d.f. for S_3 is $\nu_3 = N - (p + 1)$. Similarly, for S_1, we have $n_k - (p + 1)$ degrees of freedom for the kth term (representing the kth group) in the second member of Eq. 3.33. Hence,

$$\nu_1 = \sum_{k=1}^{K} [n_k - (p + 1)] = N - (p + 1)K$$

In getting S_2, we have estimated K additive constants (one for each group) and p regression coefficients (one for each covariate, common to all K groups). Hence

$$\nu_2 = N - K - p$$

The degrees of freedom for S_4 and S_5 are given by the differences $\nu_2 - \nu_1$ and $\nu_3 - \nu_2$, respectively, as before.

The several n.d.f.'s derived above are listed in Table 3.4. It should be noted that these νs reduce to those shown in Table 3.1 when $p = 1$.

TABLE 3.4 N.D.F.'s for the Various Residual Sums-of-Squares.

Sum-of-Squares	Defining Equation	N.D.F.
S_1	3.33	$\nu_1 = N - (p + 1)K$
S_2	3.34	$\nu_2 = N - K - p$
S_3	3.35	$\nu_3 = N - (p + 1)$
S_4	3.36	$\nu_4\ (= \nu_2 - \nu_1) = p(K - 1)$
S_5	3.38	$\nu_5\ (= \nu_3 - \nu_2) = K - 1$

When these n.d.f.'s are appropriately substituted in Eq. 3.37′ and 3.39′, we get the complete formulas for the two F-ratios needed in our tests. Thus,

$$F_\beta = \frac{S_4/p(K - 1)}{S_1/[N - (p + 1)K]} \tag{3.37}$$

with S_4 and S_1 as defined in Eqs. 3.36 and 3.33, respectively. And

$$F_\mu = \frac{S_5/(K - 1)}{S_2/(N - K - p)} \tag{3.39}$$

with S_5 and S_2 as defined in Eqs. 3.38 and 3.34, respectively.

3.4 NUMERICAL EXAMPLE

We illustrate the computations for covariance analysis with more than one covariate by means of a fictitious example involving two covariates and two groups. The various totals, means, and sums of squares and products of raw scores are shown in Table 3.5.

TABLE 3.5 Sums, Means, Sums-of-Squares, and Products (Raw Scores) of Criterion Variable X_0 and Two Covariates X_1 and X_2 for Two Groups and Total Sample.

$(\alpha =)$	$(\beta =)$	0	1	2
	$X_{\beta 1.}$	90	119	446
	$X_{\beta 2.}$	176	173	455
	$X_{\beta ..}$	266	292	901
	$\bar{X}_{\beta 1.}$	1.667	2.204	8.259
	$\bar{X}_{\beta 2.}$	3.088	3.035	7.982
	$\bar{X}_{\beta ..}$	2.396	2.631	8.117
	$\sum_i X_{01i} X_{\beta 1i}$	252	273	702
0	$\sum_i X_{02i} X_{\beta 2i}$	824	752	1324
	$\sum_k \sum_i X_{0ki} X_{\beta ki}$	1076	1025	2026
	$\sum_i X_{11i} X_{\beta 1i}$	273	435	894
1	$\sum_i X_{12i} X_{\beta 2i}$	752	831	1351
	$\sum_k \sum_i X_{1ki} X_{\beta ki}$	1025	1266	2245
	$\sum_i X_{21i} X_{\beta 1i}$	702	894	3952
2	$\sum_i X_{22i} X_{\beta 2i}$	1324	1351	3895
	$\sum_k \sum_i X_{2ki} X_{\beta ki}$	2026	2245	7847
		$n_1 = 54$	$n_2 = 57$	$N = 111$

In this table, α and β are used as generic subscripts for variable number ($\alpha = 0, 1, 2$; $\beta = 0, 1, 2$). In each block of three rows, the first two rows refer to the relevant quantities for groups 1 and 2, respectively, and the third, for the total composite sample. The first block shows the totals for X_0, X_1, and X_2, in the three columns labeled $\beta = 0$, 1, and 2, respectively. The next block shows the two group means and the grand mean for each of the three variables. The third block shows the raw-score sums of products of X_0 with each of the three variables, in the appropriate columns. Thus, the first-column entries in this block are the three sums of X_0^2; the second-column entries are sums of X_0X_1; and the third-column entries are sums of X_0X_2. The last two blocks show the raw-score sums of products of X_1 with each of the three variables, and of X_2 with each of the three variables, respectively.

Substituting in Eqs. 3.26 to 3.28 from Table 3.5, we compute the elements of the several SSCP matrices, $\mathbf{S}_1$, $\mathbf{S}_2$, $\mathbf{W}$, and $\mathbf{T}$. The results are:

$$\mathbf{S}_1 = \left[\begin{array}{c|cc} 102.00 & 74.56 & -41.33 \\ \hline 74.56 & 172.76 & -88.85 \\ -41.33 & -88.85 & 268.37 \end{array}\right]$$

$$\mathbf{S}_2 = \left[\begin{array}{c|cc} 280.56 & 217.94 & -80.91 \\ \hline 217.94 & 305.93 & -29.97 \\ -80.91 & -29.97 & 262.98 \end{array}\right]$$

$$\mathbf{W} = \left[\begin{array}{c|cc} 382.56 & 292.50 & -122.24 \\ \hline 292.50 & 478.69 & -118.82 \\ -122.24 & -118.82 & 531.35 \end{array}\right]$$

and

$$\mathbf{T} = \left[\begin{array}{c|cc} 438.56 & 325.25 & -133.15 \\ \hline 325.25 & 497.86 & -125.20 \\ -133.15 & -125.20 & 533.48 \end{array}\right]$$

The next step is to compute the various quadratic forms with the general schema

$$\mathbf{S}_{0p}\mathbf{S}_{pp}^{-1}\mathbf{S}_{p0}$$

as appearing in Eq. 3.32. These are the quantities to be subtracted from the corresponding sums-of-squares of X_0 to yield the various residual sums-of-squares. We illustrate the computation of $\mathbf{W}_{0p}\mathbf{W}_{pp}^{-1}\mathbf{W}_{p0}$ in detail and list the values of the remaining quadratic forms.

By inverting $\mathbf{W}_{pp}$ (the 2×2 matrix forming the lower right-hand portion of $\mathbf{W}$) in accordance with the rule stated in Chapter 2 (Eq. 2.16), we find

$$\mathbf{W}_{pp}^{-1} = \begin{bmatrix} 2.2120 & .4946 \\ .4946 & 1.9928 \end{bmatrix} \times 10^{-3}$$

(The factoring out of 10^{-3} is for the purpose of avoiding a string of zeros preceding the first significant digit in some of the elements. In general, especially when the sample size is large, the elements of the several inverse matrices will be quite small. It is generally good computational practice to factor out a suitable negative power of 10 so that most of the recorded diagonal elements lie between 1 and 10.)

We then form the product

$$\begin{aligned} \mathbf{W}_{0p}\mathbf{W}_{pp}^{-1} &= [292.50, \ -122.24] \begin{bmatrix} 2.2120 & .4946 \\ .4946 & 1.9928 \end{bmatrix} \times 10^{-3} \\ &= [.5866, \ -.0989] \end{aligned}$$

and finally the triple product

$$\begin{aligned} (\mathbf{W}_{0p}\mathbf{W}_{pp}^{-1})\mathbf{W}_{p0} &= [.5866, \ -.0989] \begin{bmatrix} 292.50 \\ -122.24 \end{bmatrix} \\ &= 183.67 \end{aligned}$$

As an arithmetic check, we may compute the quadratic form by the alternative method suggested in Chapter Two (p. 19). That is, we first compute the matrix product

$$\begin{aligned} \mathbf{W}_{p0}\mathbf{W}_{0p} &= \begin{bmatrix} 292.50 \\ -122.24 \end{bmatrix} [292.50, \ -122.24] \\ &= \begin{bmatrix} 85.568 & -35.755 \\ -35.755 & 14.943 \end{bmatrix} \times 10^{3} \end{aligned}$$

We then multiply each element of this matrix by the corresponding element of $\mathbf{W}_{pp}^{-1}$, and form the algebraic sum of the resulting products. Thus,

$$\begin{aligned} \mathbf{W}_{0p}\mathbf{W}_{pp}^{-1}\mathbf{W}_{p0} &= (85.568)(2.2120) + (-35.755)(.4946) \\ &\quad + (-35.755)(.4946) + (14.943)(1.9928) \\ &= 183.69 \end{aligned}$$

which equals the previously obtained value (183.67) within rounding error.

Similar computations yield values for the remaining three (or, in general, $K + 1$) quadratic forms, as follows:

$$\mathbf{S}_{1,0p}\mathbf{S}_{1,pp}^{-1}\mathbf{S}_{1,p0} = 32.21$$

$$\mathbf{S}_{2,0p}\mathbf{S}_{2,pp}^{-1}\mathbf{S}_{2,p0} = 168.91$$

and

$$\mathbf{T}_{0p}\mathbf{T}_{pp}^{-1}\mathbf{T}_{p0} = 217.75$$

We now have all the intermediate quantities needed for conducting the first significance test in our covariance analysis: that of the hypothesis of equal slopes of the population regression hyperplanes. (Actually, the last quantity listed above, $\mathbf{T}_{0p}\mathbf{T}_{pp}^{-1}\mathbf{T}_{p0}$, is not needed unless and until we conduct the second significant test.) Using Eq. 3.33, we find

$$\begin{aligned} S_1 &= \sum_{k=1}^{2} [\mathbf{S}_{k,00} - \mathbf{S}_{k,0p}\mathbf{S}_{k,\ pp}^{-1}\mathbf{S}_{k,p0}] \\ &= (102.00 - 32.21) + (280.56 - 168.91) \\ &= 69.79 + 111.65 = 181.44 \end{aligned}$$

Next, following Eq. 3.34, we compute

$$\begin{aligned} S_2 &= W_{00} - \mathbf{W}_{0p}\mathbf{W}_{pp}^{-1}\mathbf{W}_{p0} \\ &= 382.56 - 183.67 = 198.89 \end{aligned}$$

Then, we form the differences

$$S_4 = S_2 - S_1 = 198.89 - 181.44 = 17.45$$

Since W_{00} is involved in both S_2 and S_1 as an additive term, it cancels out in the difference $S_2 - S_1$, which could hence have been obtained more directly as given in Eq. 3.36:

$$\begin{aligned} S_4 &= \sum_{k=1}^{2} \mathbf{S}_{k,0p}\mathbf{S}_{k,\ pp}^{-1}\mathbf{S}_{k,p0} - \mathbf{W}_{0p}\mathbf{W}_{pp}^{-1}\mathbf{W}_{p0} \\ &= (32.21 + 168.91) - 183.67 = 17.45 \end{aligned}$$

However, it is of interest actually to compute both S_1 and S_2, because each has a meaning as one kind of residual sum-of-squares.

The appropriate F-ratio is then computed in accordance with Eq. 3.37

$$\begin{aligned} F_\beta &= \frac{S_4/\,p(K-1)}{S_1/[N - K(p+1)]} \\ &= \frac{17.45/(2)(2-1)}{181.44/[111 - (2)(2+1)]} = 5.05 \end{aligned}$$

with $\text{n.d.f.}_1 = 2$ and $\text{n.d.f.}_2 = 105$. This value is significant at the 1% level, so we would reject the hypothesis that the two population regression hyperplanes are parallel to each other. Hence, the second significance test is not valid, and we would have to use other methods to investigate whether the group means on the criterion variable differ significantly after covariance adjustments have been made on the two covariates. However, we carry out the second test below, for illustrative purposes. Substituting in Eq. 3.35 the value of $\mathbf{T}_{0p}\mathbf{T}_{pp}^{-1}\mathbf{T}_{p0}$ computed above, we find

$$S_3 = 438.56 - 217.75 = 220.81$$

With this and the previously determined value of S_2, we get, in accordance with Eq. 3.38,

$$S_5 = S_3 - S_2 = 220.81 - 198.89 = 21.92$$

The F-ratio for testing the hypothesis of equal adjusted criterion-variable means in the two populations is then computed from Eq. 3.39.

$$F_\mu = \frac{S_5/(K-1)}{S_2/(N-K-p)} = \frac{21.92/(2-1)}{198.89/(111-2-2)} = 11.78$$

which, with n.d.f.$_1$ = 1, n.d.f.$_2$ = 107, is significant at the 1% level.

Thus, if this test were a valid one (that is, if the null hypothesis concerning parallelism of the population regression hyperplanes had not been rejected), we would have concluded that the group means on X_0 are significantly different after covariance adjustments for differences on X_1 and X_2 are made. In that case we would need to compute the adjusted group means on X_0 to determine the direction of the difference.

For this purpose we construct, for each group, a multiple regression equation of X_0 and X_1 and X_2, using the within-groups regression coefficients. These are the elements of the row vector $\mathbf{W}_{0p}\mathbf{W}_{pp}^{-1}$ obtained earlier as the first step in computing the quadratic form $\mathbf{W}_{0p}\mathbf{W}_{pp}^{-1}\mathbf{W}_{p0}$. For the present example, $\mathbf{W}_{0p}\mathbf{W}_{pp}^{-1} = [.5866, -.0989]$, as shown on p. 58. The elements of this vector are the coefficients to be used in the multiple regression equations for estimating what the group means on X_0 might have been if the group means on the covariates had all been equal to the grand means. With p covariates, the general equation is

$$\bar{X}^*_{0k.} = \bar{X}_{0k.} + b_{w1}(\bar{X}_{1..} - \bar{X}_{1k.}) + b_{w2}(\bar{X}_{2..} - \bar{X}_{2k.}) + \cdots + b_{wp}(\bar{X}_{p..} - \bar{X}_{pk.}); \quad k = 1, 2, \ldots, k \tag{3.40}$$

In the present example, appropriate substitution in Eq. 3.40 yields

$$\bar{X}^*_{01.} = 1.667 + (.5866)(2.631 - 2.204) + (-.0989)(8.117 - 8.259) = 1.932$$

and

$$\bar{X}^*_{02.} = 3.088 + (.5866)(2.631 - 3.035) + (-.0989)(8.117 - 7.982) = 2.838$$

Thus, if the test were a valid one, the conclusion would be that group 2 would have had a higher criterion-variable mean that group 1, even if the two groups had had equal means on the two covariates, respectively.

EXERCISE

Fifteen subjects were randomly assigned to three treatment groups, five *S*s each, in an experiment using X_0 as dependent variable and X_1 and X_2 as covariates. The data were as shown below.

Treatment 1			*Treatment 2*			*Treatment 3*		
X_0	X_1	X_2	X_0	X_1	X_2	X_0	X_1	X_2
6	3	5	11	2	8	20	2	12
6	9	7	14	7	8	21	9	10
8	16	10	18	13	15	25	14	18
13	19	17	18	19	14	21	20	18
12	24	15	20	23	17	29	23	20

Carry out an analysis of covariance for these data. Several intermediate results are given below so that you may check your arithmetic step by step.

$$\mathbf{S}_1 = \begin{bmatrix} 44.0 & 96.0 & 67.0 \\ 96.0 & 274.8 & 154.2 \\ 67.0 & 154.2 & 104.8 \end{bmatrix} \qquad \mathbf{S}_2 = \begin{bmatrix} 52.8 & 119.2 & 57.6 \\ 119.2 & 292.8 & 130.4 \\ 57.6 & 130.4 & 69.2 \end{bmatrix}$$

$$\mathbf{S}_3 = \begin{bmatrix} 56.8 & 88.4 & 48.4 \\ 88.4 & 285.2 & 125.2 \\ 48.4 & 125.2 & 75.2 \end{bmatrix} \qquad \mathbf{T} = \begin{bmatrix} 657.73 & 281.93 & 343.13 \\ 281.93 & 857.73 & 405.53 \\ 343.13 & 405.53 & 308.93 \end{bmatrix}$$

$$S_1 = w_{00} - \sum_{k=1}^{3} \mathbf{S}_{k,0p}\mathbf{S}_{k,pp}^{-1}\mathbf{S}_{k,p0}$$

$$= 153.60 - [42.97 + 50.35 + 31.95] = 28.33$$

$$S_2 = w_{00} - \mathbf{W}_{0p}\mathbf{W}_{pp}^{-1}\mathbf{W}_{p0} = 153.60 - 122.18 = 31.42$$

$$S_3 = T_{00} - \mathbf{T}_{0p}\mathbf{T}_{pp}^{-1}\mathbf{T}_{p0} = 657.73 - 468.43 = 189.30$$

Chapter Four

Multivariate Significance Tests of Group Differences

The technique of covariance analysis with multiple *covariates*, discussed in the previous chapter, involved only one dependent variable on which the groups were compared. We now turn to "genuine" multivariate methods involving a multiplicity of *dependent* variables.

Many problems in educational and psychological research require the simultaneous use of several dependent variables. For example, the outcomes of an instructional program in the language arts may be assessed in terms of vocabulary acquisition, reading comprehension, effectiveness of verbal and written communication, ability to discriminate between good and poor literary style, and so on. Suppose that two or more such instructional programs are to be compared in terms of these several criteria. We may, of course, be interested in the differences among the programs on each of these criteria, separately, in its own right. In that case, we might conduct a series of univariate significance tests (t-tests or F-tests)—one test for each criterion. But what if the outcomes (of two programs, for example) differed only slightly on each criterion, so that none of the univariate tests detected a significant difference? Would we conclude that the two programs seem no different in toto even if the differences on the several criteria were slight but consistent? Or do we conclude that the small differences, taken together, point to a real difference? Alternatively, it may be that the difference is significant (at a given α level) for some criteria but not others. Can we take the significant and nonsignificant differences at their face values? Or do we have to concede that, given a large number of criterion variables, we would expect to find significant differences on a few of them by pure chance (one out of twenty if $\alpha = .05$, for example)?

In order to answer the foregoing questions, we need to have a method (or several methods) for studying the differences among the groups in terms of many dependent variables *considered simultaneously*. The multivariate significance tests described in this chapter are among those that serve this purpose. If we wish also to know along what dimensions do the stable differences occur, we shall need to invoke the more refined technique of *discriminant analysis*, described in Chapter Six.

4.1 THE MULTIVARIATE NORMAL DISTRIBUTION

Throughout the sequel, a basic assumption needed for strict validity of the significance tests is that the variables under study follow a *multivariate normal distribution*. For the case of two variables X_1 and X_2, the bivariate normal density function is

$$\phi(X_1, X_2) = \frac{1}{2\pi\sigma_1\sigma_2\sqrt{1-\rho^2}} \exp\left[\frac{-1}{2(1-\rho^2)}\left\{\frac{(X_1-\mu_1)^2}{\sigma_1^{\,2}} + \frac{(X_2-\mu_2)^2}{\sigma_2^{\,2}} - 2\rho\frac{(X_1-\mu_1)(X_2-\mu_2)}{\sigma_1\sigma_2}\right\}\right], \tag{4.1}$$

where μ_i and $\sigma_i^{\,2}$ are the mean and variance of X_i ($i = 1, 2$), and ρ is the correlation coefficient. The surface represented by this equation resembles a bell-shaped mound, distorted by being stretched out in one direction and compressed in the direction perpendicular to the first—the degrees of distortion being dependent on the value of ρ and the ratio σ_1/σ_2. Figure 4.1 depicts the general appearance of the bivariate normal surface for $\rho = .60$ and $\sigma_1/\sigma_2 = 1$. Since such a perspective drawing is difficult to construct and conveys little or no quantitative information, it is customary to represent a bivariate normal surface by drawing an arbitrarily selected *isodensity contour*—that is, the cross section of the surface made by a plane parallel to the (X_1, X_2) plane. The equation for such a contour curve is obtained by setting the expression in braces in Eq. 4.1 equal to a positive constant C. The smaller this constant, the larger is the altitude (representing density) along the contour, since the altitude is proportional to the negative exponential of $C/2(1 - \rho^2)$.

By use of elementary analytic geometry, it can be shown that the equation in question, namely

$$\frac{(X_1-\mu_1)^2}{\sigma_1^{\,2}} + \frac{(X_2-\mu_2)^2}{\sigma_2^{\,2}} - 2\rho\frac{(X_1-\mu_1)(X_2-\mu_2)}{\sigma_1\sigma_2} = C \tag{4.2}$$

represents an ellipse with center at the point (μ_1, μ_2), which is called the *centroid* of the bivariate population, and major or minor axis along the line

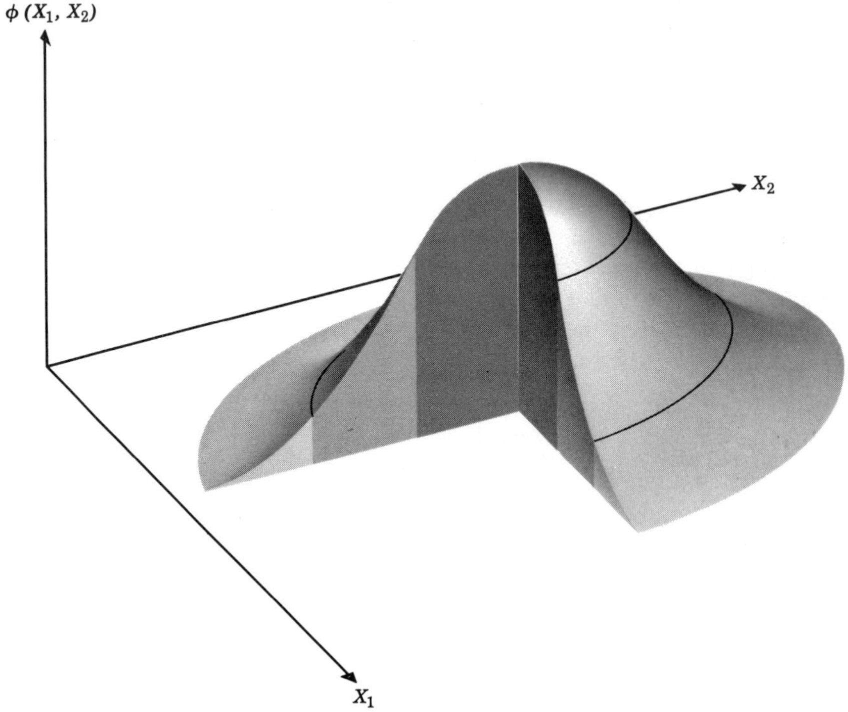

FIGURE 4.1 ***A bivariate normal density surface.***

passing through this point and making the following angle with the positive X_1-axis:

$$\theta = \begin{cases} \frac{1}{2}\ \text{Arctan}\ \dfrac{2\rho\sigma_1\sigma_2}{\sigma_1^2 - \sigma_2^2} & \text{when } \sigma_1 \neq \sigma_2 \\ 45^\circ & \text{when } \sigma_1 = \sigma_2 \end{cases}$$

(This line contains the major axis if $\rho > 0$, the minor axis if $\rho < 0$.) Since the angle θ depends only on σ_1, σ_2, and ρ, and is independent of C, it follows that taking various values of C (in other words, making cross sections of the density surface with planes at various elevations) generates a family of concentric ellipses, all with the same orientation. Thus, a representation by a series of contour lines of the bivariate normal surface depicted in Fig. 4.1 would look something like Fig. 4.2. If the reader imagines piling up a large number of elliptic discs of sizes as shown in Fig. 4.2 and thicknesses of about $\frac{1}{8}$ inch each, he will have a good idea of what a bivariate normal surface looks like.

EXAMPLE 4.1. Suppose that X_1 and X_2 follow a bivariate normal distribution with $\mu_1 = 15$, $\mu_2 = 20$, $\sigma_1 = \sigma_2 = 5$, and $\rho = .60$. Then the expression in braces in Eq. 4.1 for the density function becomes

$$(X_1 - 15)^2/25 + (X_2 - 20)^2/25 - 2(.60)(X_1 - 15)(X_2 - 20)/25$$

Setting this expression equal to any positive constant C gives the equation of an isodensity contour ellipse. Thus, treating C as a parameter, the equation

$$(X_1 - 15)^2 + (X_2 - 20)^2 - 1.2(X_1 - 15)(X_2 - 20) = 25C$$

represents a family of concentric ellipses centered at (15, 20), with major axis making a 45° angle with the X_1-axis, because $\sigma_1 = \sigma_2$ and $\rho > 0$. Figure 4.2 shows

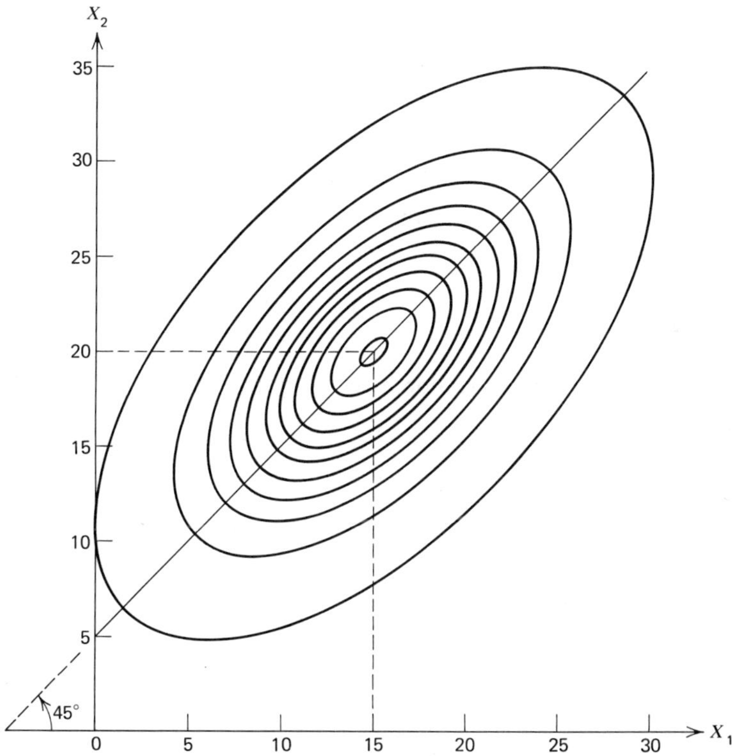

FIGURE 4.2 ***Several members of the family of ellipses $(X_1 - 15)^2 + (X_2 - 20)^2 - 1.2(X_1 - 15)(X_2 - 20) = 25C$ with selected values for the parameter C.***

several members of this family of ellipses for selected values of C, the first three of which are 5.89, 2.95, and 2.06, and the last three are .29, .13, and .01.

Before giving the equation for the multivariate normal density function for more than two variables, it is convenient to rewrite the quantities in Eq. 4.1 in matrix notation. The generalization to p variables will then be almost self-evident. We first define the *variance-covariance matrix*, or *dispersion matrix*, for a bivariate population as follows:

$$\mathbf{\Sigma}_2 = \begin{bmatrix} \sigma_1^2 & \rho\sigma_1\sigma_2 \\ \rho\sigma_2\sigma_1 & \sigma_2^2 \end{bmatrix} \tag{4.3}$$

The determinant of this matrix is

$$|\mathbf{\Sigma}_2| = \sigma_1^2\sigma_2^2(1 - \rho^2) \tag{4.4}$$

Hence, the inverse of $\mathbf{\Sigma}_2$ is given, in accordance with Eq. 2.16, by

$$\begin{aligned} \mathbf{\Sigma}_2^{-1} &= 1/\sigma_1^2\sigma_2^2(1 - \rho^2) \begin{bmatrix} \sigma_2^2 & -\rho\sigma_1\sigma_2 \\ -\rho\sigma_2\sigma_1 & \sigma_1^2 \end{bmatrix} \\ &= 1/(1 - \rho^2) \begin{bmatrix} 1/\sigma_1^2 & -\rho/\sigma_1\sigma_2 \\ -\rho/\sigma_2\sigma_1 & 1/\sigma_2^2 \end{bmatrix} \end{aligned} \tag{4.5}$$

It is now readily seen that the expression in the exponent of Eq. 4.1, apart from the factor $-1/2$, is equivalent to the quadratic form

$$[X_1 - \mu_1, X_2 - \mu_2]\, \mathbf{\Sigma}_2^{-1} \begin{bmatrix} X_1 - \mu_1 \\ X_2 - \mu_2 \end{bmatrix}$$

We let χ^2 symbolize this expression, which, on introducing

$$\mathbf{x}' = [X_1 - \mu_1, X_2 - \mu_2] \tag{4.6}$$

may be written as

$$\chi^2 = \mathbf{x}'\mathbf{\Sigma}_2^{-1}\mathbf{x} \tag{4.7}$$

Next, the constant factor $1/2\pi\sigma_1\sigma_2\sqrt{1 - \rho^2}$ of the expression for $\phi(X_1, X_2)$ may be written as $(2\pi)^{-1}|\mathbf{\Sigma}_2|^{-1/2}$, since $\sigma_1\sigma_2\sqrt{1 - \rho^2}$ is the square root of $|\mathbf{\Sigma}_2|$, as seen from Eq. 4.4.

Thus, Eq. 4.1, specifying the bivariate normal density function, may be written compactly as

$$\phi(X_1, X_2) = (2\pi)^{-1}|\mathbf{\Sigma}_2|^{-1/2} \exp(-\chi^2/2) \tag{4.8}$$

with $\mathbf{\Sigma}_2$ and χ^2 defined by Eqs. 4.3 and 4.7, respectively.

The extension to the p-variate case is now almost obvious. If we define the dispersion matrix as

$$\mathbf{\Sigma} = \begin{bmatrix} \sigma_1{}^2 & \rho_{12}\sigma_1\sigma_2 & \cdots & \rho_{1p}\sigma_1\sigma_p \\ \rho_{21}\sigma_2\sigma_1 & \sigma_2{}^2 & \cdots & \rho_{2p}\sigma_2\sigma_p \\ \vdots & & \ddots & \vdots \\ \rho_{p1}\sigma_p\sigma_1 & \rho_{p2}\sigma_p\sigma_2 & \cdots & \sigma_p{}^2 \end{bmatrix} \tag{4.9}$$

where $\sigma_i{}^2$ is the variance of X_i, and ρ_{ij} $(i \neq j)$ is the coefficient of correlation between X_i and X_j, and let

$$\chi^2 = \mathbf{x}'\mathbf{\Sigma}^{-1}\mathbf{x} \tag{4.10}$$

with

$$\mathbf{x}' = [X_1 - \mu_1, X_2 - \mu_2, \ldots, X_p - \mu_p] \tag{4.11}$$

then the p-variate normal density function is given by

$$\phi(X_1, X_2, \ldots, X_p) = K \exp(-\chi^2/2)$$

where only the constant K remains to be determined.

Examination of Eq. 4.8 alone is insufficient to permit our inferring what powers of 2π and of $|\mathbf{\Sigma}|$ are involved in K. But a comparison with the univariate normal density function,

$$\phi(X) = \frac{1}{\sqrt{2\pi}\,\sigma} \exp[-(X - \mu)^2/2\sigma^2] \tag{4.12}$$

in which the normalizing constant is

$$(2\pi)^{-1/2}(\sigma^2)^{-1/2}$$

leads to the following conjecture: the power of 2π seems to be $-1/2$ times the number of variables, while the power of $|\mathbf{\Sigma}|$ (which reduces to σ^2 in the univariate case) is $-1/2$ regardless of the number of variables. We thus infer that, for the p-variate case,

$$K = (2\pi)^{-p/2}|\mathbf{\Sigma}|^{-1/2}$$

This conjecture proves to be correct (see, for example, Anderson, 1958, p. 17), and we have, as the complete equation for a p-variate normal density function,

$$\phi(X_1, X_2, \ldots, X_p) = (2\pi)^{-p/2}|\mathbf{\Sigma}|^{-1/2} \exp(-\chi^2/2) \tag{4.13}$$

with $\mathbf{\Sigma}$ and χ^2 defined by Eqs. 4.9 and 4.10, respectively. We shall denote this distribution by the symbol $\mathrm{N}(\boldsymbol{\mu}, \mathbf{\Sigma})$, meaning a multivariate normal distribution with centroid $\boldsymbol{\mu} = [\mu_1, \mu_2, \ldots, \mu_p]'$ and dispersion matrix $\mathbf{\Sigma}$.

The close analogy between the p-variate normal density function (4.13) for $N(\boldsymbol{\mu}, \boldsymbol{\Sigma})$ and the familiar univariate normal density function (4.12) for $N(\mu, \sigma^2)$ should be noted. In particular, we point out that the expression

$$-\chi^2/2 = -(\mathbf{x}'\boldsymbol{\Sigma}^{-1}\mathbf{x})/2$$

occurring in the exponent in Eq. 4.13 is a "natural" generalization of the expression

$$-(X - \mu)^2/2\sigma^2 = -[(X - \mu)(\sigma^2)^{-1}(X - \mu)]/2$$

which occurs in the univariate case.

The reason for denoting the expression $\mathbf{x}'\boldsymbol{\Sigma}^{-1}\mathbf{x}$ as χ^2 may best be understood with reference to the special case in which the p variables are pairwise uncorrelated, that is, when $\rho_{ij} = 0$ for all $i \neq j$. For variables whose joint distribution is multivariate normal, uncorrelatedness implies statistical independence—that is, that their joint density function can be expressed as a product of their respective marginal density functions (see Exercise 1 on p. 72). In this case, the dispersion matrix given by Eq. 4.9 reduces to a diagonal matrix:

$$\boldsymbol{\Sigma} = \begin{bmatrix} \sigma_1^{\,2} & 0 & \cdots & 0 \\ 0 & \sigma_2^{\,2} & \cdots & 0 \\ \vdots & & \ddots & \vdots \\ 0 & 0 & \cdots & \sigma_p^{\,2} \end{bmatrix}$$

and Eq. 4.10 for χ^2 reduces to

$$\chi^2 = [X_1 - \mu_1, X_2 - \mu_2, \cdots, X_p - \mu_p]$$

$$\times \begin{bmatrix} \sigma_1^{\,-2} & & & 0 \\ & \sigma_2^{\,-2} & & \\ & & \ddots & \\ 0 & & & \sigma_p^{-2} \end{bmatrix} \begin{bmatrix} X_1 - \mu_1 \\ X_2 - \mu_2 \\ \vdots \\ X_p - \mu_p \end{bmatrix}$$

$$= \sum_{i=1}^{p} (X_i - \mu_i)^2/\sigma_i^{\,2} \tag{4.14}$$

Each term in this sum is the square of a unit normal variate (a z^2 in customary notation), and is hence a chi-square variate with 1 degree of freedom (d.f.). Then, using the additivity property of independent chi-square variates, we obtain the important conclusion that, when the X_i are statistically independent, the sum denoted by χ^2 is in fact a chi-square variate with p d.f.'s. In the next chapter it will be shown that when the X_i are not independently distributed, a new set of variables Y_i can be derived that *are* independently distributed, and for which the χ^2 quantity is equivalent to that for the original X_i's. Pending this demonstration, therefore, we have proved the following

Theorem. Given a p-variate normal population $\mathrm{N}(\boldsymbol{\mu}, \boldsymbol{\Sigma})$ with density function

$$\phi(X_1, X_2, \ldots, X_p) = (2\pi)^{-p/2}|\boldsymbol{\Sigma}|^{-1/2} \exp(-\chi^2/2)$$

the quantity $\chi^2 = \mathbf{x}'\boldsymbol{\Sigma}^{-1}\mathbf{x}$ in the exponent is a chi-square variate with p d.f.'s.

The importance of this theorem lies in its enabling us to determine the multivariate analogues of the various centile points of a normal distribution, *without* the aid of special tables of multivariate normal integrals. We return to the bivariate case to describe the procedure in detail.

Regions Enclosing Specified Percentages of a Bivariate Normal Population. Recall, from Eq. 4.2, that an isodensity contour ellipse of a bivariate normal distribution may be specified by choosing a suitable positive constant C, and setting

$$\frac{(X_1 - \mu_1)^2}{\sigma_1{}^2} + \frac{(X_2 - \mu_2)^2}{\sigma_2{}^2} - 2\rho\frac{(X_1 - \mu_1)(X_2 - \mu_2)}{\sigma_1\sigma_2} = C$$

In our present notation, the left-hand side of this equation is equal to $(1 - \rho^2)\mathbf{x}'\boldsymbol{\Sigma}^{-1}\mathbf{x}$, or $(1 - \rho^2)\chi^2$. Hence, an isodensity ellipse may also be specified by choosing a particular value for χ^2 in the equation

$$\mathbf{x}'\boldsymbol{\Sigma}^{-1}\mathbf{x} = \frac{1}{1 - \rho^2}\left[\frac{(X_1 - \mu_1)^2}{\sigma_1{}^2} + \frac{(X_2 - \mu_2)^2}{\sigma_2{}^2} - 2\rho\frac{(X_1 - \mu_1)(X_2 - \mu_2)}{\sigma_1\sigma_2}\right]$$

$$= \chi^2 \tag{4.15}$$

Note that this corresponds, in the univariate normal case, to specifying two points with equal ordinates (representing density) by fixing the value of z^2 (which is $\chi_1{}^2$) in

$$\frac{(X - \mu)^2}{\sigma^2} = z^2$$

For any choice of z, say z_0, the abscissas of these points are $X = \mu - z_0\sigma$ and $X = \mu + z_0\sigma$, and the interval along the X-axis between these points is such that the area under the normal curve in this interval is equal to some particular value between 0 and 1, determined by the value z_0. This area represents the probability that a random observation X drawn from $\mathrm{N}(\mu, \sigma^2)$ will fall in the interval $[\mu - z_0\sigma, \mu + z_0\sigma]$. For example, with $z^2 = (1.645)^2 = 2.706$, we have

$$p(\mu - 1.645\sigma \le X \le \mu + 1.645\sigma) = .90$$

or, if we denote the interval $[\mu - 1.645\sigma, \mu + 1.645\sigma]$ by $R_1(2.706)$, where the subscript indicates that we refer to the univariate case, and the argument is the z^2 value, we may write:

$$p(X \in R_1(2.706)) = .90$$

where the symbol "$\in$" is read "belongs to" or "is a member of".

In the same way, for the bivariate case, the ellipse specified by a particular choice of χ^2 value defines a region in the (X_1, X_2)-plane such that the point (X_1, X_2), representing a random observation from $N(\boldsymbol{\mu}, \boldsymbol{\Sigma})$, has a specified probability of falling in this region. This probability, according to the theorem stated earlier, is equal to the probability that a chi-square variate with 2 d.f.'s has a value not exceeding the value selected for χ^2. For example, from Table E.3, we find that $p(\chi^2 \leq 4.605) = .90$. Hence, if we denote by $R_2(4.605)$ the region bounded by and including the ellipse

$$\frac{1}{1-\rho^2}\left[\frac{(X_1-\mu_1)^2}{\sigma_1{}^2} + \frac{(X_2-\mu_2)^2}{\sigma_2{}^2} - 2\rho\frac{(X_1-\mu_1)(X_2-\mu_2)}{\sigma_1\sigma_2}\right] = 4.605$$

we can assert that

$$p[(X_1, X_2) \in R_2(4.605)] = .90$$

Geometrically, this means that the volume of the three-dimensional region $R_2(4.605)$ between the bivariate normal surface and the (X_1, X_2)-plane, bounded laterally by the right elliptic cylinder based on the ellipse $\mathbf{x}'\boldsymbol{\Sigma}^{-1}\mathbf{x} = 4.605$, is equal to .90. The statistical implication is that if we draw many independent observations at random from a bivariate normal population $N(\boldsymbol{\mu}, \boldsymbol{\Sigma})$ and each timc dctermine a point with coordinates (X_1, X_2) given by the paired observation, then 90% of these points will lie in the region $R_2(4.605)$, that is, inside or on the ellipse $\mathbf{x}'\boldsymbol{\Sigma}^{-1}\mathbf{x} = 4.605$.

EXAMPLE 4.2. Consider once again the bivariate normal population of Example 4.1, with $\boldsymbol{\mu}' = [15, 20]$ and

$$\boldsymbol{\Sigma} = \begin{bmatrix} 25 & 15 \\ 15 & 25 \end{bmatrix} \qquad \text{(that is, } \sigma_1 = \sigma_2 = 5,\ \rho = .60\text{)}$$

We are now interested in the expression $\mathbf{x}'\boldsymbol{\Sigma}^{-1}\mathbf{x}$ which, according to Eq. 4.15, is $1/(1 - \rho^2)$ times the expression used in Example 4.1. Hence,

$$\begin{aligned}\mathbf{x}'\boldsymbol{\Sigma}^{-1}\mathbf{x} &= [1/(1-\rho^2)](1/25)[(X_1-15)^2 + (X_2-20)^2 \\ &\qquad - 1.2(X_1-15)(X_2-20)] \\ &= (1/16)[(X_1-15)^2 + (X_2-20)^2 \\ &\qquad - 1.2(X_1-15)(X_2-20)]\end{aligned}$$

or, using deviation-score notation ($X_1 - 15 = x_1$, $X_2 - 20 = x_2$),

$$\mathbf{x}'\boldsymbol{\Sigma}^{-1}\mathbf{x} = (1/16)(x_1^2 + x_2^2 - 1.2x_1x_2)$$

Geometrically, this is equivalent to having translated the coordinate axes so that the new origin is at the population centroid. In this new reference system the 90% ellipse (that is, the ellipse inside or on which 90% of the population lies) is specified by equating the above expression to 4.605, which is the value of a chi-square variate with 2 d.f.'s that is exceeded 10% of the time:

$$x_1^2 + x_2^2 - 1.2x_1x_2 = (16)(4.605) = 73.680$$

The larger of the two ellipses shown in Fig. 4.3 is the 90% ellipse.

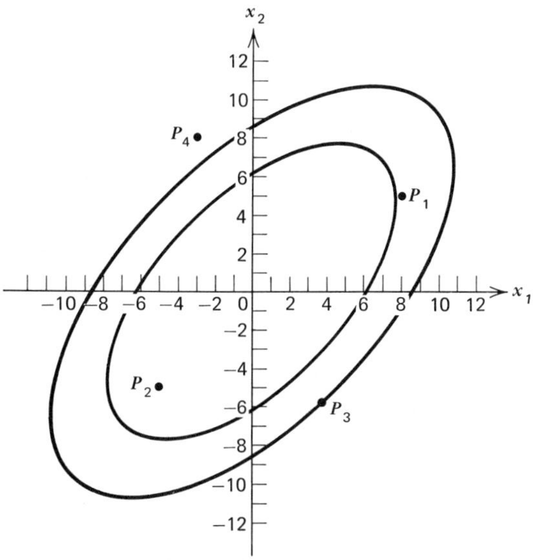

FIGURE 4.3 ***Two isodensity ellipses of the bivariate normal distribution with $\sigma_1 = \sigma = 5$, $\rho = .60$.***

The two points P_1 and P_2 shown inside the larger ellipse have coordinates (8, 5) and (−5, −5), respectively; point P_3, lying on this ellipse, has coordinates (4, −5.565); P_4, which is outside the ellipse, has coordinates (−3, 8). Computing the quantity

$$Q = \mathbf{x}'\boldsymbol{\Sigma}^{-1}\mathbf{x} = (1/16)(x_1^2 + x_2^2 - 1.2x_1x_2)$$

for each of these four points, we find the values to be as follows:

Point	Q-value
P_1 (8, 5)	2.563
P_2 (−5, −5)	1.250
P_3 (4, −5.565)	4.605
P_4 (−3, 8)	6.363

Note that $Q \leq 4.605$ for the three points that are inside or on the 90% ellipse, while $Q > 4.605$ for P_4, which lies outside this ellipse. In general, for any chosen value of χ^2, it is true that $\mathbf{x}'\mathbf{\Sigma}^{-1}\mathbf{x} \leq \chi^2$ for any point in $R_2(\chi^2)$, and $\mathbf{x}'\mathbf{\Sigma}^{-1}\mathbf{x} > \chi^2$ for any point outside $R_2(\chi^2)$. This fact was tacitly assumed, in addition to the theorem about $\mathbf{x}'\mathbf{\Sigma}^{-1}\mathbf{x}$ being a chi-square variate, when it was asserted that the various percent ellipses could be determined by selecting the appropriate values of χ^2 from a table of chi-square distributions.

When the number of variables is three or greater, Eq. 4.7 represents a family of ellipsoids or hyperellipsoids instead of ellipses, but the interpretation exactly parallels that for the bivariate case. For example, when $p = 4$, the 90% hyperellipsoidal region is bounded by the hypersurface

$$\mathbf{x}'\mathbf{\Sigma}^{-1}\mathbf{x} = 7.779$$

the value 7.779 being the 90th centile of the chi-square distribution with n.d.f. = 4. This region, $R_4(7.779)$, is such that 90% of the points in four-dimensional space representing observation quadruples (x_1, x_2, x_3, x_4) drawn from the four-variate normal population $N(\mathbf{0}, \mathbf{\Sigma})$ lie inside it. Furthermore, the value of $\mathbf{x}'\mathbf{\Sigma}^{-1}\mathbf{x}$ computed for a point with coordinates (x_1, x_2, x_3, x_4) is less than or equal to 7.779 if and only if $P \in R_4(7.779)$.

EXERCISES

1. If p variables $X_1, X_2, \ldots, X_p$ are independently normally distributed with mean μ_i and variance σ_i^2 (for each $i = 1, 2, \ldots, p$), their joint density function is given by the product of their respective univariate density functions. That is,

$$\phi(X_1, X_2, \ldots, X_p) = \prod_{i=1}^{p} \left\{ \frac{1}{\sqrt{2\pi}\,\sigma_i} \exp[-(X_i - \mu_i)^2/2\sigma_i^2] \right\}$$

Show that this is a special case of the general multivariate density function function 4.13 with

$$\mathbf{\Sigma} = \begin{bmatrix} \sigma_1^2 & & & 0 \\ & \sigma_2^2 & & \\ & & \ddots & \\ 0 & & & \sigma_p^2 \end{bmatrix}$$

2. Write the equation for the 75% ellipse of the bivariate normal population considered in Example 4.2. By making a rough plot, verify that the smaller ellipse shown in Fig. 4.3 corresponds to this. Then show that the values of $\mathbf{x}'\mathbf{\Sigma}^{-1}\mathbf{x}$ for the points indicated in Fig. 4.3 satisfy the proper magnitude relations.

3. For the same population as in Exercise 2, determine approximately what percent of the population lies in the region bounded by the contour ellipse passing through the point (5, 8).

   NOTE: When a region is specified in terms of a point on the boundary ellipse, we say that the points inside the region are "closer" to the centroid than is the given point. Here we are measuring "closeness" by a probabilistic notion of "distance", not by the ordinary Euclidean distance. Thus, if two contour ellipses differing only slightly in size are constructed, there will be many points on the smaller ellipse that are *geometrically* farther from the centroid than are some points on the larger ellipse. Nevertheless, *any* point on the smaller ellipse is said to be "closer" to the centroid in the above sense than is *any* point on the larger ellipse.

4.2 THE SAMPLING DISTRIBUTION OF SAMPLE CENTROIDS

The reader should be familiar with the fact that when independent, random samples of size N are drawn from a univariate normal population $\mathrm{N}(\mu, \sigma^2)$, the sample means are distributed as $\mathrm{N}(\mu, \sigma^2/N)$. An exact analogue to this fact holds in the multivariate case.

If a sample of size N is drawn from a p-variate normal population, there are p sample means $\bar{X}_1, \bar{X}_2, \ldots, \bar{X}_p$. Over successive, independent random samples, these means jointly follow a p-variate normal distribution with the same centroid $\boldsymbol{\mu} = [\mu_1, \mu_2, \ldots, \mu_p]'$ as the parent population, and with dispersion matrix equal to $1/N$ times the dispersion matrix $\mathbf{\Sigma}$ of the parent population. Thus, if we define a *sample centroid* as the vector

$$\bar{\mathbf{X}} = \begin{bmatrix} \bar{X}_1 \\ \bar{X}_2 \\ \vdots \\ \bar{X}_p \end{bmatrix} \tag{4.16}$$

we may summarize the foregoing in the form of a theorem.

> ***Theorem.*** Sample centroids $\bar{\mathbf{X}}$ based on independent, random samples of size N from a population $\mathrm{N}(\boldsymbol{\mu}, \mathbf{\Sigma})$ have the sampling distribution $\mathrm{N}(\boldsymbol{\mu}, \mathbf{\Sigma}/N)$.

The density function for $\mathrm{N}(\boldsymbol{\mu}, \mathbf{\Sigma}/N)$ is found, by making the appropriate substitutions in Eq. 4.13, to be

$$\phi(\bar{X}_1, \bar{X}_2, \ldots, \bar{X}_p) = (2\pi)^{-p/2} N^{p/2} |\mathbf{\Sigma}|^{-1/2} \exp[-N(\bar{\mathbf{x}}'\mathbf{\Sigma}^{-1}\bar{\mathbf{x}})/2] \tag{4.17}$$

where $\bar{\mathbf{x}} = \bar{\mathbf{X}} - \boldsymbol{\mu}$ is the deviation of the sample centroid from the population centroid.

We can apply the theorem in the previous section to the sampling distribution of $\bar{\mathbf{X}}$, and thereby determine elliptical regions within which any specified percentage of the centroids based on random samples of a given size will lie. Or, conversely, we can determine what percent of the samples of a given size will have centroids that lie in a specified elliptical region.

Suppose, for instance, that samples of size $N = 10$ are drawn from the bivariate normal population of Example 4.2. Then, the region in which 90% of the sample centroids lie is bounded by the ellipse $N\bar{\mathbf{x}}'\boldsymbol{\Sigma}^{-1}\bar{\mathbf{x}} = 4.605$, or

$$10[\bar{X}_1 - 15, \bar{X}_2 - 20] \begin{bmatrix} 25 & 15 \\ 15 & 25 \end{bmatrix}^{-1} \begin{bmatrix} \bar{X}_1 - 15 \\ \bar{X}_2 - 20 \end{bmatrix} = 4.605$$

where the constant on the right is the 90th centile of the chi-square distribution with 2 d.f.'s, as before. Dividing through by 10 (or, in general, the sample size N), the equation becomes

$$\bar{\mathbf{x}}'\boldsymbol{\Sigma}^{-1}\bar{\mathbf{x}} = .4605$$

in which the left-hand side is formally identical to that for the 90% ellipse of the parent population, but the right-hand side is smaller by a factor of $1/N$. Figure 4.4 shows the 90% ellipse of the sampling distribution of $\bar{\mathbf{x}}$ and

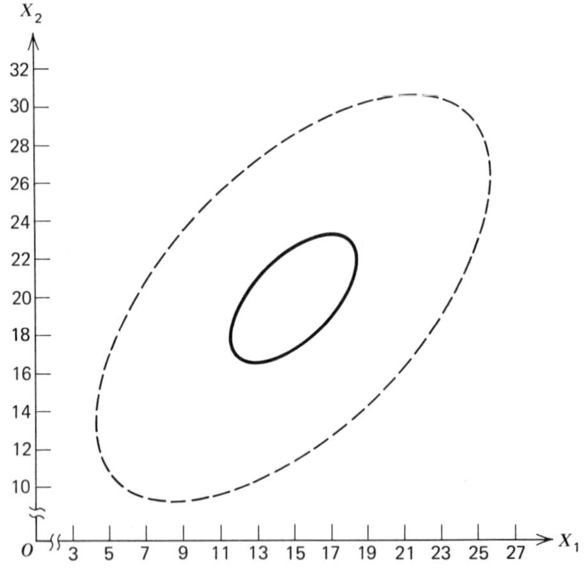

FIGURE 4.4 ***A 90% ellipse of sampling distribution of centroids of samples of size 10 from a bivariate normal population, and a 90% ellipse of the parent population.***

the corresponding ellipse for the parent population (in dotted lines). Note that the former is about one third, in linear size, of the latter; more precisely, the ratio of the linear dimensions of the two ellipses is $1 : \sqrt{10}$, or $1 : \sqrt{N}$ in general.

What we have just seen is a 90% *probability region* for sample centroids, given the centroid and dispersion matrix of a multivariate normal population. It corresponds, in the univariate case, to the symmetric 90% probability interval $[\mu - 1.64\sigma/\sqrt{N}, \mu + 1.64\sigma/\sqrt{N}]$ for means of samples of size N from $N(\mu, \sigma^2)$. Now, the reader is probably familiar with the transition from a probability interval for sample means to a confidence interval for the population mean, effected by algebraic manipulation of the inequalities enclosed by parentheses in a probability statement like

$$p(\mu - 1.64\sigma/\sqrt{N} \leq \bar{X} \leq \mu + 1.64\sigma/\sqrt{N}) = .90$$

transforming it into

$$p(\bar{X} - 1.64\sigma/\sqrt{N} \leq \mu \leq \bar{X} + 1.64\sigma/\sqrt{N}) = .90$$

In exactly the same way, the probability statement

$$p[(\bar{\mathbf{X}} - \boldsymbol{\mu})'\boldsymbol{\Sigma}^{-1}(\bar{\mathbf{X}} - \boldsymbol{\mu}) \leq 4.605/N] = .90$$

is readily transformed into

$$p[(\boldsymbol{\mu} - \bar{\mathbf{X}})'\boldsymbol{\Sigma}^{-1}(\boldsymbol{\mu} - \bar{\mathbf{X}}) \leq 4.605/N] = .90$$

and is construed as follows: Imagine that we know the dispersion matrix $\boldsymbol{\Sigma}$, but not the centroid $\boldsymbol{\mu}$, of a bivariate normal population. We draw successive random samples of size N from this population, and for each sample we construct an ellipse

$$(\mathbf{Y} - \bar{\mathbf{X}})'\boldsymbol{\Sigma}^{-1}(\mathbf{Y} - \bar{\mathbf{X}}) = 4.605/N \tag{4.18}$$

centered at the *sample* centroid $\bar{\mathbf{X}} = [\bar{X}_1, \bar{X}_2]$, using a coordinate system (Y_1, Y_2). Then, 90% of these ellipses will contain the population centroid $\boldsymbol{\mu}$ as an interior (or boundary) point.

Consequently, for any single sample of size N, the region bounded by the ellipse (4.18), with center at the particular sample centroid $\bar{\mathbf{X}}$, is called a 90% *confidence region* for the unknown population centroid $\boldsymbol{\mu}$ of the bivariate normal population $N(\boldsymbol{\mu}, \boldsymbol{\Sigma})$. More generally, a $100(1 - \alpha)$% confidence region for the centroid $\boldsymbol{\mu}$ of a p-variate normal population $N(\boldsymbol{\mu}, \boldsymbol{\Sigma})$ with known $\boldsymbol{\Sigma}$ may be determined by constructing the ellipse, with center at $\bar{\mathbf{X}}$, defined by the equation

$$(\mathbf{Y} - \bar{\mathbf{X}})'\boldsymbol{\Sigma}^{-1}(\mathbf{Y} - \bar{\mathbf{X}}) = \chi_p^2(1 - \alpha)/N \tag{4.19}$$

where $\chi_p^2(1 - \alpha)$ denotes the $100(1 - \alpha)$ centile point of the chi-square distribution with p d.f.'s, and N is the sample size.

EXERCISE

Suppose that a bivariate normal population has a dispersion matrix

$$\begin{bmatrix} 25 & 12 \\ 12 & 16 \end{bmatrix}$$

and that a random sample of 20 observations yielded a centroid [18, 15]. Construct the elliptical 95% confidence region for $\boldsymbol{\mu}$, centered around the sample centroid.

4.3 SIGNIFICANCE TEST: ONE-SAMPLE PROBLEM

Although the purposes differ, conducting a significance test and establishing a confidence region are formally identical in rationale: the test statistic takes a value that leads to rejection, at a given α level, of a hypothesis about a parameter if and only if the $100(1 - \alpha)\%$ confidence region fails to include the hypothesized parameter value.

To illustrate with a familiar univariate example, suppose that the variance of a normal population is known to be equal to 144, and that the mean is unknown but hypothesized to be equal to 100. If a sample of 36 observations from this population yields a mean of 104, the test statistic has the value

$$z = \frac{104 - 100}{12/\sqrt{36}} = 2$$

which leads to a rejection of the hypothesis that $\mu = 100$ at the .05 level (against a two-sided alternative). At the same time, the 95% confidence interval for μ, based on this sample, is

$$[104 - 1.96(12/\sqrt{36}),\ 104 + 1.96(12/\sqrt{36})]$$

or

$$[100.08,\ 107.92]$$

which does not include the hypothesized value, $\mu = 100$.

In the same way, a hypothesis concerning the centroid of a multivariate normal population *with known dispersion matrix* $\boldsymbol{\Sigma}$ can be tested by substituting the hypothesized population centroid $\boldsymbol{\mu}_0$ and the observed sample centroid, respectively, for $\mathbf{Y}$ and $\overline{\mathbf{X}}$ in the left-hand side of Eq. 4.19. If the quadratic form thus computed has a value exceeding $\chi_p^2(1 - \alpha)/N$, then the hypothesis $\boldsymbol{\mu} = \boldsymbol{\mu}_0$ is rejected at the α level of significance. The basis for this decision is that, from the argument presented in Section 4.1, we know that

$$(\boldsymbol{\mu}_0 - \overline{\mathbf{X}})'\boldsymbol{\Sigma}^{-1}(\boldsymbol{\mu}_0 - \overline{\mathbf{X}}) \quad >, =, \text{or} < \quad \chi_p^2(1 - \alpha)/N$$

according to whether the point $\boldsymbol{\mu}_0$ lies outside, on, or inside the ellipse (4.19) centered at $\overline{\mathbf{X}}$, respectively. In practice, it is more convenient to use

$$Q = N(\boldsymbol{\mu}_0 - \overline{\mathbf{X}})'\boldsymbol{\Sigma}^{-1}(\boldsymbol{\mu}_0 - \overline{\mathbf{X}}) \tag{4.20}$$

as the test statistic, and reject the hypothesis that $\boldsymbol{\mu} = \boldsymbol{\mu}_0$ at the α level if and only if $Q > \chi_p^2(1 - \alpha)$.

EXAMPLE 4.3. Suppose that a sample of 25 observations is drawn from a bivariate normal population with unknown centroid $\boldsymbol{\mu}$ and dispersion matrix

$$\boldsymbol{\Sigma} = \begin{bmatrix} 16 & 8 \\ 8 & 9 \end{bmatrix}$$

If the sample centroid is found to be $\overline{\mathbf{X}} = [15.4, 9.9]$, test the hypothesis that $\boldsymbol{\mu} = [17, 10]$ at the 5% significance level.

Since the inverse of the population dispersion matrix is

$$\boldsymbol{\Sigma}^{-1} = (1/80)\begin{bmatrix} 9 & -8 \\ -8 & 16 \end{bmatrix}$$

the test statistic Q of Eq. 4.20 has the value

$$Q = (25)(1/80)[17 - 15.4, 10 - 9.9]\begin{bmatrix} 9 & -8 \\ -8 & 16 \end{bmatrix} \times \begin{bmatrix} 17 - 15.4 \\ 10 - 9.9 \end{bmatrix} = 6.45$$

From Table E.3, we find that the 95th centile of the χ^2 distribution with 2 d.f.'s is 5.991. Thus, $Q > \chi_2^2(.95)$, so the hypothesis that $\boldsymbol{\mu} = [17, 10]$ is rejected.

Population Dispersion Matrix Unknown. The foregoing discussions assumed the population dispersion matrix to be known. In practice, however, when a hypothesis concerning $\boldsymbol{\mu}$ is to be tested, $\boldsymbol{\Sigma}$ is usually also unknown and has to be estimated from the sample. In the corresponding univariate case, the appropriate test statistic is the familiar

$$t = \frac{\overline{X} - \mu_0}{s/\sqrt{N}}$$

where s is the square root of the unbiased estimate $\sum(X - \bar{X})^2/(N - 1)$ of σ^2. This statistic, as is well known, follows Student's t-distribution with $N - 1$ d.f.'s, instead of N(0, 1), which is the distribution of

$$z = \frac{(\bar{X} - \mu_0)}{\sigma/\sqrt{N}}$$

when σ is known.

The multivariate counterpart of t, or rather its square,

$$t^2 = \frac{(\bar{X} - \mu_0)^2}{s^2/N} = N(\bar{X} - \mu_0)(s^2)^{-1}(\bar{X} - \mu_0)$$

is obtained by replacing the scalar quantities $\bar{X} - \mu_0$ and s^2 in the last expression by their matrix analogues $\bar{\mathbf{X}} - \boldsymbol{\mu}_0$ and $\mathbf{S}/(N - 1)$, respectively ($\mathbf{S}$ being the sample SSCP matrix). The result is, of course, very similar to the Q of Eq. 4.20: the only difference is that $\boldsymbol{\Sigma}$ is replaced by its unbiased estimate $\mathbf{S}/(N - 1)$. This statistic was denoted T^2 by Hotelling (1931), who first studied its sampling distribution. We shall use a subscript 1 to indicate that it is the T^2 for a one-sample problem. Thus,

$$T_1{}^2 = N(N - 1)(\bar{\mathbf{X}} - \boldsymbol{\mu}_0)'\mathbf{S}^{-1}(\bar{\mathbf{X}} - \boldsymbol{\mu}_0) \tag{4.21}$$

What Hotelling showed was that the following multiple of $T_1{}^2$ has an F-distribution:

$$\frac{N - p}{(N - 1)p} T_1^2 = F^p_{N-p} \tag{4.22}$$

where the superscript p and the subscript $N - p$ indicate the numerator and denominator d.f.'s of the F statistic. (Note that for $p = 1$, that is, the univariate case, Eq. 4.22 reduces to the familiar relation, $t^2 = F^1_{N-1}$.)

EXAMPLE 4.4. The centroid and SSCP matrix for a sample of 22 observations from a bivariate normal population were:

$$\bar{X}' = [32.6, 33.5] \quad \text{and} \quad \mathbf{S} = \begin{bmatrix} 47.25 & 42.02 \\ 42.02 & 111.09 \end{bmatrix}$$

Test the hypothesis that $\boldsymbol{\mu}_0 = [31, 32]'$ at the 1% level of significance.

The inverse of the given SSCP matrix is found to be

$$\mathbf{S}^{-1} = \begin{bmatrix} 3.1891 & -1.2063 \\ -1.2063 & 1.3564 \end{bmatrix} \times 10^{-2}$$

From Eqs. 4.21 and 4.22, the quantity distributed as an F-variate is

$$\frac{N - p}{(N - 1)p} T_1^2 = \frac{N(N - p)}{p} (\bar{\mathbf{X}} - \boldsymbol{\mu}_0)\, \mathbf{S}^{-1}(\bar{\mathbf{X}} - \boldsymbol{\mu}_0)$$

$$= [(22)(20)/2][1.6, 1.5]$$

$$\times \begin{bmatrix} 3.1891 & -1.2063 \\ -1.2063 & 1.3564 \end{bmatrix} \begin{bmatrix} 1.6 \\ 1.5 \end{bmatrix} \times 10^{-2}$$

$$= 11.9$$

From Table E.4, we find the 99th centile of the F distribution with $p = 2$ and $N - p = 20$ d.f.'s to be 5.85. Since the observed F value exceeds this, we may reject the hypothesis $\boldsymbol{\mu}_0 = [31, 32]'$ at the 1% level.

The one-sample $T_1{}^2$-test finds an important application in testing the significance of the difference between the centroids of matched samples (where each member of one sample is paired with a member of the other sample in some meaningful fashion), or the centroids of the same sample in a test-retest design. Just as in the univariate case, the multivariate matched-samples problem is reduced to a one-sample problem by using the differences between the several paired observations as the dependent variables.

Thus, in the p-variate case, the ith pair of observation vectors yields a difference vector

$$\mathbf{d}_i' = [X_{1i}^{(1)} - X_{1i}^{(2)}, X_{2i}^{(1)} - X_{2i}^{(2)}, \ldots, X_{pi}^{(1)} - X_{pi}^{(2)}] \tag{4.23}$$

and the mean difference vector, or the centroid of the difference scores, is given by

$$\bar{\mathbf{d}}' = \left(\sum_{i=1}^{N} \mathbf{d}_i'\right)/N = [\bar{X}_1^{(1)} - \bar{X}_1^{(2)}, \bar{X}_2^{(1)} - \bar{X}_2^{(2)}, \ldots, \bar{X}_p^{(1)} - \bar{X}_p^{(2)}] \tag{4.24}$$

which is the same as the difference between the centroid of the two matched samples.

The hypothesis to be tested is that the population centroid of difference scores,

$$\boldsymbol{\mu}_d' = [\mu_1^{(1)} - \mu_1^{(2)}, \mu_2^{(1)} - \mu_2^{(2)}, \ldots, \mu_p^{(1)} - \mu_p^{(2)}]$$

is equal to the null vector $[0, 0, \ldots, 0]$. Hence, the appropriate $T_1{}^2$-statistic for the matched-samples problem is given by

$$T_d{}^2 = N(N - 1)\bar{\mathbf{d}}'\mathbf{S}_d{}^{-1}\bar{\mathbf{d}} \tag{4.25}$$

where $\mathbf{S}_d$ is the sample SSCP matrix of the p difference scores. The test is carried out by computing expression (4.22) and comparing the result with the appropriate centile value of the F^p_{N-p}-distribution.

EXAMPLE 4.5. A researcher at a school for the deaf gave several motor tests to the resident students, and also tested a group of hearing children, paired child for child on the basis of sex, age, and height with the deaf children. Scores for 10 deaf girls and their hearing counterparts on a test of grip (X_1) and a test of balance (X_2) were as shown below. Using Hotelling's $T_1{}^2$-test, test the significance of the difference between the centroids of the deaf and hearing groups at $\alpha = .01$.

Pair No.		1	2	3	4	5	6	7	8	9	10
X_1	D	25	22	28	35	37	48	49	54	65	57
	H	26	22	29	39	34	51	42	54	77	68
X_2	D	2.0	2.0	2.7	2.7	3.0	1.7	2.0	2.0	2.7	1.0
	H	2.3	1.0	3.7	3.3	10.0	4.3	4.7	7.0	3.3	1.7

The difference scores (H − D) for each pair on each of the two variables are:

d_1	1	0	1	4	−3	3	−7	0	12	11	$\Sigma d_1 = 22$
d_2	0.3	−1.0	1.0	.06	7.0	2.6	2.7	5.0	0.6	0.7	$\Sigma d_2 = 19.5$

The centroid of the difference scores, the SSCP matrix, and its inverse are (as the reader should verify as an exercise) as follows:

$$\bar{\mathbf{d}} = \begin{bmatrix} 2.20 \\ 1.95 \end{bmatrix}, \qquad \mathbf{S}_d = \begin{bmatrix} 301.600 & -56.400 \\ -56.400 & 53.325 \end{bmatrix}$$

and

$$\mathbf{S}_d^{-1} = \begin{bmatrix} 4.133 & 4.371 \\ 4.371 & 23.376 \end{bmatrix} \times 10^{-3}$$

Therefore, from Eq. 4.25 we compute

$$T_d{}^2 = (10)(9)[2.20, 1.95] \begin{bmatrix} 4.133 & 4.371 \\ 4.371 & 23.376 \end{bmatrix} \begin{bmatrix} 2.20 \\ 1.95 \end{bmatrix} \times 10^{-3} = (90)(146.39) \times 10^{-3}$$

and, from Eq. 4.22,

$$F_8^2 = \frac{8}{(9)(2)}(90)(146.39) \times 10^{-3}$$
$$= (40)(146.39) \times 10^{-3} = 5.86$$

which falls short of the 99th centile of the F-distribution with 2 and 8 d.f.'s (which is 8.65). Therefore, the null hypothesis that the population centroid of the difference scores is [0, 0] is not rejected.

4.4 SIGNIFICANCE TEST: TWO-SAMPLE PROBLEM

The multivariate analogue of the familiar t-ratio for testing the significance of the difference between two independent means (that is, means based on two unrelated samples) may also be written out by making appropriate replacements of scalars by vectors or matrices in the expression for t^2; namely,

$$t^2 = \frac{(\bar{X}_1 - \bar{X}_2)^2}{s^2(1/n_1 + 1/n_2)} = (\bar{X}_1 - \bar{X}_2)\left[s^2\frac{n_1 + n_2}{n_1 n_2}\right]^{-1}(\bar{X}_1 - \bar{X}_2)$$

where

$$s^2 = \frac{\sum x_1{}^2 + \sum x_2{}^2}{n_1 + n_2 - 2}$$

is the pooled within-groups estimate of the assumed common variance σ^2 of the two populations.

To obtain the multivariate counterpart of s^2, we divide the within-groups SSCP matrix $\mathbf{W}$ (defined in Eq. 3.30) by its n.d.f., $n_1 + n_2 - 2$. Thus, the required analogue of s^2 is

$$\frac{\mathbf{W}}{n_1 + n_2 - 2}$$

The difference $\bar{X}_1 - \bar{X}_2$ between the two means is, of course, replaced by the difference between the two centroid vectors, that is, $\bar{\mathbf{X}}^{(1)} - \bar{\mathbf{X}}^{(2)}$, in the multivariate case. On making these replacements in the expression above for t^2, we obtain Hotelling's T^2-statistic for the two-sample problem, which we here denote by $T_2{}^2$:

$$T_2^2 = (\bar{\mathbf{X}}^{(1)} - \bar{\mathbf{X}}^{(2)})'\left[\frac{\mathbf{W}}{n_1 + n_2 - 2}\,\frac{n_1 + n_2}{n_1 n_2}\right]^{-1}(\bar{\mathbf{X}}^{(1)} - \bar{\mathbf{X}}^{(2)})$$

$$= \frac{n_1 n_2(n_1 + n_2 - 2)}{n_1 + n_2}(\bar{\mathbf{X}}^{(1)} - \bar{\mathbf{X}}^{(2)})'\mathbf{W}^{-1}(\bar{\mathbf{X}}^{(1)} - \bar{\mathbf{X}}^{(2)}) \qquad (4.26)$$

Just as in the case of the one-sample T_1^2-statistic, T_2^2 is derivable from a quantity like Q (Eq. 4.20), which in turn derives from considering a confidence region for the difference between the centroids of two multivariate normal populations with the same dispersion matrix $\boldsymbol{\Sigma}$. After obtaining the appropriate form of Q for the two-sample case, one has merely to replace $\boldsymbol{\Sigma}$ by its pooled within-groups estimate, $\mathbf{W}/(n_1 + n_2 - 2)$. We shall not go into the derivation here, but merely point out that the sampling distribution of the difference between pairs of sample centroids based on pairs of independent samples drawn from $\mathrm{N}(\boldsymbol{\mu}_1, \boldsymbol{\Sigma})$ and $\mathrm{N}(\boldsymbol{\mu}_2, \boldsymbol{\Sigma})$, respectively, is given by the multivariate normal distribution

$$\mathrm{N}\left[\boldsymbol{\mu}_1 - \boldsymbol{\mu}_2, \boldsymbol{\Sigma}\left(\frac{1}{n_1} + \frac{1}{n_2}\right)\right]$$

What Hotelling (1931) showed was that the following multiple of the T_2^2, thus derived, is distributed as an F-variate with p and $n_1 + n_2 - p - 1$ d.f.'s:

$$\frac{n_1 + n_2 - p - 1}{(n_1 + n_2 - 2)p} T_2^2 = F^p_{n_1+n_2-p-1} \qquad (4.27)$$

(Note, again, that for the univariate case, with $p = 1$, this reduces to the relation $t^2 = F^1_{n_1+n_2-2}$.)

EXAMPLE 4.6. A miniature, fictitious example will suffice to illustrate the computations, and will also indicate the difference between a multivariate significance test and a sequence of univariate significance tests. Suppose that two treatment groups, in an experiment using the randomized-groups design, were measured on two criterion variables X_1 and X_2, and that the scores were as follows:

	Group 1		Group 2	
	X_1	X_2	X_1	X_2
	3	6	2	11
	9	6	7	14
	16	8	13	18
	19	13	19	18
	24	12	23	20
Group Totals	71	45	64	81
Group Means	14.2	9.0	12.8	16.2

If univariate t-tests are conducted for the two criterion variables separately, it will be found (as the reader should verify) that:

$$t < 1 \qquad \text{for} \quad X_1$$
$$t = -3.27 \qquad \text{for} \quad X_2 \ (.01 < p < .02)$$

Would you conclude that the two groups are significantly different?

To conduct the T_2^2-test, we first compute the SSCP matrices for the two groups:

$$\mathbf{S}_1 = \begin{bmatrix} 274.8 & 96.0 \\ 96.0 & 44.0 \end{bmatrix} \quad \text{and} \quad \mathbf{S}_2 = \begin{bmatrix} 292.8 & 119.2 \\ 119.2 & 52.8 \end{bmatrix}$$

Hence,

$$\mathbf{W} = \mathbf{S}_1 + \mathbf{S}_2 = \begin{bmatrix} 567.6 & 215.2 \\ 215.2 & 96.8 \end{bmatrix}$$

and

$$\mathbf{W}^{-1} = \begin{bmatrix} 1.121 & -2.493 \\ -2.493 & 6.575 \end{bmatrix} \times 10^{-2}$$

The vector of centroid difference is:

$$\begin{aligned} \bar{\mathbf{X}}^{(1)'} - \bar{\mathbf{X}}^{(2)'} &= [14.2 - 12.8 \qquad 9.0 - 16.2] \\ &= [1.4 \quad -7.2] \end{aligned}$$

From the foregoing intermediate results, we compute T_2^2 in accordance with Eq. 4.26, thus:

$$\begin{aligned} T_2^2 &= [(5)(5)(8)/(10)][1.4, \ -7.2] \begin{bmatrix} 1.121 & -2.493 \\ -2.493 & 6.575 \end{bmatrix} \\ &\quad \times \begin{bmatrix} 1.4 \\ -7.2 \end{bmatrix} \times 10^{-2} = 78.66 \end{aligned}$$

Hence, the corresponding F-statistic is, by Eq. 4.27, computed as:

$$F_7^2 = [7/(8)(2)](78.66) = 34.41$$

which far exceeds the 99th centile of the F-distribution with 2 and 7 d.f.'s. The two group centroids are therefore significantly different beyond the 1% level—a conclusion that is not between the respective results for the two univariate t-tests, it should be noted.

4.5 SIGNIFICANCE TEST: K-SAMPLE PROBLEM

The multivariate significance tests discussed in the foregoing hinged primarily on using the dispersion matrix $\mathbf{\Sigma}$ as a multivariate analogue of the variance σ^2 of a univariate distribution. There is another extension of the concept of variance to multivariate distributions, and this is the *determinant* $|\mathbf{\Sigma}|$ of the dispersion matrix, which is called the *generalized variance.*

We digress briefly to give a geometric interpretation of the generalized variance, which emerges when we use the *N-space* (short for N-dimensional person space) representation of multivariate observations—that is, a representation in which the coordinate axes correspond to individuals instead of variates, as they do in test space. Thus, each point (or vector from the origin to the point) in N-space represents a separate test; its coordinates (or components of the vector) are the scores earned by N individuals on that test. Each such vector is called a *test vector.* This representation has two important properties. First, when the scores are in deviations from the mean ($X - \bar{X} = x$), the length of a test vector is

$$|\mathbf{x}| = \sqrt{\sum_{i=1}^{N} x_i^2} = \sqrt{N-1}\, s$$

that is, $\sqrt{N-1}$ times the standard deviation of that test. Second, the cosine of the angle between any two test vectors (likewise in deviation form) is equal to the product–moment correlation between these tests.

With this background, let us consider the sample counterpart of $|\mathbf{\Sigma}|$, namely $|\mathbf{S}/(N-1)|$. For simplicity, we treat the bivariate case, but the result is immediately generalizable to the p-variate case, and is equally applicable to the population dispersion.

The sample dispersion matrix may be written as:

$$\mathbf{S}/(N-1) = \frac{1}{N-1}\,\mathbf{x}'\mathbf{x} = \begin{bmatrix} s_1^2 & s_1 s_2 r \\ s_2 s_1 r & s_2^2 \end{bmatrix}$$

Hence, the generalized sample variance for the bivariate case is:

$$|\mathbf{S}/(N-1)| = s_1^2 s_2^2 (1 - r^2) = (s_1 s_2)^2 (1 - \cos^2\theta) = (s_1 s_2 \sin\theta)^2$$

where θ is the angle between the test vectors in deviation-score form. But, as shown above, each standard deviation is $1/\sqrt{N-1}$ times the length of the corresponding test vector. Therefore,

$$s_1 s_2 \sin\theta = \frac{|\mathbf{x}_1|}{\sqrt{N-1}} \times \frac{|\mathbf{x}_2|}{\sqrt{N-1}} \sin\theta$$

is equal to the area of the parallelogram formed by the "rescaled" test vectors $\mathbf{x}_1/\sqrt{N-1}$ and $\mathbf{x}_2/\sqrt{N-1}$ in N-space, as seen from Fig. 4.5. Thus, the generalized variance is the square of this area. In the p-variate case, the generalized variance is the square of the p-dimensional "volume" of the parallelotope formed by the rescaled test vectors

$$\mathbf{x}_i/\sqrt{N-1} \qquad (i = 1, 2, \ldots, p)$$

in the N-dimensional person space.

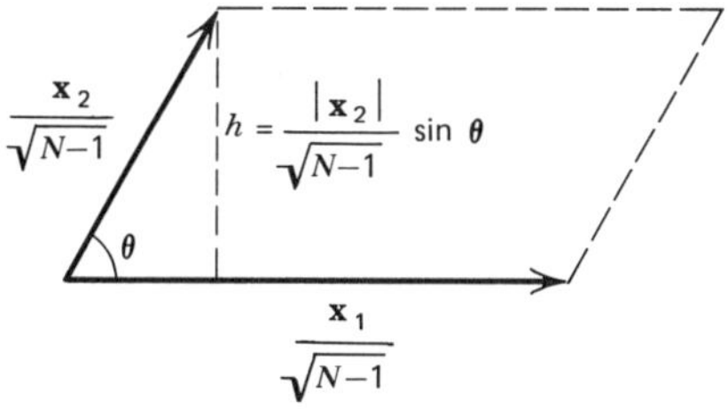

FIGURE 4.5 ***Generalized variance as a squared area (for p = 2).***

We thus see that the two multivariate analogues of variance, $\mathbf{\Sigma}$ and $|\mathbf{\Sigma}|$, have their geometric interpretations in the p-dimensional test-space (as the matrix of isodensity ellipsoids) and the N-dimensional person space, respectively. It is the latter analogue of variance that plays a prominent role in the Λ criterion, due to Wilks (1932), for testing the significance of the overall difference among several sample centroids, constituting a multivariate extension of the F-ratio test in simple analysis of variance.

The rationale for the Λ criterion stems from the likelihood-ratio principle (see Anderson, 1958, pp. 187–191), but here we shall confine ourselves to showing that Λ reduces to a function of the familiar F-ratio when $p = 1$. Greater insight into the nature of the Λ criterion will be gained in Chapter Six, after the necessary mathematical background has been developed in Chapter Five. It will then be seen that the Λ criterion is applicable to many other situations besides that of a multivariate extension of one-way ANOVA.

For the present, we confine our attention to the multivariate significance test for the K-sample problem. In this context—the simplest case for which Wilks' likelihood-ratio criterion is used—Λ is defined as follows:

$$\Lambda = \frac{|\mathbf{W}|}{|\mathbf{T}|} \tag{4.28}$$

where $\mathbf{T}$ is the total-sample SSCP matrix, and $\mathbf{W}$ is the within-groups SSCP matrix, defined in Eq. 3.30 as the sum of the separate SSCP matrices of the K groups; that is,

$$\mathbf{W} = \mathbf{S}_1 + \mathbf{S}_2 + \cdots + \mathbf{S}_K$$

To see what Λ reduces to when $p = 1$, we note that in this case $|\mathbf{W}| = SS_w$, and $|\mathbf{T}| = SS_t = SS_w + SS_b$. Hence,

$$\Lambda_{(p=1)} = \frac{SS_w}{SS_w + SS_b} = \frac{1}{1 + (SS_b/SS_w)}$$

But the customary F-ratio for simple ANOVA is given by

$$F = \frac{SS_b/(K - 1)}{SS_w/(N - K)} = \frac{SS_b}{SS_w} \frac{N - K}{K - 1}$$

where K is the number of groups, and $N = n_1 + n_2 + \cdots + n_K$ is the total sample size. Consequently,

$$\frac{SS_b}{SS_w} = \frac{K - 1}{N - K} \cdot F$$

which, when substituted in the preceding expression for $\Lambda_{(p=1)}$, yields

$$\Lambda_{(p=1)} = \frac{1}{1 + [(K - 1)/(N - K)]F}$$

One fact immediately evident from this relationship, in the univariate case, between Λ and F is that the former is an *inversely* related measure of disparity among the several group means. The larger this disparity is (relative to within-group variability), the larger is F, and hence the smaller is Λ. This observation holds true in the multivariate case as well: the greater the disparity among the several group centroids (relative to the within-groups generalized variance), the smaller the value of Λ.

The sampling distribution of Λ under the null hypothesis (that is, in the present context, when the K population centroids are equal) was recently obtained by Schatzoff (1964, 1966a). His results showed that exact numerical computations are feasible only when p is an even number or K is an odd number. For other cases (that is, p is odd *and* K is even), linear interpolation between tabled values for adjacent p or K was shown to give reasonably accurate approximations.

For reasons of tabular economy Schatzoff's tables were not constructed to give the centile points of Λ itself, but those of a certain logarithmic function of Λ. This function is

$$\begin{aligned} V &= -[N - 1 - (p + K)/2] \ln \Lambda \\ &= -2.3026[N - 1 - (p + K)/2] \log \Lambda \end{aligned} \tag{4.29}$$

which was shown by Bartlett (1947) to be distributed approximately as a chi-square with $p(K - 1)$ degrees of freedom, provided $N - 1 - (p + K)/2$ is large. Actually, what Schatzoff's tables explicitly display are "correction factors" for converting selected centiles of $\chi^2_{p(K-1)}$ to the corresponding centiles of the statistic V. One multiplies the appropriate chi-square value (shown at the foot of each column) by the proper entry in his tables to obtain the critical value of V for significance at the specified α level. Table 4.1 below, adapted from one of Schatzoff's sequence of tables, shows the critical values of V at $\alpha = .05$, .025, and .01 for different sample sizes (N) when $p = 10$ and $K = 7$.

TABLE 4.1 95th, 97.5th, and 99th Centiles of V with $p = 10$, $K = 7$ for Selected Values of N (Sample Size).

N \ α	.05	.025	.01
20	91.26	96.71	103.32
22	86.75	91.71	97.75
24	84.45	89.21	94.92
26	83.11	87.71	93.24
28	82.24	86.72	92.09
30	81.61	86.05	91.38
32	81.14	85.47	90.76
34	80.74	85.13	90.41
36	80.50	84.88	90.06
40	80.19	84.47	89.71
46	79.87	84.13	89.26
56	79.56	83.80	88.91
76	79.32	83.55	88.64
100	79.21	83.44	88.53
136	79.16	83.38	88.47
180	79.14	83.34	88.43
∞	79.08	83.30	88.38

The values in the bottom row ($N \to \infty$) are exactly equal to the corresponding centiles of $\chi^2_{p(K-1)}$, because the correction factors (which always decrease monotonically with increasing N) converge to unity as the sample size becomes indefinitely large, no matter what values of p, K, and α are considered.

From Table 4.1, we see that, with $p = 10$ and $K = 7$, the chi-square approximation to V is correct two significant digits when N is about 100 or more, and correct to three digits when N exceeds 180. The same degree of accuracy is achieved with smaller N's when the number of variables and number of groups are smaller. For instance, with $p = K = 5$, a total sample size of 70 (that is, an average of 14 members per sample group) is sufficient to assure three-digit accuracy. In view of the difficulty of providing extensive tables for the exact distribution of V (since it depends on the three quantities, p, K, and N), the main practical contribution of Schatzoff's theoretically important work may well lie in its having demonstrated, as just exemplified, that the chi-square approximation to Bartlett's V is reasonably good for moderately large sample sizes. Another implication is that, since Schatzoff's correction factors are always greater than 1 for finite N, using the chi-square approximation to V always offers a conservative test: if the value of V as computed from Eq. 4.29 is smaller than the $100(1 - \alpha)$th centile of $\chi^2_{p(K-1)}$, we can be sure that the exact test will not lead to rejection of the null hypothesis at the $100\alpha\%$ level of significance. It is when the value of V slightly exceeds the specified centile of $\chi^2_{p(K-1)}$ that one would do well to consult Schatzoff's tables in order to see whether the V is significant at that level.

There is another function of Λ that seems to offer a better approximate test than does Bartlett's V (without Schatzoff's correction), and is therefore useful when Schatzoff's tables are not available, or when $p(K - 1)$ exceeds 70—which, due to limitations in computer capacity, is the maximum value of $p(K - 1)$ for which Schatzoff's tables are currently in existence. This function, due to Rao (1952), is given by the formula,

$$R = \frac{1 - \Lambda^{1/s}}{\Lambda^{1/s}} \frac{ms - p(K - 1)/2 + 1}{p(K - 1)} \tag{4.30}$$

where

$$m = N - 1 - (p + K)/2 \qquad \text{and} \qquad s = \sqrt{\frac{p^2(K - 1)^2 - 4}{p^2 + (K - 1)^2 - 5}}$$

The statistic R is distributed approximately as an F-variate with $\nu_1 = p(K - 1)$ and $\nu_2 = ms - p(K - 1)/2 + 1$ degrees of freedom. When ν_2 is not an integer, we may either interpolate in the F-table or simply use the integer closest to ν_2 as the n.d.f. for the denominator.

It should be pointed out that Bartlett's V and Rao's R each has its advantages. Besides leading to an exact test with the aid of Schatzoff's tables, V has the advantage that, being a logarithmic function of Λ, it can be expressed as a sum of several terms—each of which is itself an approximate chi-square variate—when Λ is expressed as a product of several factors, as it will turn out to be the case in more advanced applications of Λ (see Chapter Six). On the other hand, the R-statistic, in addition to permitting a closer approximate test, has the property that it reduces to an *exact* F-variate when $p = 1$ or 2, *or* when $K = 2$ or 3. These special cases (which were known long before the R-statistic was developed) are shown in Table 4.2 below, adapted from Rao (1952, p. 260).

TABLE 4.2 Functions of Λ Distributed as *F*-Ratios for Special Values of *p* and *K*.

Any p,	$K = 2$	$\dfrac{1-\Lambda}{\Lambda}\,\dfrac{N-p-1}{p}$	$= F^{p}_{N-p-1}$
	$K = 3$	$\dfrac{1-\Lambda^{1/2}}{\Lambda^{1/2}}\,\dfrac{N-p-2}{p}$	$= F^{2p}_{2(N-p-2)}$
Any K,	$p = 1$	$\dfrac{1-\Lambda}{\Lambda}\,\dfrac{N-K}{K-1}$	$= F^{K-1}_{N-K}$
	$p = 2$	$\dfrac{1-\Lambda^{1/2}}{\Lambda^{1/2}}\,\dfrac{N-K-1}{K-1}$	$= F^{2(K-1)}_{2(N-K-1)}$

Let us verify, for $K = 3$, the reduction of R to the tabled F-variate. Making the appropriate substitutions in the expressions for m and s, following Eq. 4.30, we obtain:

$$m = (2N - p - 5)/2 \qquad s = \sqrt{\frac{p^2(2)^2 - 4}{p^2 + (2)^2 - 5}} = 2$$

and

$$\frac{ms - p(K-1)/2 + 1}{p(K-1)} = \frac{2N - p - 5 - p + 1}{2p} = \frac{N - p - 2}{p}$$

Therefore,

$$R = \frac{1 - \Lambda^{1/2}}{\Lambda^{1/2}}\,\frac{N - p - 2}{p}$$

which is precisely the function of Λ displayed in Table 4.2 for $K = 3$.

It is also of interest to note that for $K = 2$ the function in Table 4.2 has the same distribution as

$$T_2{}^2\,\frac{N - p - 1}{(N - 2)p}$$

In fact, it can be shown (Section 6.6) that

$$\frac{1 - \Lambda}{\Lambda} = \frac{T^2}{N - 2}$$

in the two-group case; that is, Wilks' Λ and Hotelling's T^2 are functionally related in this case.

EXAMPLE 4.7.[1] Employees in three job categories of an airline company were administered an Activity Preference Questionnaire consisting of three bipolar scales: X_1 = Outdoor–Indoor preferences; X_2 = Convivial–Solitary preferences; X_3 = Conservative–Liberal preferences. (A high score on each scale indicates a preponderance of choices of activities of the first-named type over those of the second-named type.) The means of the three groups of employees on the three scales were as follows:

		Means on:		
(k)	n_k	X_1	X_2	X_3
1. Passenger Agents	85	12.59	24.22	9.02
2. Mechanics	93	18.54	21.14	10.14
3. Operations Control Men	66	15.58	15.45	13.24

The within-groups SSCP matrix **W** (that is, the sum of the SSCP matrices $\mathbf{S}_1$, $\mathbf{S}_2$, and $\mathbf{S}_3$, for the three groups taken separately), and the total SSCP matrix **T** (with the three groups merged into a single conglomerate sample) were found to be as shown below.

$$\mathbf{W} = \begin{bmatrix} 3967.8301 & 351.6142 & 76.6342 \\ 351.6142 & 4406.2517 & 235.4365 \\ 76.6342 & 235.4365 & 2683.3164 \end{bmatrix}$$

and

$$\mathbf{T} = \begin{bmatrix} 5540.5742 & -421.4364 & 350.2556 \\ -421.4364 & 7295.5710 & -1170.5590 \\ 350.2556 & -1170.5590 & 3374.9232 \end{bmatrix}$$

Given the above data, we wish to test whether the three group centroids are significantly different from one another; that is, to determine whether the data warrant our rejection of the null hypothesis that the centroids of the three populations, from which our samples were drawn, are all equal.

[1]Source of contrived data: Rulon et al. (1967), by permission of the publishers.

To find the value of Wilks' Λ, we compute the determinants of $\mathbf{W}$ and $\mathbf{T}$, as required by Eq. 4.28. The results are:

$$|\mathbf{W}| = 46.348 \times 10^9 \qquad \text{and} \qquad |\mathbf{T}| = 127.679 \times 10^9$$

The likelihood-ratio Λ is then given by the ratio

$$\Lambda = |\mathbf{W}|/|\mathbf{T}| = 46.348/127.679 = .36301$$

Since the number of groups is three in this example, an exact F-test is possible, based on the Λ just computed. From the appropriate entry of Table 4.2, we see that

$$\frac{1 - \Lambda^{1/2}}{\Lambda^{1/2}} \frac{N - p - 2}{p} = \frac{1 - \sqrt{.36301}}{\sqrt{.36301}} \frac{244 - 3 - 2}{3}$$
$$= 52.57$$

may be referred to the F-distribution with $2p = 6$ d.f.'s in the numerator and $2(N - p - 2) = 478$ (or, for practical purposes, infinite) d.f.'s in the denominator. The observed value 52.57 far exceeds any tabled centile point of the F^6_∞-distribution, so we reject the null hypothesis that that three population centroids are identical.

Purely for illustrative purposes, let us also compute the approximate chi-square statistic V, given in Eq. 4.29, for the above Λ value:

$$V = -2.3026[244 - 1 - (3 + 3)/2] \log (.36301)$$
$$= (-2.3026)(240)(-.44008) = 243.20$$

which, as a chi-square with $p(K - 1) = 6$ d.f.'s, far exceeds the 99.9th centile. Since a chi-square value divided by its n.d.f. is an F-variate with the same n.d.f. for the numerator and infinite d.f.'s for the denominator, it is possible to make a check on the degree of approximation offered by the V statistic in this example.

$$V/6 = 243.20/6 = 40.53$$

is to be compared with the exact F value obtained above. The approximation is seen to be only moderate.

EXERCISES

1. Ten pairs of college freshmen with matching IQs were used in a problem-solving experiment, members of each pair being assigned at random to

the experimental and control groups. The task was to solve two sets of riddle-like problems, one set being entirely verbal and the other set involving some numerical reasoning. The problems in the two sets had similar logical structures, so that some practice effect could be anticipated even in the control group. The experimental group was given instructions that should facilitate performance, especially on the second set of problems.

The scores earned by the ten experimental *S*s and their matched controls on the two sets of problems were as follows:

Exp.	X_1	19	28	30	31	33	34	41	35	45	53
	X_2	26	33	37	34	41	36	45	40	42	56
Control	X_1	19	27	31	30	34	34	39	36	42	50
	X_2	18	31	36	31	36	37	39	41	43	52

Test the significance of the difference between the experimental and control group centroids at the five percent level of significance.

2. A group of twenty drug addicts and a group of thirty chronic alcoholics were given two psychological tests *A* and *B*, each yielding standardized scores on a 10-point scale. The means of the two groups on the two tests, and the within-groups SSCP matrix **W** were as follows:

	A	B	(n_k)
Drug addicts	5.8	2.9	(20)
Alcoholics	6.7	2.3	(30)

$$\mathbf{W} = \begin{bmatrix} 95.4 & 11.6 \\ 11.6 & 29.2 \end{bmatrix}$$

Test the significance of the difference between the drug-addict and alcoholic group centroids, using $\alpha = .01$.

3. Sixty subjects were assigned at random to three experimental conditions, twenty *S*s per group, in a learning experiment. The criterion performance was scored in two ways: speed (X_1) and accuracy (X_2). The means of the three groups on the two criterion measures, the within-groups SSCP matrix **W**, and the total SSCP matrix **T** were as shown below.

	$\bar{X}_1$	$\bar{X}_2$
Group 1	38.5	19.6
Group 2	34.0	22.4
Group 3	30.1	16.2

$$\mathbf{W} = \begin{bmatrix} 982.8 & 205.2 \\ 205.2 & 643.5 \end{bmatrix} \qquad \mathbf{T} = \begin{bmatrix} 1689.6 & 472.8 \\ 472.8 & 1029.1 \end{bmatrix}$$

Test the significance of the overall difference among the three group centroids, using $\alpha = .01$.

[Note: if the reader is puzzled to find that knowledge of the groups' centroids is apparently not needed in carrying out the significance test, he should observe that this problem could have been solved with only the centroids and the **W** matrix given. Matrix **T** can then be computed without its being given. Although the procedure anticipates material discussed in Chapter Six, the reader may find it informative and challenging to try it at this point. The first step is to compute the between-groups SSCP matrix **B** in accordance with Equations 6.3 and 6.4. The total SSCP matrix is then obtained as $\mathbf{W} + \mathbf{B} = \mathbf{T}$, which is the multivariate counterpart of the relation $SS_w + SS_b = SS_t$ in univariate, one-way analysis of variance.]

Chapter Five

More Matrix Algebra: Linear Transformations, Axis Rotation, and Eigenvalue Problems

Reference was made in the preceding chapter to the fact (Theorem on p. 69) that the quantity $\chi^2 = \mathbf{x}'\mathbf{\Sigma}^{-1}\mathbf{x}$ in the exponent of the p-variate normal density function is a chi-square variate with p degrees of freedom. This fact was proved only for the special case when the p random variables X_i are independently distributed, and hence uncorrelated. The proof for the general case of p-variate normal populations with nonzero correlation coefficients hinges on the fact that a set of p independently normally distributed variables can be constructed from linear combinations of the original p variables. In this chapter we develop matrix methods that enable us, among other things, to determine these linear combinations. To understand how this works, it is helpful to relate the algebraic process of forming linear combinations to the geometric operation of rotating the axes of a coordinate system.

5.1 LINEAR TRANSFORMATIONS AND AXIS ROTATION

In discussing the p-variate normal distribution, we associated a p-dimensional vector $[X_1, X_2, \ldots, X_p]$ with a point in p-dimensional space having the coordinates $(X_1, X_2, \ldots, X_p)$. Let us go back to the simplest case of $p = 2$ to study the relationship between linear combinations of variables and axis rotation.

A person with scores 8 and 5, respectively, on two tests X_1 and X_2 may be represented by a vector [8, 5]. This vector, in turn, can be geometrically represented by an arrow from the origin O of a Cartesian coordinate system in the (X_1, X_2)-plane to the point $P(8, 5)$, as in Fig. 5.1.

Now let us rotate the axes 20° in the counterclockwise direction, and call the resulting coordinate system (OY_1, OY_2), as shown in Fig. 5.2. The coordinates of point P in the new system may be found by letting

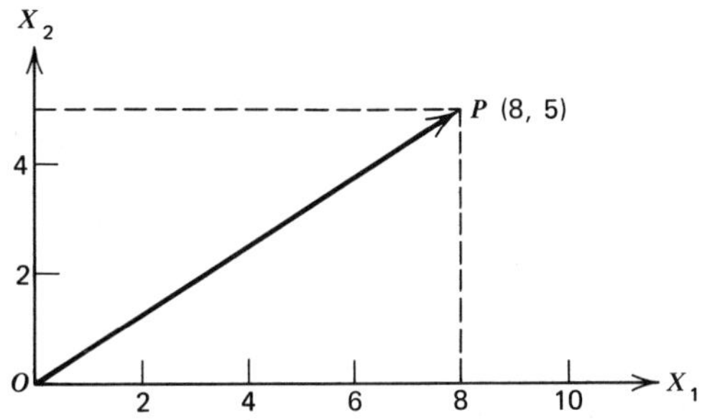

FIGURE 5.1 ***Vector representing a person with scores 8 and 5 on Tests X_1 and X_2.***

$X_1 = 8$, $X_2 = 5$, and $\theta = 20°$ in the axis-rotation formulas

$$\begin{cases} Y_1 = (\cos\theta)X_1 \quad + (\sin\theta)X_2 \\ Y_2 = (-\sin\theta)X_1 + (\cos\theta)X_2 \end{cases} \tag{5.1}$$

Substituting the appropriate trigonometric function values for $\theta = 20°$, the results are:

$$\begin{cases} Y_1 = (.940)(8) \quad + (.342)(5) = 9.230 \\ Y_2 = (-.342)(8) + (.940)(5) = 1.964 \end{cases}$$

Thus, in the new coordinate system, the vector OP has the representation [9.230, 1.964], as may be graphically verified by referring to Fig. 5.2.

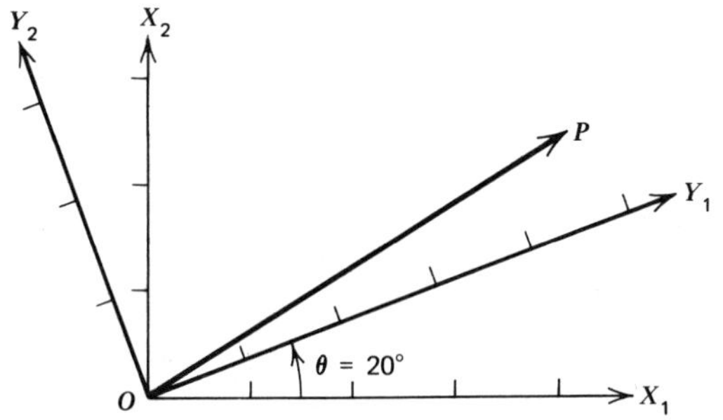

FIGURE 5.2 ***Vector OP shown with reference to the original coordinate system (OX_1, OX_2) and a rotated system (OY_1, OY_2).***

What is more to the point for our present purposes is to note that each of the transformation equations (5.1) expresses one of the new coordinates of point P as a linear combination of the original coordinates X_1 and X_2. Alternatively, we may regard these equations as defining two new *variables* Y_1 and Y_2 in terms of linear combinations of the original variables X_1 and X_2. We thus see that, when we perform a rotation of axes, we are in effect defining a new set of variables, each of which is a linear combination of the original variables.

Is it the case, conversely, that *any* linear combination of a given set of variables expresses the result of rotating one of the axes of a coordinate system through some angle? Not quite so. The coefficients of the linear combination must satisfy a certain condition before that linear combination qualifies as a "rotated axis." This condition may be inferred by examining the coefficients in the expressions for Y_1 and Y_2 in Eqs. 5.1. Note that, from the trigonometric identities

$$(\cos\theta)^2 + (\sin\theta)^2 = 1$$

and

$$(-\sin\theta)^2 + (\cos\theta)^2 = 1$$

the *squares* of the coefficients add up to unity in each case. That this is a sufficient condition for a linear combination to represent a rotated axis may be established by the following observation:

> If $\quad Y = aX_1 + bX_2 \quad$ and $\quad a^2 + b^2 = 1$
>
> then an angle θ can always be found such that $\cos\theta = a$ and $\sin\theta = b$

Thus, a linear combination of two variables, X_1 and X_2, formed by using coefficients whose squares sum to unity can always be represented by an axis accessible by some rotation in the (X_1, X_2)-plane from the X_1 axis.

EXAMPLE 5.1. Let $Y = .6X_1 + .8X_2$. Then, since $(.6)^2 + (.8)^2 = 1$, an angle θ may be found such that $\cos\theta = .6$ and $\sin\theta = .8$. Determine such an angle and draw an axis OY superimposed on a facsimile of Fig. 5.1. Also, verify that the foot of the perpendicular dropped from point P to line OY is $(.6)(8) + (.8)(5) = 8.8$ units away from the origin O. (A more concise way of expressing this fact is to say that *the projection of vector OP onto axis OY has length* 8.8.)

From a table of trigonometric functions, we find $\sin 53°8' = .80003$. So we may take $\theta = 53°8'$. Alternatively, since $\tan\theta = \sin\theta/\cos\theta = .8/.6 = 4/3$, we may determine a point Q whose ordinate and abscissa stand in the ratio 4:3. The line through the origin O and such a point Q gives the desired Y-axis, without our actually having to find angle θ numerically.

Next, given two linear combinations of X_1 and X_2 that are both representable as rotated axes, what further condition among the coefficients must hold in order that the two new axes be perpendicular to each other? The answer is again exemplified in Eqs. 5.1:

$$(\cos\theta)(-\sin\theta) + (\sin\theta)(\cos\theta) = 0$$

To generalize, if the coefficients of two linear combinations

$$Y = aX_1 + bX_2 \qquad \text{and} \qquad Y' = a'X_1 + b'X_2$$

are such that

$$a^2 + b^2 = a'^2 + b'^2 = 1 \tag{5.2}$$

and the further condition,

$$aa' + bb' = 0 \tag{5.3}$$

holds, then the axes OY and OY', representing the two linear combinations, are perpendicular, or *orthogonal*, to each other. In other words, if both Eqs. 5.2 and 5.3 are satisfied by the coefficients of two linear combinations of a pair of variables X_1 and X_2, then the axes OY and OY' together constitute a coordinate system resulting from a *rigid* (or angle-preserving) rotation of the original rectangular coordinate system (OX_1, OX_2). (A slight qualification of this statement, which we do not wish to introduce at this point, will subsequently be made.)

EXAMPLE 5.2. If $Y = .6X_1 + .8X_2$ and $Y' = -.8X_1 + .6X_2$, we have

$$(.6)^2 + (.8)^2 = (-.8)^2 + (.6)^2 = 1$$

and

$$(.6)(-.8) + (.8)(.6) = 0$$

Therefore, axes OY and OY' are mutually orthogonal. On the other hand, let

$$Y'' = .707X_1 - .707X_2$$

Since

$$(.707)^2 + (-707)^2 = 1$$

Y'' is also representable by an axis in the (OX_1, OX_2)-plane. However,

$$(.6)(.707) + (.8)(-.707) \neq 0$$

so axes OY and OY'' are *not* mutually orthogonal. The foregoing statements may be verified by determining angles θ, θ', and θ'' such that $\cos\theta = .6$, $\cos\theta' = -.8$, and $\cos\theta'' = .707$, and drawing axes OY, OY', and OY'', making these angles, respectively, with axis OX_1 of the original system.

The preceding discussion of linear combinations and axis rotation in a plane may be extended to spaces of three or more dimensions by introducing a redundant but more symmetrical notation for the angles among various pairs of axes. That is, instead of using, in Eqs. 5.1, only the angle θ between axes OX_1 and OY_1 of Fig. 5.2, we introduce four angles, θ_{11}, θ_{12}, θ_{21}, and θ_{22}, as shown in Fig. 5.3, where θ_{ij} denotes the angle from the (old) X_i axis to the (new) Y_j axis.

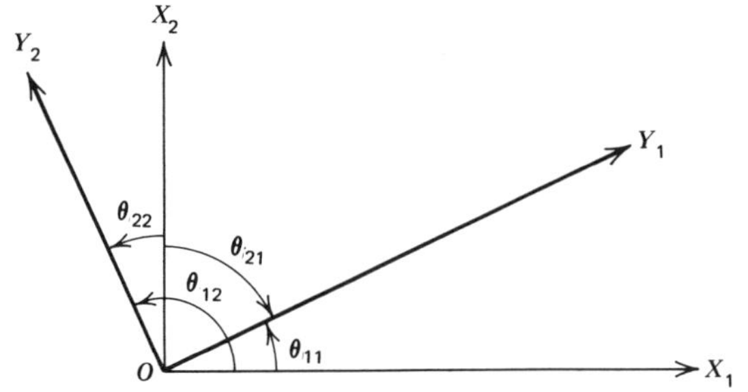

FIGURE 5.3 ***Rigid axis rotation designated by four related angles*** θ_{ij}.

The relation between each of these angles and the previous θ is as follows:

$$\theta_{11} = \theta \qquad \theta_{12} = \theta + 90^\circ$$
$$\theta_{21} = \theta - 90^\circ \qquad \theta_{22} = \theta$$

Consequently, by use of trigonometric formulas for functions of complementary angles, we see that

$$\sin\theta = \sin(\theta_{21} + 90^\circ) = \cos\theta_{21}$$

and

$$-\sin\theta = -\sin(\theta_{12} - 90^\circ) = \sin(90^\circ - \theta_{12}) = \cos\theta_{12}$$

Substituting these equivalents for $\sin\theta$ and $-\sin\theta$, we may rewrite Eqs. 5.1 so that all the coefficients are cosines, thus:

$$\begin{aligned} Y_1 &= (\cos\theta_{11})X_1 + (\cos\theta_{21})X_2 \\ Y_2 &= (\cos\theta_{12})X_1 + (\cos\theta_{22})X_2 \end{aligned} \qquad (5.4'')$$

This pair of equations has another feature that facilitates generalization to higher dimensional spaces. Not only are all the coefficients of the form $\cos\theta_{ij}$, but in each case the i agrees with the subscript of X of which $\cos\theta_{ij}$ is the coefficient, and the j agrees with the subscript of Y in whose expression the coefficient appears.

The rule just stated allows us to write the appropriate set of equations corresponding to rigid rotations of axes in spaces of any dimensionality. Thus, for three dimensions, we have:

$$\begin{aligned} Y_1 &= (\cos\theta_{11})X_1 + (\cos\theta_{21})X_2 + (\cos\theta_{31})X_3 \\ Y_2 &= (\cos\theta_{12})X_1 + (\cos\theta_{22})X_2 + (\cos\theta_{32})X_3 \\ Y_3 &= (\cos\theta_{13})X_1 + (\cos\theta_{23})X_2 + (\cos\theta_{33})X_3 \end{aligned} \tag{5.5''}$$

where θ_{ij} is again the angle from the ith axis of the original system to the jth axis of the new system, as depicted in Fig. 5.4.

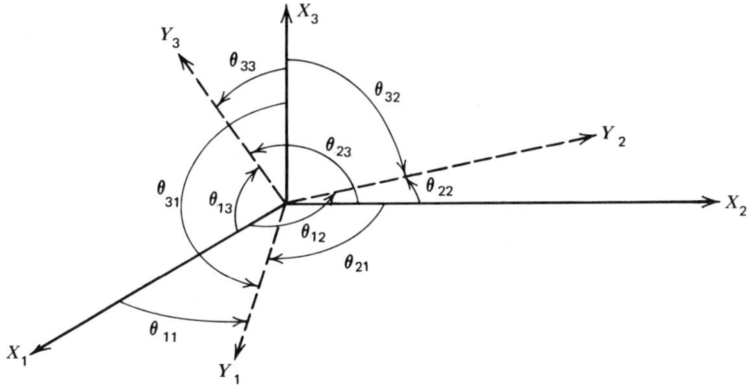

FIGURE 5.4 ***Rigid rotation in space designated by nine related angles*** θ_{ij}.

The transformation equations may be written more compactly by introducing the abbreviation

$$v_{ij} = \cos\theta_{ij}$$

Equations 5.4″ and 5.5″, for two and three dimensions, respectively, become, in this notation,

$$\begin{cases} Y_1 = v_{11}X_1 + v_{21}X_2 \\ Y_2 = v_{12}X_1 + v_{22}X_2 \end{cases} \tag{5.4'}$$

$$\begin{cases} Y_1 = v_{11}X_1 + v_{21}X_2 + v_{31}X_3 \\ Y_2 = v_{12}X_1 + v_{22}X_2 + v_{32}X_3 \\ Y_3 = v_{13}X_1 + v_{23}X_2 + v_{33}X_3 \end{cases} \tag{5.5'}$$

The systematicness and compactness of notation in Eqs. 5.4′ and 5.5′ are not gained without some sacrifice in completeness. What was immediately clear in Eqs. 5.1, using $\sin\theta$ and $\cos\theta$, must now be stated as separate conditions on the coefficients v_{ij} before Eqs. 5.4′ are unambiguously recognizable as those for a rigid rotation of a rectangular coordinate system in a plane. The conditions, which were stated in different notation in Eqs. 5.2 and 5.3, now become:

$$\begin{cases} v_{11}^2 + v_{21}^2 = v_{12}^2 + v_{22}^2 = 1 \\ v_{11}v_{12} + v_{21}v_{22} = 0 \end{cases} \tag{5.6}$$

In the three-dimensional case, in order to assure that Eqs. 5.5′ refer to a rigid rotation of axes, the conditions to be stated are six in number:

$$\begin{cases} v_{11}^2 + v_{21}^2 + v_{31}^2 = v_{12}^2 + v_{22}^2 + v_{32}^2 \\ \qquad = v_{13}^2 + v_{23}^2 + v_{33}^2 = 1 \\ v_{11}v_{12} + v_{21}v_{22} + v_{31}v_{32} = v_{11}v_{13} + v_{21}v_{23} + v_{31}v_{33} \\ \qquad = v_{12}v_{13} + v_{22}v_{23} + v_{32}v_{33} = 0 \end{cases} \tag{5.7}$$

The number of conditions rapidly increases with the dimensionality p of the space. There is one condition requiring that the sum of the squares of the coefficients appearing in each transformation equation be equal to unity (totalling p conditions); and one condition requiring that the sum of cross products between coefficients in each *pair* of transformation equations be equal to zero (totalling $p(p - 1)/2$ conditions). Thus, for a p-dimensional space, the total number of conditions to be satisfied by the p^2 coefficients v_{ij} in the transformation equations is $p(p + 1)/2$. However, a useful notational convention allows us to express all $p(p + 1)/2$ conditions in the form of a single equation type. This is to use the symbol δ_{jk}, known as *Kronecker's delta*, defined as follows:

$$\delta_{jk} = \begin{cases} 1 & \text{if } j = k \\ 0 & \text{if } j \neq k \end{cases} \tag{5.8}$$

It may be readily verified that, using this symbol, the sets of Eqs. 5.6 and 5.7 are each embraced by the single equation type,

$$\sum_{i=1}^{p} v_{ij}v_{ik} = \delta_{jk} \qquad (j = 1, 2, \ldots, p;\ k = j, j + 1, \ldots, p) \tag{5.9}$$

An easier way of stating the $p(p + 1)/2$ conditions to be satisfied by the coefficients in the set of p transformation equations representing a rigid rotation of a rectangular coordinate system in p-dimensional space results from writing the set of equations in matrix form. Equations 5.4′ and 5.5′ then become

$$[Y_1, Y_2] = [X_1, X_2]\begin{bmatrix} v_{11} & v_{12} \\ v_{21} & v_{22} \end{bmatrix} \tag{5.4}$$

and

$$[Y_1, Y_2, Y_3] = [X_1, X_2, X_3]\begin{bmatrix} v_{11} & v_{12} & v_{13} \\ v_{21} & v_{22} & v_{23} \\ v_{31} & v_{32} & v_{33} \end{bmatrix} \tag{5.5}$$

respectively. The obvious extension to p-dimensional space is

$$[Y_1, Y_2, \ldots, Y_p] = [X_1, X_2, \ldots, X_p] \begin{bmatrix} v_{11} & v_{12} & \cdots & v_{1p} \\ v_{21} & v_{22} & \cdots & v_{2p} \\ \vdots & \vdots & \ddots & \vdots \\ v_{p1} & v_{p2} & \cdots & v_{pp} \end{bmatrix}$$

or, in symbolic matrix notation,

$$\mathbf{Y}' = \mathbf{X}'\mathbf{V} \tag{5.10}$$

where

$$\mathbf{V} = (v_{ij})$$

We digress briefly to mention that any transformation of the form of Eq. 5.10, with no restriction on the $p \times p$ matrix $\mathbf{V}$, is called a *linear transformation*, since all quantities involved appear only in the first degree. The matrix $\mathbf{V}$ is called the *transformation matrix*, and we say that the vector $\mathbf{X}$ is transformed linearly into vector $\mathbf{Y}$ by matrix $\mathbf{V}$. An important property of linear transformations is that, if $\mathbf{X}_1'$ and $\mathbf{X}_2'$ are two p-dimensional row vectors and $\mathbf{V}$ is a $p \times p$ transformation matrix, then it follows from the distributive law of matrix multiplication (Eq. 2.10b) that

$$(\mathbf{X}_1' + \mathbf{X}_2')\mathbf{V} = \mathbf{X}_1'\mathbf{V} + \mathbf{X}_2'\mathbf{V}$$

(that is, the linear transform of the sum of two vectors is equal to the sum of their respective linear transforms.)

Returning to our main present concern, we inquire (in the terminology just introduced) what condition the transformation matrix $\mathbf{V}$ must satisfy in order that the linear transformation (5.10) represent a rigid rotation. Since the elements v_{ij} must satisfy Eqs. 5.9, it follows that the matrix $\mathbf{V}$ must satisfy the condition

$$\mathbf{V}'\mathbf{V} = \mathbf{I} \tag{5.11'}$$

as the reader may easily verify for the simple cases when $p = 2$ or 3. (More generally, observe that those members of the set of Eqs. 5.9 for which $j = k$ assert that the diagonal elements of $\mathbf{V}'\mathbf{V}$ are all equal to unity, and those members with $j \neq k$ assert that every off-diagonal element of $\mathbf{V}'\mathbf{V}$ is equal to zero.)

In the foregoing discussions, we have acted as though the stated conditions on the coefficients of linear combinations—presented in various forms, Eqs. 5.2, 5.3, 5.9, and finally 5.11′—were both necessary and sufficient for ensuring that transformation Eq. 5.10 represent a rigid rotation. It is now time to introduce a qualification that we held in abeyance to simplify our discussion. The fact is, condition (5.11′) is necessary, but not quite sufficient by itself to restrict (5.10) to rigid rotations in the true sense of the term.

To see why this is so, we again examine the two-dimensional case. The linear transformation

$$[Y_1, Y_2] = [X_1, X_2] \begin{bmatrix} \cos\theta & \sin\theta \\ \sin\theta & -\cos\theta \end{bmatrix}$$

is an instance of (5.10) with

$$\mathbf{V} = \begin{bmatrix} \cos\theta & \sin\theta \\ \sin\theta & -\cos\theta \end{bmatrix}$$

which satisfies condition (5.11′):

$$\mathbf{V}'\mathbf{V} = \begin{bmatrix} \cos\theta & \sin\theta \\ \sin\theta & -\cos\theta \end{bmatrix} \begin{bmatrix} \cos\theta & \sin\theta \\ \sin\theta & -\cos\theta \end{bmatrix} = \begin{bmatrix} 1 & 0 \\ 0 & 1 \end{bmatrix}$$

However, a comparison with Eq. 5.1, which may be written in the form

$$[Y_1, Y_2] = [X_1, X_2] \begin{bmatrix} \cos\theta & -\sin\theta \\ \sin\theta & \cos\theta \end{bmatrix}$$

reveals that our present transformation matrix $\mathbf{V}$ differs from that of Eq. 5.1 in the signs of the second-column elements, and only in this respect. That is, the first column of the two matrices are identical, but the second column of $\mathbf{V}$ is -1 times the second column of the transformation matrix of 5.1—which unquestionably represents a rigid rotation. Yet, $\mathbf{V}$ satisfies condition (5.11′), as we just saw.

What, then, is the nature of the transformation effected by our present $\mathbf{V}$, and what further condition besides (5.11′) must we impose on a transformation matrix before (5.10) definitely represents a rigid rotation and nothing else? To distinguish between the linear combinations displayed in Eq. 5.1 and those generated by $\mathbf{V}$, let us denote the latter by Y_1' and Y_2'. Thus,

$$\begin{cases} Y_1' = (\cos\theta)X_1 + (\sin\theta)X_2 \\ Y_2' = (\sin\theta)X_1 + (-\cos\theta)X_2 \end{cases}$$

whereas

$$\begin{cases} Y_1 = (\cos\theta)X_1 + (\sin\theta)X_2 \\ Y_2 = (-\sin\theta)X_1 + (\cos\theta)X_2 \end{cases}$$

It is then evident that $Y_1' = Y_1$, and $Y_2' = -Y_2$. Consequently, if Eq. 5.1 represents a rotation by an angle θ, then our present transformation by $\mathbf{V}$, leading to Y_1' and Y_2', must represent the same rotation followed by a *reversal* in the positive sense of the Y_2 axis. Such a reversal in direction is called a *reflection*. Thus, the transformation (5.1) represents a "pure" rigid rotation, while the transformation by $\mathbf{V}$ represents a rigid rotation combined with a reflection of the Y_2 axis. Figure 5.5 depicts these two transformations.

What property of the transformation matrix makes the difference between pure rotation and rotation-plus-reflection? From Eq. 5.11′ and the

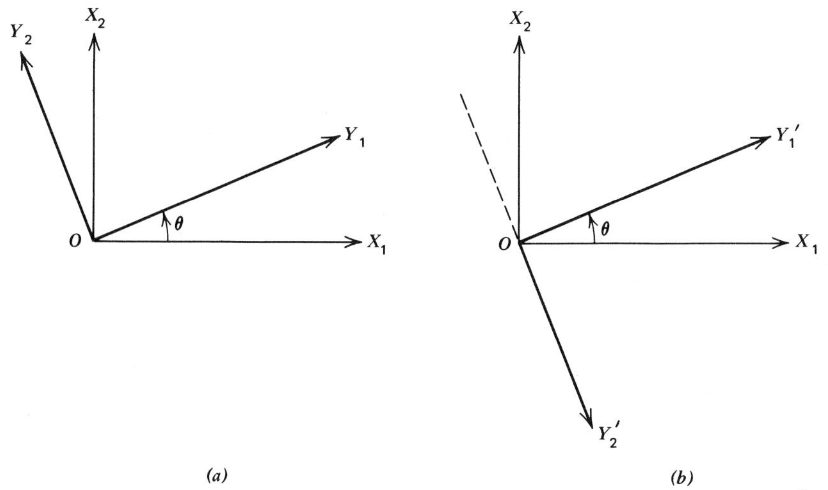

FIGURE 5.5 **(a)** ***Rigid rotation through an angle*** θ***, and*** **(b)** ***rigid rotation and reflection of second axis.***

fact that the determinant of the product of two square matrices is equal to the product of their respective determinants (see Eq. A.6, Appendix A), it follows that

$$|\mathbf{V}'\mathbf{V}| = |\mathbf{V}'||\mathbf{V}| = |\mathbf{V}|^2 = |\mathbf{I}| = 1$$

Hence,

$$|\mathbf{V}| = \pm 1$$

That is, any matrix satisfying Eq. 5.11′ has the property that its determinant is equal to either 1 or -1. On evaluating the determinant of the transformation matrix of 5.1, we find that

$$\begin{vmatrix} \cos\theta & -\sin\theta \\ \sin\theta & \cos\theta \end{vmatrix} = 1$$

Hence, our present $|\mathbf{V}|$, whose second column is the negative of that of the above, must be equal to -1, as the reader may verify directly.

The generality of the above argument for any 2×2 matrix satisfying (5.11′) is obvious. Any transformation matrix representing pure rigid rotation must be of the form of Eq. 5.1:

$$\begin{bmatrix} \cos\theta & -\sin\theta \\ \sin\theta & \cos\theta \end{bmatrix}$$

for some angle θ. Then, any 2×2 matrix

$$\begin{bmatrix} a & a' \\ b & b' \end{bmatrix}$$

whose elements satisfy Eqs. 5.2 and 5.3, that is,

$$\begin{cases} a^2 + b^2 = a'^2 + b'^2 = 1 \\ aa' + bb' = 0 \end{cases}$$

(which are together equivalent to Eq. 5.11′), must either be of the above form (as implied in the oversimplified argument on p. 97), *or* differ from it only in change of signs in one or both columns. The latter cases are possible because replacing a by $-a$ and b by $-b$, exclusively or in addition to replacing a' by $-a'$ and b' by $-b'$, will not affect the conditions (5.2) and (5.3). However, if both columns differed in signs from those of the transformation matrix in (5.1), the new matrix would be equal to

$$\begin{bmatrix} \cos(\theta + 180^\circ) & -\sin(\theta + 180^\circ) \\ \sin(\theta + 180^\circ) & \cos(\theta + 180^\circ) \end{bmatrix}$$

which represents a pure rotation by an angle $\theta + 180^\circ$, and its determinant would equal 1. Thus, only when the transformation matrix involves sign changes in just one of the columns will it represent something other than a pure rotation; namely rotation plus reflection of one axis. And in this case the determinant is equal to -1. We thus conclude that, among 2×2 transformation matrices satisfying (5.11′), those whose determinants are equal to 1 represent pure rigid rotation, while those with determinants equal to -1 represent rigid rotation *and* reflection of one axis.

The foregoing arguments may be extended to three-dimensional space. The transformation matrix

$$\mathbf{V} = \begin{vmatrix} v_{11} & v_{12} & v_{13} \\ v_{21} & v_{22} & v_{23} \\ v_{31} & v_{32} & v_{33} \end{vmatrix}$$

of a rigid rotation may be shown to satisfy $|\mathbf{V}| = 1$ in addition to Eq. 5.11′, while a matrix with sign changes in one or *all three* columns (that is, an odd number of columns) from the above will have a determinant equal to -1. Such a transformation matrix represents rotation plus a reflection of one axis or of all three axes. The reader should convince himself—by using his thumb, forefinger, and middle finger spread out at right angles to one another in both hands—that no rotation alone can lead from one coordinate system (the one represented by the left hand, say) to another in which one or all three axes (fingers) are pointed in opposite directions from the first. But, if two of the columns of $\mathbf{V}$ above are subjected to sign changes, the resulting matrix will again have a determinant equal to 1. Correspondingly, a pure rotation will suffice to go from one coordinate system to another in which two of the axes are respectively pointed in opposite directions from the former.

For transformations in spaces of dimensionality greater than three, concrete geometric argument no longer avails. Nevertheless, we still speak of rotations followed by an even number of axis reflections as "pure rotations," and those followed by an odd number of axis reflections as "rotation-plus-reflection." The transformation matrices for pure rotations have determinants equal to 1, and those for rotation-plus-reflection have determinants equal to -1. We thus arrive at the necessary and sufficient condition for a linear transformation of the form of Eq. 5.10 to represent a rigid rotation by conjoining the condition $|\mathbf{V}| = 1$ to Eq. 5.11′; that is,

$$\mathbf{V}'\mathbf{V} = \mathbf{I} \qquad \text{and} \qquad |\mathbf{V}| = 1 \tag{5.11}$$

As a further point of terminology, a square matrix that satisfies conditions (5.11) is called an *orthogonal matrix*. By contrast, a square matrix that satisfies (5.11′) but not (5.11)—that is, one whose determinant is equal to -1—is sometimes referred to as an *improper orthogonal matrix*. We mention in passing that both "proper" and improper orthogonal matrices have the convenient property that their inverse is simply their transpose:

$$\mathbf{V}' = \mathbf{V}^{-1} \tag{5.12}$$

which follows from postmultiplying both sides of Eq. 5.11′ by $\mathbf{V}^{-1}$.

Using the terminology just introduced, we may summarize the entire development of this section as follows: A linear transformation $\mathbf{Y}' = \mathbf{X}'\mathbf{V}$ represents a rigid rotation of a Cartesian coordinate system if and only if $\mathbf{V}$ is an orthogonal matrix; if $\mathbf{V}$ is an improper orthogonal matrix, the transformation represents a rigid rotation followed by reflections of an odd number of coordinate axes.

EXERCISE

Prove that two rigid rotations performed in succession may be replaced by a single rigid rotation. That is, if $\mathbf{Y}' = \mathbf{X}'\mathbf{V}$ and $\mathbf{Z}' = \mathbf{Y}'\mathbf{U}$ are linear transformations representing rigid rotations, then $\mathbf{Z}'$ may be obtained directly from $\mathbf{X}'$ by a rigid rotation.

5.2 THE EFFECT OF A LINEAR TRANSFORMATION ON AN SSCP MATRIX

Having established the relationship between a rotation of axes and a linear transformation, we turn now to the question. "How does an SSCP matrix transform when the basic variables are subjected to a linear transformation?" To state the question more operationally, we want to investigate the following problem:

Given the SSCP matrix $\mathbf{S}(X)$ for a set of variables $X_1, X_2, \ldots, X_p$, how do we find the SSCP matrix $\mathbf{S}(Y)$ for a set of variables $Y_1, Y_2, \ldots, Y_q$ $(q \leq p)$, each of which is defined as a linear combination of the X's?

Note that the number of new (transformed) variables may be equal to or less than the number of original variables. In fact, in some applications we shall be concerned with just one linear combination of the X's, in which case $\mathbf{S}(Y)$ will of course reduce to a single sum-of-squares.

The first step in determining $\mathbf{S}(Y)$ is to express the means of the Y's in terms of the means of the X's and the transformation matrix. Consider just one linear combination Y_1 of the X's:

$$Y_1 = v_{11}X_1 + v_{21}X_2 + \cdots + v_{p1}X_p$$

If we denote the score of the αth individual on X_i ($i = 1, 2, \ldots, p$) by $X_{\alpha i}$, then his Y_1 score is

$$Y_{\alpha 1} = v_{11}X_{\alpha 1} + v_{21}X_{\alpha 2} + \cdots + v_{p1}X_{\alpha p}$$

For a sample of N individuals, the Y_1 mean is

$$\begin{aligned}\overline{Y}_1 = \frac{1}{N}\sum_{\alpha=1}^{N} Y_{\alpha 1} &= \frac{1}{N}\sum_{\alpha=1}^{N} (v_{11}X_{\alpha 1} + v_{21}X_{\alpha 2} + \cdots + v_{p1}X_{\alpha p}) \\ &= v_{11}\left(\frac{1}{N}\sum_{\alpha=1}^{N} X_{\alpha 1}\right) + v_{21}\left(\frac{1}{N}\sum_{\alpha=1}^{N} X_{\alpha 2}\right) + \cdots \\ &\qquad + v_{p1}\left(\frac{1}{N}\sum_{\alpha=1}^{N} X_{\alpha p}\right) \\ &= v_{11}\overline{X}_1 + v_{21}\overline{X}_2 + \cdots + v_{p1}\overline{X}_p\end{aligned}$$

where $\overline{X}_i$ denotes the mean of X_i for the sample. In other words, the mean of a linear combination of several variables is given by the same linear combination of the means of the original variables.

If several linear combinations $Y_1, Y_2, \ldots, Y_q$ of the X's are considered simultaneously, then each Y mean is expressible as the appropriate linear combination of the X means, and hence the Y centroid $[\overline{Y}_1, \overline{Y}_2, \ldots, \overline{Y}_q]$ is related to the X centroid as follows:

$$[\overline{Y}_1, \overline{Y}_2, \ldots, \overline{Y}_q] = [\overline{X}_1, \overline{X}_2, \ldots, \overline{X}_p]\begin{bmatrix} v_{11} & v_{12} & \cdots & v_{1q} \\ v_{21} & v_{22} & \cdots & v_{2q} \\ \vdots & \vdots & & \vdots \\ v_{p1} & v_{p2} & \cdots & v_{pq}\end{bmatrix}$$

where the elements of the jth column of the rectangular matrix (v_{ij}) are the coefficients for the jth linear combination Y_j. Introducing the abbreviations $\overline{\mathbf{X}}'$ and $\overline{\mathbf{Y}}'$ for the X and Y centroids and $\mathbf{V}$ for the transformation matrix (v_{ij}), the above equation may be written as

$$\overline{\mathbf{Y}}' = \overline{\mathbf{X}}'\mathbf{V} \tag{5.13}$$

which is identical in form to the linear transformation Eq. 5.10. Of course, the **V** here need not be an orthogonal matrix (nor even a square matrix unless $q = p$) if the Y's are simply an arbitrary set of linear combinations of the X's. However, the case when the Y's correspond to a set of coordinate axes resulting from a rigid rotation is included in Eq. 5.13, just as in Eq. 5.10. To represent this special case, we need merely to specify that **V** is an orthogonal matrix.

Next, let us apply Eq. 5.10 to each of N row vectors

$$\begin{aligned}\mathbf{X}_1' &= [X_{11}, X_{12}, \ldots, X_{1p}]\\ \mathbf{X}_2' &= [X_{21}, X_{22}, \ldots, X_{2p}]\\ &\vdots\\ \mathbf{X}_N' &= [X_{N1}, X_{N2}, \ldots, X_{Np}]\end{aligned}$$

representing the scores of N individuals on p variables. The resulting transformed row vectors

$$\begin{aligned}\mathbf{Y}_1' &= \mathbf{X}_1'\mathbf{V}\\ \mathbf{Y}_2' &= \mathbf{X}_2'\mathbf{V}\\ &\vdots\\ \mathbf{Y}_N' &= \mathbf{X}_N'\mathbf{V}\end{aligned}$$

will then represent the same N individuals' scores on p (or, more generally, $q \leq p$) linear combinations $Y_1, Y_2, \ldots, Y_p$ (or Y_q) of the X's.

Now suppose we collect the original N row vectors into a single score matrix, which we denote by **X** as in Chapter Two; that is,

$$\mathbf{X} = \begin{bmatrix} X_{11} & X_{12} & \cdots & X_{1p}\\ X_{21} & X_{22} & \cdots & X_{2p}\\ \vdots & \vdots & & \vdots\\ X_{N1} & X_{N2} & \cdots & X_{Np}\end{bmatrix}$$

It then follows (as the reader should verify) that **XV** will be the $N \times p$ (or $N \times q$) transformed score matrix consisting of the row vectors $\mathbf{Y}_1', \mathbf{Y}_2', \ldots, \mathbf{Y}_N'$. If we denote this matrix by **Y**, we have

$$\mathbf{Y} = \mathbf{XV} \tag{5.10a}$$

which is formally identical to Eq. 5.10 except for the lack of primes on **X** and **Y**. This difference may seem confusing at first, but it serves to remind us that in Eq. 5.10 a row vector $\mathbf{X}'$ is being transformed, while in Eq. 5.10a an $N \times p$ matrix **X** (each row of which is an instance of the earlier $\mathbf{X}'$) is being transformed.[1]

[1]The reason why the row vectors in Eq. 5.10 were not denoted by primed small letters $\mathbf{x}'$ and $\mathbf{y}'$ is, of course, that these symbols are reserved for vectors of deviation scores. We thus made an exception to the general rule of denoting vectors by small letters.

Similarly, by writing N identical rows

$$\bar{\mathbf{X}}' = [\bar{X}_1, \bar{X}_2, \ldots, \bar{X}_p]$$

in the form of an $N \times p$ matrix, we obtain the mean-score matrix of the X's, which was also introduced in Chapter Two and denoted by $\bar{\mathbf{X}}$ (without a prime). It then follows that

$$\bar{\mathbf{Y}} = \bar{\mathbf{X}}\mathbf{V} \tag{5.13a}$$

is the corresponding mean-score matrix of the Y's, consisting of N identical rows $\bar{\mathbf{Y}}'$, each having as its elements the values of the means $\bar{Y}_1$, $\bar{Y}_2, \ldots, \bar{Y}_p$ (or $\bar{Y}_q$).

According to Eq. 2.7, an SSCP matrix may be expressed in terms of the score matrix and the mean-score matrix as

$$\mathbf{S}(X) = \mathbf{X}'\mathbf{X} - \bar{\mathbf{X}}'\bar{\mathbf{X}} \tag{5.14}$$

Correspondingly, for the transformed variables, the SSCP matrix is

$$\mathbf{S}(Y) = \mathbf{Y}'\mathbf{Y} - \bar{\mathbf{Y}}'\bar{\mathbf{Y}} \tag{5.15}$$

We now substitute in this equation the expressions for $\mathbf{Y}$ and $\bar{\mathbf{Y}}$ in terms of $\mathbf{X}$, $\bar{\mathbf{X}}$, and $\mathbf{V}$ from Eqs. 5.10a and 5.13a, and obtain

$$\begin{aligned}
\mathbf{S}(Y) &= (\mathbf{XV})'(\mathbf{XV}) - (\bar{\mathbf{X}}\mathbf{V})'(\bar{\mathbf{X}}\mathbf{V}) \\
&= (\mathbf{V}'\mathbf{X}')(\mathbf{XV}) - (\mathbf{V}'\bar{\mathbf{X}}')(\bar{\mathbf{X}}\mathbf{V}) \\
&= \mathbf{V}'(\mathbf{X}'\mathbf{X}')\mathbf{V} - \mathbf{V}'(\bar{\mathbf{X}}'\bar{\mathbf{X}})\mathbf{V} \\
&= \mathbf{V}'(\mathbf{X}'\mathbf{X} - \bar{\mathbf{X}}'\bar{\mathbf{X}})\mathbf{V}
\end{aligned}$$

where we have used the associative law in going from the second to the third expression on the right, and the distributive law (twice) in going from the third to the last expression. Now, the quantity in parentheses in this last expression is, according to Eq. 5.14, none other than $\mathbf{S}(X)$, the SSCP matrix for the original variables. We thus have the following important relation between the SSCP matrices for the original and the transformed variables:

$$\mathbf{S}(Y) = \mathbf{V}'\mathbf{S}(X)\mathbf{V} \tag{5.16}$$

EXAMPLE 5.3. A small numerical example may clarify the procedures developed above. Suppose that a group of 10 college students was given a Personal Preferences Questionnaire which yields, among other things, scores on two scales: X_1 = Outdoor Scale; and X_2 = Gregariousness Scale.

The score matrix was:

$$\mathbf{X} = \begin{bmatrix} 21 & 16 \\ 22 & 23 \\ 18 & 17 \\ 25 & 22 \\ 10 & 21 \\ 19 & 20 \\ 25 & 18 \\ 23 & 17 \\ 16 & 15 \\ 20 & 17 \end{bmatrix}$$

Hence, the mean-score matrix is

$$\bar{\mathbf{X}} = \begin{bmatrix} 19.7 & 18.6 \\ 19.7 & 18.6 \\ \vdots & \vdots \\ 19.7 & 18.6 \end{bmatrix} \qquad \text{(10 identical rows)}$$

From these, the SSCP matrix is computed, in accordance with Eq. 5.14, as

$$\begin{aligned} \mathrm{S}(X) &= \mathbf{X}'\mathbf{X} - \bar{\mathbf{X}}'\bar{\mathbf{X}} \\ &= \begin{bmatrix} 4049 & 3673 \\ 3673 & 3526 \end{bmatrix} - \begin{bmatrix} 3880.9 & 3664.2 \\ 3664.2 & 3459.6 \end{bmatrix} \\ &= \begin{bmatrix} 168.1 & 8.8 \\ 8.8 & 66.4 \end{bmatrix} \end{aligned}$$

Subsequently, a researcher defines two derived scores that are of special interest to him. They are:

$$Y_1 = X_1 + X_2$$

(indicating overall "strength" of preferences)

and

$$Y_2 = X_1 - X_2$$

(indicating excess of Outdoor over Gregarious preferences)

He wishes to compute the SSCP matrix for these linearly transformed variables without having to start from scratch; that is, he wants to utilize the fact that the SSCP matrix $\mathrm{S}(X)$ for the original variables is already known.

The transformation matrix here is

$$\mathbf{V} = \begin{bmatrix} 1 & 1 \\ 1 & -1 \end{bmatrix}$$

(Note that this matrix satisfies Eq. 5.3 but not 5.2. Hence $\mathbf{Y}' = \mathbf{X}'\mathbf{V}$ does not represent a rotation. However, we can construct an orthogonal matrix based on $\mathbf{V}$ by normalizing each of its columns to unity—that is, by dividing the elements of its first column by

$$\pm\sqrt{(1)^2 + (1)^2} = \pm\sqrt{2}$$

and those of the second column by

$$\pm\sqrt{(1)^2 + (-1)^2} = \pm\sqrt{2}$$

The signs of the divisors must be chosen so that the resulting matrix, say $\mathbf{V}_0$, has determinant equal to 1. Using $\sqrt{2}$ as the first-column divisor and $-\sqrt{2}$ as the second, we obtain

$$\mathbf{V}_0 = \begin{bmatrix} 1/\sqrt{2} & -1/\sqrt{2} \\ 1/\sqrt{2} & 1/\sqrt{2} \end{bmatrix}$$

which is orthogonal. What is the angle θ of the rotation represented by the transformation $Y' = X'V_0$?)

Going back to the main problem, we can compute the SSCP matrix for the Y's in accordance with Eq. 5.16, thus:

$$\begin{aligned} \mathbf{S}(Y) &= \mathbf{V}'\mathbf{S}(X)\mathbf{V} \\ &= \begin{bmatrix} 1 & 1 \\ 1 & -1 \end{bmatrix} \begin{bmatrix} 168.1 & 8.8 \\ 8.8 & 66.4 \end{bmatrix} \begin{bmatrix} 1 & 1 \\ 1 & -1 \end{bmatrix} \\ &= \begin{bmatrix} 176.9 & 75.2 \\ 159.3 & -57.6 \end{bmatrix} \begin{bmatrix} 1 & 1 \\ 1 & -1 \end{bmatrix} \\ &= \begin{bmatrix} 252.1 & 101.7 \\ 101.7 & 216.9 \end{bmatrix} \end{aligned}$$

The reader should verify this result by computing $\mathbf{S}(Y)$ from scratch. That is, he should first compute the score matrix and mean-score matrix for Y_1 and Y_2 by using Eqs. 5.10a and 5.13a, and then obtain $\mathbf{S}(Y)$ in accordance with Eq. 5.15.

5.3 VARIANCE-MAXIMIZING ROTATIONS

We have seen how axis rotation is expressed matrix-algebraically, and how an SSCP matrix transforms under axis rotation, or under linear transformations more generally. We are now in a position to investigate the problem of how to determine a rotated axis or system of axes possessing certain desirable properties in terms of the transformed SSCP matrix.

One such desirable property is rooted in a general tenet of all scientific endeavor—the canon of parsimony—which seeks to explain as much as possible of observed systematic variation in terms of as few variables as possible. Carried to the logical extreme and translated into multivariate statistical language, this tenet leads one to pose the following task:

Given a set of p-variate observations, determine a rotated axis such that the variable thereby defined as a linear combination of the original p variables has maximum variance.

Intuitive Considerations for Multivariate Normal Distributions. The task posed above can obviously be set and carried out irrespective of the nature of the multivariate distribution followed by the observations. However, if the distribution happens to be multivariate normal, there is a simple geometric interpretation of the axis being sought. We describe this in detail with reference to the bivariate normal distribution.

The bivariate normal density function, Eq. 4.1, may be written in deviation-score form as:

$$\phi(x_1, x_2) = \frac{1}{2\pi\sigma_1\sigma_2\sqrt{1-\rho^2}} \exp\left[\frac{-1}{2(1-\rho^2)}\left\{\frac{x_1^2}{\sigma_1^2} + \frac{x_2^2}{\sigma_2^2} - 2\rho\frac{x_1x_2}{\sigma_1\sigma_1}\right\}\right] \tag{5.17}$$

where $x_1 = X_1 - \mu_1$ and $x_2 = X_2 - \mu_2$. As described in Chapter Four, the density surface is shaped like a somewhat distorted bell such that the cross sections by planes parallel to the (x_1, x_2)-plane are all ellipses, several of which are shown in Fig. 5.6. Since the density function is in deviation-

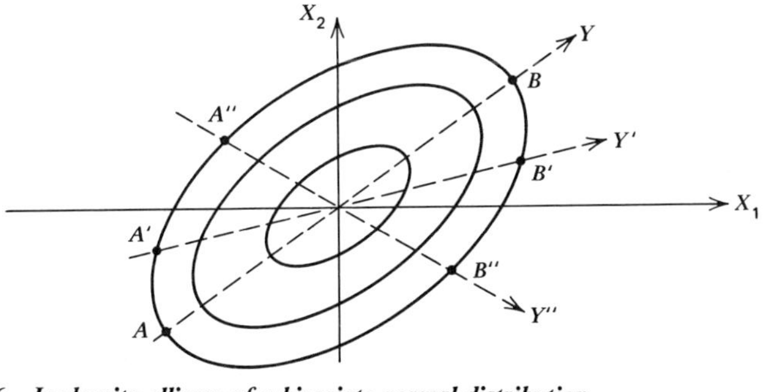

FIGURE 5.6 ***Isodensity ellipses of a bivariate normal distribution.***

score form, the center of these isodensity ellipses coincides with the origin of the coordinate system.

Now imagine the cross sections of the density surface made by various planes containing the density axis, that is, planes perpendicular to the (x_1, x_2)-plane and passing through the origin O. It can be shown by analytic geometry that these cross sections all have the shape of univariate normal density curves. Considering that the intersection of each such plane with the (x_1, x_2)-plane is a straight line indicating a possible position of a rotated axis, it should be intuitively plausible that the cross-sectional curve represents the univariate normal density function of the new variable corresponding to the rotated axis. Actually, the height of the curve has to be proportionally adjusted so that the area under it becomes unity before the curve is exactly that of the density function. However, this adjustment does not affect the "width" of the curve, which is a measure of the variance of the transformed variable.

Consequently, the relative magnitudes of the variances of transformed variables defined by different rotated axes may be assessed by comparing the lengths of the line segments (like $A''B''$, $A'B'$, and AB in Fig. 5.6) that are cut off from the several axes by a given isodensity ellipse of the original bivariate normal surface. For instance, if the largest ellipse in Fig. 5.6 represents the 90% ellipse of the bivariate distribution, then segment $A''B''$ is the symmetric 90% interval of the normal curve which represents the distribution of the transformed variable defined by axis Oy''. Similar remarks hold for segments $A'B'$, AB, and any other diameter of the ellipse in question. Since all the transformed variables have normal distributions, the longer the 90% interval, the larger the variance.

On the basis of the foregoing intuitive reasoning, it is clear that the axis defining the variable with maximum variance must coincide with the major axis of any isodensity ellipse. (In Fig. 5.6, Oy represents this axis.) This, then, is the special geometric interpretation that obtains for the variance-maximizing rotated axis when the original variables follow a bivariate normal distribution. The argument can be extended to higher dimensions, and we have the following theorem.

Theorem. Given p variables following a p-variate normal distribution, the axis defining the linearly transformed variable with maximum variance is the major axis of the isodensity hyperellipsoid

$$\mathbf{x}'\boldsymbol{\Sigma}^{-1}\mathbf{x} = C$$

where $\mathbf{x}' = [X_1 - \mu_1, X_2 - \mu_2, \ldots, X_p - \mu_p]$, $\boldsymbol{\Sigma}$ is the variance-covariance matrix, and C is an arbitrary positive constant.

Before developing a general method for determining the variance-maximizing axis for any set of multivariate data, let us utilize the formula for locating the major axis of an ellipse, cited in Chapter Four, to determine the desired axis in the case of a bivariate normal distribution. We use Example 5.3 of the preceding section to illustrate the procedure. (In so doing we are temporarily pretending that the variance-covariance matrix obtainable by dividing the SSCP matrix of that example by 10, is that of a bivariate normal distribution; we shall subsequently see that such a pretense can be dispensed with.)

For the stated numerical example, the variance-covariance matrix is:

$$\mathbf{\Sigma} = \begin{bmatrix} 16.81 & .88 \\ .88 & 6.64 \end{bmatrix}$$

Any isodensity ellipse of the bivariate normal distribution with this variance-covariance matrix, therefore, has the equation,

$$(1/16.81)x_1^2 + (1/6.64)x_2^2 - [(2)(.88)/(16.81)(6.64)]x_1x_2 = C$$

where C is some positive constant. (Note that the coefficient of x_1x_2 here is equivalent to the coefficient $-2\rho/\sigma_1\sigma_2$ given in Eq. 5.17 because $\rho\sigma_1\sigma_2/\sigma_1^2\sigma_2^2 = \rho/\sigma_1\sigma_2$.)

Therefore, the angle θ from the x_1 axis to the major axis (since $\rho > 0$) of the ellipse is the smallest positive angle satisfying the equation

$$\tan 2\theta = \frac{2\rho\sigma_1\sigma_2}{\sigma_1^2 - \sigma_2^2} = (2)(.88)/(16.81 - 6.64) = .17306$$

Consulting a table of trigonometric functions, we find $2\theta = 9°49'$, and hence $\theta = 4°54.5'$. Evidently, the major axis of the isodensity ellipse slopes only very slightly away from the x_1 axis—a fact that could have been anticipated by noting that the correlation coefficient is very small,

$$\rho = .88/\sqrt{(16.81)(6.64)} = .0833$$

To verify that the axis just located is indeed the variance-maximizing axis, let us compute the variance of the transformed variable thereby defined and compare it with the variances of several other transformed variables corresponding to other rotated axes. We use a special case of Eq. 5.16 in which the transformation matrix $\mathbf{V}$ is a two-dimensional column vector with elements $\cos\theta$ and $\sin\theta$. Again consulting a table of trigonometric functions, we find

$$\cos 4°54.5' = .996332$$

and

$$\sin 4°54.5' = .085572$$

Therefore, the variance of the new variable y defined by the rotation

$$y = .996332x_1 + .085572x_2$$

is given by

$$\begin{aligned}\sigma_y^2 &= [.996332 \quad .085572] \begin{bmatrix} 16.81 & .88 \\ .88 & 6.64 \end{bmatrix} \begin{bmatrix} .996332 \\ .085572 \end{bmatrix} \\ &= 16.8856\end{aligned}$$

Let us take two other variables y' and y'' defined by axes slightly off from and on opposite sides of Oy, by choosing $\theta = 4°$ and $\theta = 6°$ as the angles of rotation. Carrying out the same operations as we just did, but using the transformation matrices

$$\begin{bmatrix} \cos 4° \\ \sin 4° \end{bmatrix} \quad \text{and} \quad \begin{bmatrix} \cos 6° \\ \sin 6° \end{bmatrix}$$

we find:

$$\begin{aligned}\sigma_{y'}^2 &= [.99756 \quad .06976] \begin{bmatrix} 16.81 & .88 \\ .88 & 6.64 \end{bmatrix} \begin{bmatrix} .99756 \\ .06976 \end{bmatrix} \\ &= 16.8830\end{aligned}$$

and

$$\begin{aligned}\sigma_{y''}^2 &= [.99452 \quad .10453] \begin{bmatrix} 16.81 & .88 \\ .88 & 6.64 \end{bmatrix} \begin{bmatrix} .99452 \\ .10553 \end{bmatrix} \\ &= 16.8818\end{aligned}$$

It is thus seen that σ_y^2 is greater than either $\sigma_{y'}^2$ or $\sigma_{y''}^2$. It will be instructive for the reader to compute the variances for several other rotated axes, with angles of rotation 45°, 60°, 90°, 120°, 135°, and so on. It should be found that the variance steadily decreases until the angle of rotation is about 90°, and then increases again. The reader may be able to guess just what angle of rotation would yield the minimum variance. More will be said about the significance of the "variance-minimizing" axis when we discuss the variance-maximizing problem more generally.

EXERCISES

The following exercises refer to a bivariate normal distribution with variance-covariance matrix

$$\boldsymbol{\Sigma} = \begin{bmatrix} 23 & -12 \\ -12 & 16 \end{bmatrix}$$

1. Plot the graph of an arbitrary isodensity ellipse of the bivariate normal distribution specified above, taking the origin at the (unspecified) centroid.

2. Using the tan 2θ equation, locate the major axis of the ellipse you drew above.
3. Compute the variance of the transformed variable represented by the major axis located above. Compare this value with the variances of variables represented by rotated axes 10° away on the two sides of the major axis.
4. Knowing that the minor axis (that is, shortest diameter) of an ellipse is perpendicular to the major axis, determine (from your result in Exercise 2) the angle of rotation from Ox_1 to the minor axis.
5. Repeat Exercise 3 for the transformed variable represented by the minor axis, including comparisons with variances along two nearby axes.

General Formulation of Variance-Maximizing Rotation Problems. We shall now develop a more general approach to the problem of determining the rotated axis with maximum variance—one that does not depend on the identifying of this axis with the major axis of an isodensity ellipsoid. Conceptually, the approach is quite straightforward, even though its execution requires rather involved techniques. We have, in Eq. 5.16, a means for computing the sum-of-squares of a linearly transformed variable defined by an arbitrary $p \times 1$ transformation matrix $\mathbf{V}$. Since the latter is a column vector, we will henceforth denote it by $\mathbf{v}$, reserving the uppercase $\mathbf{V}$ for $p \times q$ transformation matrices with $q > 1$. The formula for computing the sum-of-squares $\sum y^2$ of a transformed variable y therefore becomes

$$\sum y^2 = \mathbf{v}'\mathbf{S}(X)\mathbf{v} \tag{5.18}$$

And, since we are concerned only with those transformed variables that are defined by axis rotation, we impose the restriction

$$\mathbf{v}'\mathbf{v} = 1 \tag{5.19}$$

which is a special case of Eq. 5.11′, on the vector of coefficients $\mathbf{v}$.

Our problem is thus reduced to the following:

Given an SSCP matrix $\mathbf{S}(X)$, determine a vector $\mathbf{v}$ such that maximizes $\mathbf{v}'\mathbf{S}(X)\mathbf{v}$, subject to the condition $\mathbf{v}'\mathbf{v} = 1$.

A new interpretation of the restriction Eq. 5.19 emerges in this mathematical formulation of the problem. Although we have hitherto associated this restriction with axis rotation, it is now seen that the maximizing problem itself is meaningless without specifying the norm $(\mathbf{v}'\mathbf{v})^{1/2}$ of the transformation vector $\mathbf{v}$. For the quantity $\mathbf{v}'\mathbf{S}(X)\mathbf{v}$ can be made as large as we please by taking numbers with large absolute values as the elements of $\mathbf{v}$, if the norm of $\mathbf{v}$ were left arbitrary. For instance, the two linear combinations $y = .6x_1 + .8x_2$ and $y' = 1.2x_1 + 1.6x_2$ have the same *relative* weights for x_1 and x_2, and are in this sense equivalent. But the variance of y' would be four times as large as that of y for any set of x_1 and x_2 values because

$$y' = 2(.6x_1 + .8x_2) = 2y$$

Similarly, a third linear combination

$$y'' = 1.8x_1 + 2.4x_2 \; (= 3y)$$

would have a variance nine times that of y. Thus, the maximizing problem makes sense only if we exclude from consideration various multiples of a vector of coefficients. This is most readily done by specifying that the only vectors to be considered will be those with a fixed length or norm. The fixed length is customarily taken as 1, because then the linear transformation will correspond to a "pure" rotation. Other lengths for the transformation vector would correspond to rotation and *change of scale*.

To return to the problem at hand, our task is to find a set of weights $\mathbf{v}' = [v_1, v_2, \ldots, v_p]$ for constructing a linear combination

$$y = v_1x_1 + v_2x_2 + \cdots + v_px_p$$

such that the quantity

$$\Sigma y^2 = \mathbf{v}'\mathbf{S}(X)\mathbf{v}$$

will be as large as possible, under the restriction that

$$\mathbf{v}'\mathbf{v} = \sum_{i=1}^{p} v_i^2 = 1$$

A convenient method for solving such a maximization problem with side conditions is that of *Lagrange multipliers*, discussed in most calculus textbooks. In general terms, the method states that if a function $f(v_1, v_2, \ldots, v_p)$ of several variables is to be maximized (or minimized) under the side condition $g(v_1, \ldots, v_p) = 0$, this can be accomplished by constructing a new function

$$F = f(v_1, v_2, \ldots, v_p) - \lambda g(v_1, v_2, \ldots, v_p)$$

(where λ is a new unknown, called the Lagrange multiplier), maximizing (or minimizing) this new function without any restriction on the variables, and then later imposing the condition $g(v_1, v_2, \ldots, v_p) = 0$. Applied to our problem, the new function to be maximized without restriction on the elements of $\mathbf{v}$ is

$$F = \mathbf{v}'\mathbf{S}(X)\mathbf{v} - \lambda(\mathbf{v}'\mathbf{v} - 1) \tag{5.20}$$

A necessary condition to be satisfied by $\mathbf{v}$ in order that F be maximized can be obtained by finding what is called the *symbolic partial derivative* of F with respect to $\mathbf{v}$ and setting this equal to the null vector $\mathbf{0}$. As explained in

Appendix C, symbolic partial differentiation by a vector is nothing more than a compact way of summarizing the results of partial differentiation by each element of the vector in turn. Thus, applying Eq. C.1s of Appendix C, we find that the symbolic partial derivative of F with respect to $\mathbf{v}$ is given by

$$\frac{\partial F}{\partial \mathbf{v}} = 2\mathbf{S}(X)\mathbf{v} - 2\lambda\mathbf{v}$$

which is a p-dimensional column vector. Setting this expression equal to the p-dimensional null vector $\mathbf{0}$ (a column consisting of p zeros), and then dividing through by 2 and factoring out $\mathbf{v}$, we obtain

$$(\mathbf{S}(X) - \lambda\mathbf{I})\mathbf{v} = \mathbf{0} \tag{5.21}$$

as the necessary condition to be satisfied by $\mathbf{v}$ in order that $\mathbf{v}'\mathbf{S}(X)\mathbf{v}$ be maximized, subject to the condition $\mathbf{v}'\mathbf{v} = 1$.

It will presently be seen that there are, in general, several vectors $\mathbf{v}$ that satisfy Eq. 5.21, but that a readily identifiable one of them will always be the required maximizing vector. It also happens that equations of the general form of Eq. 5.21 frequently turn up in connection with the several multivariate statistical techniques to be discussed in the sequel. The rationale for solving an equation of this type is, therefore, well worth our careful study—even though recourse to a computer solution becomes a practical necessity when the order of the matrix $\mathbf{S}(X)$ is larger than 3 or 4. However, for slightly larger matrices (for example, of order up to 6 or 7), an iterative method described in Appendix D is practicable.

5.4 SOLUTION OF EQUATION FOR VARIANCE-MAXIMIZING ROTATIONS: EIGENVALUE PROBLEMS

Since equations of the form of (5.21) arise in other connections besides variance-maximizing rotations, we shall replace the $\mathbf{S}(X)$ there by an arbitrary square matrix $\mathbf{A}$, and discuss the solution of the matrix equation

$$(\mathbf{A} - \lambda\mathbf{I})\mathbf{v} = \mathbf{0} \tag{5.22}$$

or, to indicate the elements of the matrix and vectors,

$$\begin{bmatrix} a_{11} - \lambda & a_{12} & \cdots & a_{1p} \\ a_{21} & a_{22} - \lambda & \cdots & a_{2p} \\ \vdots & \vdots & \ddots & \vdots \\ a_{p1} & a_{p2} & \cdots & a_{pp} - \lambda \end{bmatrix} \begin{bmatrix} v_1 \\ v_2 \\ \vdots \\ v_p \end{bmatrix} = \begin{bmatrix} 0 \\ 0 \\ \vdots \\ 0 \end{bmatrix}$$

In most, but not all, applications $\mathbf{A}$ will be a symmetrical matrix and will have a further property known as positive semi-definiteness, to be discussed later.

Equation 5.22 differs in two notable respects from the set of normal equations encountered in multiple regression analysis and given, in extended and symbolic forms, respectively, as Eqs. 2.25 and 2.27. First, the vector of constants on the right-hand side is a null vector. Such a set of equations is called a set of *homogeneous equations*. Second, the matrix of coefficients on the left-hand side, $\mathbf{A} - \lambda\mathbf{I}$, itself involves an unknown scalar quantity λ.

An obvious feature of homogeneous equations is that setting all the unknowns equal to zero will always satisfy the equations, because $(\mathbf{A} - \lambda\mathbf{I})\mathbf{0} = \mathbf{0}$, regardless of what $\mathbf{A}$ and λ are. Therefore, $\mathbf{v} = \mathbf{0}$ is called a *trivial solution* of any set of homogeneous equations. The important question is, when does a set of homogeneous equations possess a nontrivial solution, that is, a solution other than the trivial one?

Suppose, for a moment, that an appropriate value for λ had somehow been determined, so that the coefficient matrix $\mathbf{A} - \lambda\mathbf{I}$ is completely known. Now, if $\mathbf{A} - \lambda\mathbf{I}$ is nonsingular, and hence possesses an inverse, then premultiplying both members of Eq. 5.22 by $(\mathbf{A} - \lambda\mathbf{I})^{-1}$ will yield

$$\mathbf{v} = (\mathbf{A} - \lambda\mathbf{I})^{-1}\mathbf{0} = \mathbf{0}$$

that is, the trivial solution, as the *only* solution of the equation. We therefore conclude that in order for a set of homogeneous equations to possess a nontrivial solution, the matrix $\mathbf{A} - \lambda\mathbf{I}$ must *not* have an inverse. In other words, the value of λ must be such that $\mathbf{A} - \lambda\mathbf{I}$ becomes a singular matrix. That is, λ must satisfy the equation

$$|\mathbf{A} - \lambda\mathbf{I}| = 0 \tag{5.23}$$

which is called the *characteristic equation* of the matrix $\mathbf{A}$.

Thus, the first step in solving a matrix equation of the form (5.22) is to determine the roots of the characteristic equation, which are the only possible values of λ for yielding nontrivial solutions for $\mathbf{v}$. These λ values are called the *characteristic roots* or *eigenvalues* of $\mathbf{A}$. The characteristic equation is a polynomial equation in λ, of degree equal to the order of $\mathbf{A}$. Hence, if $\mathbf{A}$ is a $p \times p$ matrix, its characteristic equation will have p roots, although it is possible that not all of them are distinct, and that some of them are zero. (If $\mathbf{A}$ is symmetric, all its eigenvalues are real numbers.)

EXAMPLE 5.4. Find the eigenvalues of the variance-covariance matrix considered in the preceding section in connection with the variance-maximizing rotation problem. That matrix was

$$\mathbf{\Sigma} = \begin{bmatrix} 16.81 & .88 \\ .88 & 6.64 \end{bmatrix}$$

Its characteristic equation, $|\mathbf{\Sigma} - \lambda \mathbf{I}| = 0$, is

$$\begin{vmatrix} 16.81 - \lambda & .88 \\ .88 & 6.64 - \lambda \end{vmatrix} = 0$$

which, on expanding the determinant, becomes

$$\lambda^2 - 23.45\lambda + 110.844 = 0$$

Using the formula for the roots of a quadratic, we find

$$\lambda_1 = 16.8856 \qquad \text{and} \qquad \lambda_2 = 6.5644$$

as the eigenvalues of $\mathbf{\Sigma}$.

EXAMPLE 5.5. Find the eigenvalues of the matrix

$$\mathbf{A} = \begin{bmatrix} 136 & 104 & 94 \\ 104 & 106 & 71 \\ 94 & 71 & 65 \end{bmatrix}$$

The characteristic equation of $\mathbf{A}$ is

$$\begin{vmatrix} 136 - \lambda & 104 & 94 \\ 104 & 106 - \lambda & 71 \\ 94 & 71 & 65 - \lambda \end{vmatrix} = 0$$

Expanding the determinant either by Sarrus' rule or by cofactors along any row or column, this equation reduces to

$$-\lambda^3 + 307\lambda^2 + 1437\lambda = 0$$

a cubic with no constant term on the left-hand side. Thus, one of its roots is 0, and the other two are those of the quadratic,

$$\lambda^2 - 307\lambda - 1437 = 0$$

which are 311.6115 and -4.6115. The three eigenvalues of $\mathbf{A}$, in descending order of magnitude, are: $\lambda_1 = 311.6115$, $\lambda_2 = 0$, $\lambda_3 = -4.6115$.

EXERCISE

Find the eigenvalues of the matrix

$$\mathbf{B} = \begin{bmatrix} 7 & 0 & 1 \\ 0 & 7 & 2 \\ 1 & 2 & 3 \end{bmatrix}$$

Continuing with the solution of Eq. 5.22,

$$(\mathbf{A} - \lambda \mathbf{I})\mathbf{v} = \mathbf{0}$$

we note that there will be one vector solution $\mathbf{v}_i$ corresponding to each eigenvalue λ_i of **A**. The vector $\mathbf{v}_i$ is called the *characteristic vector* or *eigenvector* of **A** associated with the characteristic root or eigenvalue λ_i. Each eigenvector is determined as follows:

Step 1. For the given eigenvalue λ_i, write out the matrix $\mathbf{A} - \lambda_i\mathbf{I}$ by subtracting λ_i from each diagonal element of **A**.

Step 2. Compute the adjoint $\mathbf{adj}(\mathbf{A} - \lambda_i\mathbf{I})$ of the matrix $(\mathbf{A} - \lambda_i\mathbf{I})$ in accordance with the steps indicated in Chapter Two. It should be found that the columns of $\mathbf{adj}(\mathbf{A} - \lambda_i\mathbf{I})$ are all proportional to one another.

Step 3. Divide the elements of any column of $\mathbf{adj}(\mathbf{A} - \lambda_i\mathbf{I})$ by the square root of the sum of the squares of these elements. The resulting numbers are the elements of $\mathbf{v}_i$, with $\mathbf{v}_i'\mathbf{v}_i = 1$.

To see the rationale of the procedure described above, let us introduce the abbreviation

$$\mathbf{A} - \lambda_i\mathbf{I} = \mathbf{H}$$

and denote the cofactor of the (j, k)-element of **H** by H_{jk}. Considering, for simplicity, the case when **A** is of order 3, the adjoint $\mathbf{adj}(\mathbf{A} - \lambda_i\mathbf{I})$ of $\mathbf{A} - \lambda_i\mathbf{I}$ is then given by

$$\mathbf{adj}(\mathbf{H}) = \begin{bmatrix} H_{11} & H_{21} & H_{31} \\ H_{12} & H_{22} & H_{32} \\ H_{13} & H_{23} & H_{33} \end{bmatrix}$$

In accordance with *Step 3* above, we take

$$\mathbf{v}_i' = c[H_{11}, H_{12}, H_{13}]$$

where

$$c = 1/\sqrt{H_{11}^2 + H_{12}^2 + H_{13}^2}$$

is a scalar multiplier to make $\mathbf{v}_i'\mathbf{v}_i = 1$.

To show that this $\mathbf{v}_i$ satisfies Eq. 5.22, we carry out the multiplication,

$$(\mathbf{A} - \lambda_i\mathbf{I})\mathbf{v}_i = \begin{bmatrix} h_{11} & h_{12} & h_{13} \\ h_{21} & h_{22} & h_{23} \\ h_{31} & h_{32} & h_{33} \end{bmatrix} \begin{bmatrix} H_{11} \\ H_{12} \\ H_{13} \end{bmatrix} \cdot c$$

$$= c \begin{bmatrix} h_{11}H_{11} + h_{12}H_{12} + h_{13}H_{13} \\ h_{21}H_{11} + h_{22}H_{12} + h_{23}H_{13} \\ h_{31}H_{11} + h_{32}H_{12} + h_{33}H_{13} \end{bmatrix}$$

The first element of the resulting product vector (apart from the factor c) is none other than the expansion of $|\mathbf{H}|$ along its first row. Hence the value of this element is $|\mathbf{H}| = |\mathbf{A} - \lambda_i\mathbf{I}| = 0$, because λ_i is a root of the characteristic equation. Each of the other two elements of the product vector $\mathbf{Hv}_i$ represents a linear combination of the cofactors H_{11}, H_{12}, H_{13}, using elements of a *non-corresponding* row of $|\mathbf{H}|$ as coefficients. Hence, both these elements are zero, by virtue of Property 6 of determinants, stated in Appendix A. Thus, all three elements of $\mathbf{Hv}_i$ are 0; that is,

$$\mathbf{Hv}_i = (\mathbf{A} - \lambda_i\mathbf{I})\mathbf{v}_i = \mathbf{0}$$

as we set out to prove.

To illustrate the procedure, we use the example given in the exercise for finding eigenvalues. The reader should have found $\lambda_1 = 8$, $\lambda_2 = 7$, and $\lambda_3 = 2$ as the three eigenvalues of the matrix

$$\mathbf{B} = \begin{bmatrix} 7 & 0 & 1 \\ 0 & 7 & 2 \\ 1 & 2 & 3 \end{bmatrix}$$

Step 1. To find the eigenvector $\mathbf{v}_1$ corresponding to $\lambda_1 = 8$, we first write out the matrix

$$\mathbf{B} - 8\mathbf{I} = \begin{bmatrix} -1 & 0 & 1 \\ 0 & -1 & 2 \\ 1 & 2 & -5 \end{bmatrix}$$

Step 2. Compute the adjoint of the above matrix. The result is:

$$\mathbf{adj}(\mathbf{B} - 8\mathbf{I}) = \begin{bmatrix} 1 & 2 & 1 \\ 2 & 4 & 2 \\ 1 & 2 & 1 \end{bmatrix}$$

Note that all three columns are proportional, which offers a partial check on the calculations. (In this example, the first and third columns are identical, but this will not be true in general.) Before carrying out the last step, the reader should verify that

$$(\mathbf{B} - 8\mathbf{I}) \begin{bmatrix} 1 \\ 2 \\ 1 \end{bmatrix} = \mathbf{0} \qquad \text{or that} \qquad \mathbf{B} \begin{bmatrix} 1 \\ 2 \\ 1 \end{bmatrix} = 8 \begin{bmatrix} 1 \\ 2 \\ 1 \end{bmatrix}$$

Step 3. In order to have $\mathbf{v}_1$ satisfy the unit-norm condition, $\mathbf{v}_1'\mathbf{v}_1 = 1$, we divide each element of [1, 2, 1] by $\sqrt{1^2 + 2^2 + 1^2} = \sqrt{6}$, and obtain

$$\mathbf{v}_1' = [.4082 \quad .8165 \quad .4082]$$

EXERCISE

Compute the eigenvectors $\mathbf{v}_2$ and $\mathbf{v}_3$ associated with the other two eigenvalues λ_2 and λ_3 in the example above.

Next, in order to compare the results obtained by two different methods, we compute the eigenvectors of the variance-covariance matrix

$$\mathbf{\Sigma} = \begin{bmatrix} 16.81 & .88 \\ .88 & 6.64 \end{bmatrix}$$

based on the SSCP matrix used for the rotation problem of the preceding section. We have, in Example 5.4 above, already found the eigenvalues of this matrix to be

$$\lambda_1 = 16.8856 \qquad \text{and} \qquad \lambda_2 = 6.5644$$

Hence,

$$\mathbf{\Sigma} - \lambda_1\mathbf{I} = \begin{bmatrix} -.0756 & .88 \\ .88 & -10.2456 \end{bmatrix}$$

and

$$\mathbf{adj}(\mathbf{\Sigma} - \lambda_1\mathbf{I}) = \begin{bmatrix} -10.2456 & -.88 \\ -.88 & -.0756 \end{bmatrix}$$

$$\mathbf{v}_1 = \begin{bmatrix} -10.2456 \\ -.88 \end{bmatrix} \div \sqrt{(-10.2456)^2 + (-.88)^2} = \begin{bmatrix} -.99633 \\ -.08558 \end{bmatrix}$$

Similarly,

$$\mathbf{adj}(\mathbf{\Sigma} - \lambda_2\mathbf{I}) = \begin{bmatrix} .0756 & -.88 \\ -.88 & 10.2456 \end{bmatrix}$$

whence

$$\mathbf{v}_2 = \begin{bmatrix} .0756 \\ -.88 \end{bmatrix} \div \sqrt{(.0756)^2 + (-.88)^2} = \begin{bmatrix} .08558 \\ -.99633 \end{bmatrix}$$

Two important points are exemplified in the above results. First, apart from the signs, the elements of $\mathbf{v}_1$ agree (within rounding error) with the values of $\cos\theta$ and $\sin\theta$ previously obtained for the variance-maximizing rotation. The signs depend merely on whether the angle from the old x_1-axis is taken as terminating at Oy^+ (the positive direction of the new axis) or at Oy^- (the negative direction), as shown in Fig. 5.7, below, and the facts that $\cos(\theta + 180°) = -\cos\theta$ and $\sin(\theta + 180°) = -\sin\theta$.

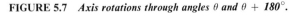

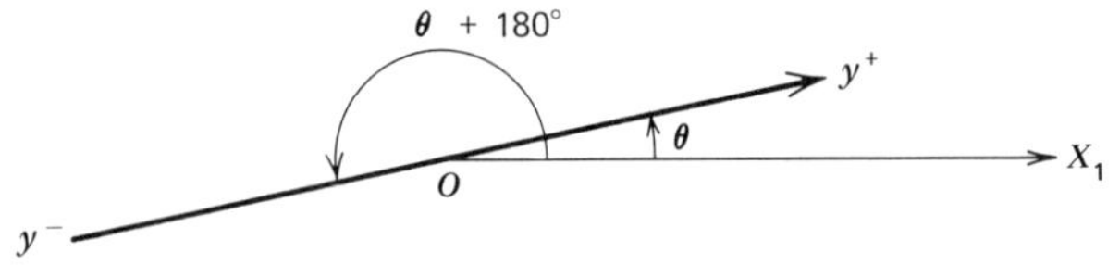

FIGURE 5.7 ***Axis rotations through angles θ and θ + 180°.***

We thus see that the assumption of a bivariate (in general, p-variate) normal distribution, made in the previous subsection for determining the variance-maximizing rotation in terms of the major axis of an ellipse, is actually gratuitous. The only difference is that, when the distribution is indeed multivariate normal, the ellipse whose major axis is determined in the variance-maximizing process is an isodensity ellipse of the distribution, and hence the intuitive geometric argument presented earlier makes sense. In other cases, the ellipse is of no particular statistical relevance, and the general approach of the current subsection provides the only adequate rationale.

The second point to be noted in the illustrative example above is that the two eigenvectors $\mathbf{v}_1$ and $\mathbf{v}_2$ satisfy the orthogonality condition, $\mathbf{v}_1'\mathbf{v}_2 = 0$, as may easily be verified. (A general proof will be given in the next section.) And, since both vectors were determined so as to have unit norms, it follows that the 2×2 transformation matrix $\mathbf{V} = [\mathbf{v}_1 \mid \mathbf{v}_2]$, with $\mathbf{v}_1$ and $\mathbf{v}_2$ as its columns, is an orthogonal matrix—which corresponds to a rigid rotation. Hence, the axis defined by $\mathbf{v}_2$ is perpendicular (or orthogonal) to the axis defined by $\mathbf{v}_1$, which we know to be the major axis of an ellipse $\mathbf{x}'\mathbf{\Sigma}^{-1}\mathbf{x} = C$. Consequently, $\mathbf{v}_2$ corresponds to the minor axis of this ellipse, which (following the geometric argument given earlier) is the variance-*minimizing* axis.

It may seem paradoxical that, even though we set out to determine the variance-maximizing axis, we got both the maximizing *and* the minimizing axes as solutions of the same matrix equation, $(\mathbf{\Sigma} - \lambda\mathbf{I})\mathbf{v} = \mathbf{0}$. The reason is that this equation is only a *necessary* condition for the variance-maximizing transformation; its solution may yield either the maximum or minimum (or even a stationary point) of the function we set out to maximize.

In the bivariate case, $\mathbf{v}_2$ indeed corresponds to the axis along which the variance is at a minimum. But when there are three or more variables, and the variance-covariance matrix has three or more eigenvectors $\mathbf{v}_i$, it turns out that all except the one associated with the smallest eigenvalue define axes along which the variances are *conditionally maximal* in the following sense: If we denote the eigenvectors as $\mathbf{v}_1, \mathbf{v}_2, \mathbf{v}_3, \ldots$, in descending order of magnitude of their associated eigenvalues $\lambda_1, \lambda_2, \lambda_3, \ldots$, and label the corresponding transformed variables $Y_1, Y_2, Y_3, \ldots$, respectively, then Y_1 has the largest possible variance; Y_2 has the largest variance among all linearly transformed variables *that are uncorrelated with* Y_1; Y_3 has the largest variance among all linearly transformed variables *that are uncorrelated with both* Y_1 *and* Y_2; and so on. (Where it is understood, of course, that by "linearly transformed" we mean so transformed by a vector of unit length.)

It may be noted in passing that even the last transformed variable Y_p, although possessing minimum variance, does have the above property of "conditional maximizing" of variance. The fact is that, once the first $p - 1$ transformed variables are determined, there is only *one* linearly transformed variable that is uncorrelated with all of these, and Y_p is this variable. Hence, Y_p vacuously satisfies the statement that it has the largest variance among those variables that are uncorrelated with $Y_1, Y_2, \ldots, Y_{p-1}$. In this respect, the bivariate case is no exception: although Y_2 has minimum variance, it is, at the same time, the linearly transformed variable with maximum variance among those uncorrelated with Y_1, since Y_2 is the *only* such variable.

The reader may have anticipated, when we numbered the transformed variables $Y_1, Y_2, Y_3, \ldots$, in descending order of magnitude of the corresponding eigenvalues $\lambda_1, \lambda_2, \lambda_3, \ldots$, that the latter must have something to do with the variances of the former. We now show that, for each i, λ_i is precisely equal to the variance of Y_i—even though it was originally introduced artificially as a Lagrange multiplier related to the side condition $\mathbf{v}_i'\mathbf{v}_i = 1$ in our maximizing problem.

Dividing both sides of Eq. 5.18 by N, we see that

$$\mathrm{Var}(Y_i) = \frac{1}{N}\mathbf{v}_i'\mathbf{S}(X)\mathbf{v}_i = \mathbf{v}_i'\mathbf{\Sigma}\mathbf{v}_i \tag{5.24}$$

But, from Eq. 5.22 with $\mathbf{A}$ replaced by $\mathbf{\Sigma}$, we know that, for each i,

$$(\mathbf{\Sigma} - \lambda_i\mathbf{I})\mathbf{v}_i = \mathbf{0}$$

or, equivalently, that

$$\mathbf{\Sigma}\mathbf{v}_i = \lambda_i\mathbf{v}_i \tag{5.25}$$

Substituting the right-hand expression here for $\mathbf{\Sigma}\mathbf{v}_i$ in Eq. 5.24, we have:

$$\begin{aligned}\mathrm{Var}(Y_i) &= \mathbf{v}_i'(\lambda_i\mathbf{v}_i)\\ &= \lambda_i(\mathbf{v}_i'\mathbf{v}_i)\end{aligned}$$

(because λ_i is a scalar), or

$$\mathrm{Var}\,(Y_i) = \lambda_i \tag{5.26}$$

(because $\mathbf{v}_i'\mathbf{v}_i = 1$, by construction), as we set out to prove.

We conclude this section by introducing the term *principal-axes rotation* to replace our earlier "variance-maximizing rotation." Since by solving the eigenvalue problem $(\mathbf{\Sigma} - \lambda\mathbf{I})\mathbf{v} = \mathbf{0}$ we obtain the principal axes of an ellipsoid (or hyperellipsoid), $\mathbf{x}'\mathbf{\Sigma}^{-1}\mathbf{x} = C$. The first principal axis is the usual major axis (the longest diameter); the second principal axis is the longest of all the diameters that are orthogonal to the first; the third is the longest of all the diameters that are orthogonal to both the first and the second; and so on.

EXERCISES

Given a set of three variables with variance-covariance matrix

$$\mathbf{\Sigma} = \begin{bmatrix} 128.7 & 61.4 & -21.0 \\ 61.4 & 56.9 & -28.3 \\ -21.0 & -28.3 & 63.5 \end{bmatrix}$$

1. Determine the (orthogonal) transformation matrix for the principal-axes rotation.
2. Using Eq. 5.24, verify that the variance of each of the transformed variables is equal to the corresponding eigenvalue of $\boldsymbol{\Sigma}$, found in the process of determining the transformation matrix above.

5.5 SOME PROPERTIES OF MATRICES RELATED TO EIGENVALUES AND EIGENVECTORS

There are many interesting and useful relationships between the eigenvalues and eigenvectors of a matrix and certain characteristics of the latter, some of which were informally mentioned in the previous section. We here state a number of these properties in a sequence of theorems, some with complete proofs, and others with proof-outlines or intuitive supporting arguments (when the proofs are unduly long or too advanced).

Theorem 1. The sum $\sum \lambda_i$ of the eigenvalues of a matrix $\mathbf{A}$ is equal to the sum $\sum a_{ii}$ of the latter's diagonal elements. This sum is called the *trace* of $\mathbf{A}$, denoted tr($\mathbf{A}$).

Outline of proof. The eigenvalues λ_i of $\mathbf{A}$ are, by definition, the roots of its characteristic equation,

$$\begin{vmatrix} a_{11} - \lambda & a_{12} & \cdots & a_{1p} \\ a_{21} & a_{22} - \lambda & \cdots & a_{2p} \\ \vdots & \vdots & \ddots & \vdots \\ a_{p1} & a_{p2} & \cdots & a_{pp} - \lambda \end{vmatrix} = 0$$

which is a polynomial equation of degree p in λ. In the theory of equations (see, for example, Richardson, 1966, p. 363), it is shown that if $\lambda_1, \lambda_2, \ldots, \lambda_p$ are the roots of a polynomial equation

$$c_0\lambda^p + c_1\lambda^{p-1} + c_2\lambda^{p-2} + \cdots + c_p = 0$$

then

$$\sum \lambda_i = -(c_1/c_0)$$

For our characteristic equation, it can be shown by expanding the determinant that

$$c_0 = (-1)^p \qquad \text{and} \qquad c_1 = (-1)^{p-1}(a_{11} + a_{22} + \cdots + a_{pp})$$

It therefore follows that

$$\sum_{i=1}^{p} \lambda_i = \sum_{i=1}^{p} a_{ii} \tag{5.27}$$

as was to be proved.

The implication of this theorem for the case in which **A** is a variance-covariance matrix (like the $\boldsymbol{\Sigma}$ of the last example in the previous section) is worthy of special note. Then the λ_i is, as shown in Eq. 5.26, the variance of the ith transformed variable Y_i; and a_{ii} is, of course, the variance of the ith original variable X_i. Equation 5.27 therefore implies that

$$\sum_{i=1}^{p} \mathrm{Var}(Y_i) = \sum_{i=1}^{p} \mathrm{Var}(X_i) \tag{5.28}$$

In words, the "aggregate variance" of the whole set of variables remains unchanged from before to after the principal-axes transformation[2] $\mathbf{Y}' = \mathbf{X}'\mathbf{V}$, where the columns of $\mathbf{V}$ are the eigenvectors $\mathbf{v}_i$ of $\boldsymbol{\Sigma}$. Thus, the effect of a principal-axes rotation is to "reshuffle" the aggregate variance so that the lion's share is allotted to the first transformed variable; the next largest share to a second transformed variable that is uncorrelated with the first, and so on.

Theorem 2. The product $\lambda_1, \lambda_2, \ldots, \lambda_p$ (which is abbreviated $\prod_{i=1}^{p} \lambda_i$) of the eigenvalues of a matrix **A** is equal to the value of the determinant $|\mathbf{A}.|$

Outline of proof: Borrowing again from the theory of equations, we have the relation

$$\prod_{i=1}^{p} \lambda_i = (-1)^p (c_p/c_0)$$

where c_p is the constant term, and c_0 the coefficient of λ^p in the characteristic equation, as in the proof outline for Theorem 1. But the constant term of $|\mathbf{A} - \lambda\mathbf{I}|$ is readily found by setting $\lambda = 0$ in this expression: $c_p = |\mathbf{A} - 0\mathbf{I}| = |\mathbf{A}|$. From this and the fact that $c_0 = (-1)^p$, stated before, it immediately follows that

$$\prod_{i=1}^{p} \lambda_i = |\mathbf{A}| \tag{5.29}$$

An important corollary to this theorem is that a matrix has one or more eigenvalues equal to zero if and only if the matrix is singular. This is immediately obvious from Eq. 5.29. Furthermore, the number of zero eigenvalues

[2] Actually, the property of leaving the aggregate variance unchanged is true not only of the principal-axes rotation, but of *all* rigid rotations. This may be seen from the following intuitive consideration: An arbitrary rigid rotation from the X_i's to a new set of variables $\{Y_i'\}$, say, may be followed by a principal-axes rotation from $\{Y_i'\}$ to $\{Y_i\}$. But this set $\{Y_i\}$ must be the same set as obtainable directly by the principal-axes rotation from $\{X_i\}$ (because two successive rigid rotations can always be replaced by a single rigid rotation). Therefore, $\sum \mathrm{Var}(Y_i') = \sum \mathrm{Var}(Y_i)$ *and* $\sum \mathrm{Var}(X_i) = \sum \mathrm{Var}(Y_i)$, from which it follows that $\sum \mathrm{Var}(Y_i') = \sum \mathrm{Var}(X_i)$.

has an important bearing on the nature of **A**—especially when the latter is a variance-covariance matrix. Even though the presence of just one eigenvalue equal to zero is sufficient to assure that **A** is singular, we somehow feel that it is "more singular" with a greater number of zero eigenvalues. A precise sense in which we can speak of various degrees of singularity will subsequently be described.

Theorem 3. Two eigenvectors $\mathbf{v}_i$ and $\mathbf{v}_j$ associated with two distinct eigenvalues λ_i and λ_j of a *symmetric* matrix are mutually orthogonal; that is, $\mathbf{v}_i'\mathbf{v}_j = 0$.

Proof: An instance of this theorem was already verified for a numerical example. The general proof is as follows.

By hypothesis, we have

$$\mathbf{A}\mathbf{v}_i = \lambda_i \mathbf{v}_i \tag{a}$$

and

$$\mathbf{A}\mathbf{v}_j = \lambda_j \mathbf{v}_j \qquad \text{with} \qquad \lambda_i \neq \lambda_j \tag{b}$$

Taking the transposes of the two members of Eq. a, we have:

$$\mathbf{v}_i'\mathbf{A} = \lambda_i \mathbf{v}_i' \tag{a$'$}$$

(since $\mathbf{A}' = \mathbf{A}$ by hypothesis).

Premultiplying both sides of Eq. b by $\mathbf{v}_i'$ yields

$$\mathbf{v}_i'\mathbf{A}\mathbf{v}_j = \lambda_j \mathbf{v}_i'\mathbf{v}_j \tag{c}$$

in the left member of which we may, from Eq. a′, substitute $\lambda_i \mathbf{v}_i'$ in place of $\mathbf{v}_i'\mathbf{A}$ to obtain

$$\lambda_i \mathbf{v}_i'\mathbf{v}_j = \lambda_j \mathbf{v}_i'\mathbf{v}_j$$

or

$$(\lambda_i - \lambda_j)(\mathbf{v}_i'\mathbf{v}_j) = 0$$

Since $\lambda_i - \lambda_j \neq 0$ by hypothesis, it follows that $\mathbf{v}_i'\mathbf{v}_j = 0$. Q.E.D.

Next, instead of considering the eigenvectors in pairs, we treat them all simultaneously by using $\mathbf{v}_1, \mathbf{v}_2, \ldots$ as the successive columns of matrix **V**. If none of the eigenvalues is zero, then **V** will be a square matrix of order equal to that of **A**. Hence, when **A** is the variance-covariance matrix $\mathbf{\Sigma}$ of a given set of p variables $X_1, X_2, \ldots, X_p$, **V** is the orthogonal transformation matrix representing a rigid rotation from the X's to the principal axes.

Since $\mathbf{\Sigma v}_i = \lambda_i \mathbf{v}_i$ for each i, it follows that, if we define a diagonal matrix

$$\mathbf{\Lambda} = \begin{bmatrix} \lambda_1 & & & 0 \\ & \lambda_2 & & \\ & & \ddots & \\ 0 & & & \lambda_p \end{bmatrix} \tag{5.30}$$

we shall have

$$\mathbf{\Sigma V} = \mathbf{V\Lambda} \tag{5.31}$$

because the columns of $\mathbf{V}$ are the successive $\mathbf{v}_i$'s, and postmultiplication by a diagonal matrix multiplies each column of $\mathbf{V}$ by the corresponding diagonal element.

Now if we premultiply both members of Eq. 5.31 by $\mathbf{V}'$, we obtain

$$\mathbf{V'\Sigma V} = \mathbf{V'V\Lambda} = \mathbf{\Lambda} \tag{5.32}$$

(where, in going from the second to the last expression, we utilized the fact that $\mathbf{V'V} = \mathbf{I}$). But, according to a result readily obtainable by dividing both sides of Eq. 5.16 by N, left-most member of Eq. 5.32 is none other than the variance-covariance matrix of the transformed variables Y_i corresponding to the principal axes. Thus, Eq. 5.32 tells us that the covariances of all pairs (Y_i, Y_j) [with $i \neq j$] of the transformed variables are equal to zero (since they are given by the off-diagonal elements of $\mathbf{\Lambda}$).

This is the basis for the assertion made earlier that Y_2 is uncorrelated with Y_i, Y_3 is uncorrelated with both Y_1 and Y_2, and so forth. We summarize the foregoing in a theorem.

Theorem 4. Given a set of variables $X_1, X_2, \ldots, X_p$, with a nonsingular variance-covariance matrix $\mathbf{\Sigma}$, we can always derive a set of uncorrelated variables $Y_1, Y_2, \ldots, Y_p$ by a set of linear transformations corresponding to the principal-axes rotation; that is, the rigid rotation whose transformation matrix $\mathbf{V}$ has, as its columns, the p eigenvectors of $\mathbf{\Sigma}$. The variance-covariance matrix of this new set of variables is the diagonal matrix $\mathbf{\Lambda} = \mathbf{V'\Sigma V}$, whose diagonal elements are the p eigenvalues of $\mathbf{\Sigma}$.

This theorem may be called the fundamental theorem of *principal components analysis*, which is the name given to the procedure of performing a principal-axes rotation. The resulting transformed variables $Y_1, Y_2, \ldots, Y_p$ are known as the *principal components*, and determining their variance-covariance matrix $\mathbf{\Lambda}$ as in Eq. 5.32 is referred to as "diagonalizing the matrix $\mathbf{\Sigma}$."

One reason why Theorem 4 is of prime importance is that it enables us to prove the fact, basic to multivariate significance tests, that the quadratic form $\chi^2 = \mathbf{x'\Sigma^{-1}x}$ in the exponent of the p-variate normal density function is a chi-square variate with p degrees of freedom. This fact was already proved in Chapter Four for the special case when $X_1, X_2, \ldots, X_p$ are *independently* normally distributed. In order to extend the proof to the general case of p variables following a multivariate normal distribution with nonzero correlations, we need only show that the quantity $\mathbf{x}'\mathbf{\Sigma}^{-1}\mathbf{x}$ remains invariant under a principal-axes rotation whereby we can obtain a set of uncorrelated variables (that is, the principal components)—because, in the case of normally distributed variables, uncorrelatedness implies independence. Actually, the invariance property just cited holds for *any* rigid rotation, as shown by the following theorem.

Theorem 5. Given a set of variables $X_1, X_2, \ldots, X_p$ with a nonsingular variance-covariance matrix $\mathbf{\Sigma}_x$, a new set of variables $Y_1, Y_2, \ldots, Y_p$ is defined by the transformation $\mathbf{Y}' = \mathbf{X}'\mathbf{V}$, where $\mathbf{V}$ is an orthogonal matrix. If the variance-covariance matrix of the Y's is denoted by $\mathbf{\Sigma}_y$, the following relation holds between the two variance-covariance matrices:

$$\mathbf{y}'\mathbf{\Sigma}_y^{-1}\mathbf{y} = \mathbf{x}'\mathbf{\Sigma}_x^{-1}\mathbf{x} \tag{5.33}$$

where $\mathbf{x} = \mathbf{X} - \overline{\mathbf{X}}$ and $\mathbf{y} = \mathbf{Y} - \overline{\mathbf{Y}}$. In words, we say that the quadratic form $\mathbf{x}'\mathbf{\Sigma}^{-1}\mathbf{x}$ is invariant under a rigid rotation.

Proof: The stated transformation holds also for deviation scores.

$$\mathbf{y}' = \mathbf{x}'\mathbf{V} \tag{a}$$

From Eq. 5.16, we know that $\mathbf{\Sigma}_y = \mathbf{V}'\mathbf{\Sigma}_x\mathbf{V}$, whence $\mathbf{\Sigma}_y^{-1} = \mathbf{V}^{-1}\mathbf{\Sigma}_x^{-1}(\mathbf{V}')^{-1}$ which, by virtue of the fact that $\mathbf{V}$ is an orthogonal matrix (so that $\mathbf{V}' = \mathbf{V}^{-1}$), reduces to

$$\mathbf{\Sigma}_y^{-1} = \mathbf{V}'\mathbf{\Sigma}_x^{-1}\mathbf{V} \tag{b}$$

Substituting a and b in the left member of the equation to be proved, we have:

$$\begin{aligned}\mathbf{y}'\mathbf{\Sigma}_y^{-1}\mathbf{y} &= (\mathbf{x}'\mathbf{V})(\mathbf{V}'\mathbf{\Sigma}_x^{-1}\mathbf{V})(\mathbf{V}'\mathbf{x}) \\ &= \mathbf{x}'(\mathbf{V}\mathbf{V}')\mathbf{\Sigma}_x^{-1}(\mathbf{V}\mathbf{V}')\mathbf{x} \\ &= \mathbf{x}'\mathbf{\Sigma}_x^{-1}\mathbf{x} \qquad \text{(because } \mathbf{V}\mathbf{V}' = \mathbf{I}\text{)}\end{aligned}$$

which is Eq. 5.33.

When this equation is applied to the case in which $\mathbf{V}$ is the transformation matrix for the principal-axes rotation, we may further invoke Theorem 4 ($\mathbf{\Sigma}_y = \mathbf{\Lambda}$, a diagonal matrix), and conclude that

$$\mathbf{x}'\mathbf{\Sigma}_x^{-1}\mathbf{x} = \mathbf{y}\,\mathbf{\Lambda}^{-1}\mathbf{y}$$

The right-hand expression is the quadratic form for a set of uncorrelated variables, which was already shown to be a chi-square variate with p degrees of freedom when the Y's are normally distributed. This completes the proof of the assertion made in Chapter Four and cited at the outset of this chapter.

Degenerate Joint Distributions. In Theorems 4 and 5, it was assumed that the variance-covariance matrix of the initial set of variables was nonsingular. We now investigate the case when this assumption does not hold. In this case we already know, from Theorem 2, that one or more of the eigenvalues of $\mathbf{\Sigma}$ are equal to zero.

Suppose that just $r(< p)$ of the eigenvalues are *not* zero. It then follows that we can determine r principal components $Y_1, Y_2, \ldots, Y_r$ (corresponding to the r eigenvectors $\mathbf{v}_1, \mathbf{v}_2, \ldots, \mathbf{v}_r$ associated with the nonzero eigenvalues of $\mathbf{\Sigma}$) such that

$$\sum_{i=1}^{r} \mathrm{Var}(Y_i) = \sum_{i=1}^{r} \lambda_i = \mathrm{tr}(\mathbf{\Sigma})$$

the step from the second to the third expressions following from Theorem 1. In other words, the aggregate variance of the original p variables $X_1, X_2, \ldots, X_p$ is completely accounted for by a smaller number r of transformed variables. The remaining $p - r$ principal components account for no variance at all. Clearly, this means that the swarm of points

$$(X_{\alpha 1}, X_{\alpha 2}, \ldots, X_{\alpha p}) \qquad (\alpha = 1, 2, \ldots, N)$$

representing the N multivariate observations in the original p-dimensional space is actually confined to a subspace of $r(< p)$ dimensions embedded in the total space. Such a "pseudo p-dimensional" swarm of points, or the joint distribution they represent, is said to be *degenerate*.

Ordinarily, the presence of a degeneracy would not be manifest in the original space with the coordinate axes representing $X_1, X_2, \ldots, X_p$ because the confining subspace would not be orthogonal to any of the axes, and hence all p variables would have nonzero variance. But when the principal-axes rotation is performed, the degeneracy (if it exists) is made explicit by the systematic reallocation of variance by the conditional variance-maximizing process described earlier. An example in three dimensional space will clarify the point.

EXAMPLE 5.6. Show that the joint distribution, whose variance-covariance matrix is given below, is a degenerate one.

$$\mathbf{\Sigma} = \begin{bmatrix} 4 & -7 & 8 \\ -7 & 13 & -17 \\ 8 & -17 & 28 \end{bmatrix}$$

The characteristic equation $|\mathbf{\Sigma} - \lambda\mathbf{I}| = 0$ reduces to $\lambda(\lambda - 42)(\lambda - 3) = 0$. Hence, $\lambda_1 = 42$, $\lambda_2 = 3$, $\lambda_3 = 0$.

Therefore $\lambda_1 + \lambda_2 = 45 = \text{tr}(\boldsymbol{\Sigma})$, indicating that the first two principal components Y_1 and Y_2 account for all the variance in the purportedly three-dimensional scatter represented by $\boldsymbol{\Sigma}$. That is, all the observation points lie in a plane parallel to the (Y_1, Y_2)-plane, so there is no spread in the direction perpendicular to this plane.

In the degenerate case, we need only to consider the r eigenvectors associated with the nonzero eigenvalues in order to effect the principal-axes rotation. Thus, the transformation matrix $\mathbf{V}$ will be of order $p \times r$, with $\mathbf{v}_1, \mathbf{v}_2, \ldots, \mathbf{v}_r$ as its columns. For the numerical example above, we have

$$\mathbf{V} = \begin{bmatrix} .26726 & .57735 \\ -.53452 & -.57735 \\ .80178 & -.57735 \end{bmatrix}$$

and $\boldsymbol{\Sigma}$ may be diagonalized into

$$\mathbf{V}'\boldsymbol{\Sigma}\mathbf{V} = \begin{bmatrix} 42 & 0 \\ 0 & 3 \end{bmatrix}$$

(The reader should verify these results as an exercise.) More generally, the diagonalization of the variance-covariance matrix of a degenerate joint distribution is represented by

$$\mathbf{V}'\boldsymbol{\Sigma}\mathbf{V} = \boldsymbol{\Lambda} \qquad \text{(an } r \times r \text{ diagonal matrix with all diagonal elements nonzero)} \tag{5.34}$$

The Rank of a Matrix. In the foregoing discussion of matrices with one or more zero eigenvalues, we saw that (when the matrix in question is a variance-covariance matrix) the number of nonzero eigenvalues corresponds to the dimensionality of the subspace to which the scatter of observation points is confined. This number, designated r above, is called the *rank* of the (square) matrix.

A more general definition of the rank of a matrix, applicable to rectangular matrices as well, involves the notion of *linear independence* of a set of vectors.

Definition. A set of vectors $\mathbf{v}_1, \mathbf{v}_2, \ldots, \mathbf{v}_k$ is said to be *linearly dependent* if some k scalars $a_1, a_2, \ldots, a_k$, not all of which zero, can be found such that $a_1\mathbf{v}_1 + a_2\mathbf{v}_2 + \cdots + a_k\mathbf{v}_k = \mathbf{0}$.

If, on the other hand, $\sum_{i=1}^{k} a_i\mathbf{v}_i = \mathbf{0}$ only when $a_1 = a_2 = \cdots = a_k = 0$, the set of vectors is said to be *linearly independent*.

The practical import of this definition is that when a set of vectors is linearly dependent, at least one member of the set can be expressed as a linear combination of the remaining members; whereas if a set is linearly independent, no member of the set can be expressed as a linear combination of the others.

EXAMPLE 5.7. The set
$\mathbf{v}_1' = [1, 2, -1] \quad \mathbf{v}_2' = [3, -1, 0] \quad \mathbf{v}_3' = [-3, 15, -6]$
is linearly dependent because $2\mathbf{v}_1' + (-1)\mathbf{v}_2' + (-1/3)\mathbf{v}_3' = [2, 4, -2] + [-3, 1, 0] + [1, -5, 2] = [0, 0, 0]$. Here $a_1 = 2$, $a_2 = -1$, $a_3 = -1/3$, *none* of which is zero, and each member of the set can be expressed as a linear combination of the other two vectors.

EXAMPLE 5.8. The set
$\mathbf{v}_1' = [2, 4, -1] \quad \mathbf{v}_2' = [3, 1, 2] \quad \mathbf{v}_3' = [-1, -2, 1/2]$
is linearly dependent because $1.\mathbf{v}_1' + 0.\mathbf{v}_2' + 2\mathbf{v}_3' = \mathbf{0}'$. (The reader should verify this.) Here $a_1 = 1$, $a_2 = 0$, $a_3 = 2$, just two of which are nonzero. In this case only $\mathbf{v}_1$ and $\mathbf{v}_3$ can be expressed as linear combinations of the other two vectors; $\mathbf{v}_2$ cannot be so expressed.

EXAMPLE 5.9. The set
$\mathbf{v}_1' = [d_1, 0, 0]$, $\mathbf{v}_2' = [0, d_2, 0]$, $\mathbf{v}_3' = [0, 0, d_3]$, where d_1, d_2, and d_3 are arbitrary nonzero numbers, is linearly *in*dependent. This is because $a_1\mathbf{v}_1' + a_2\mathbf{v}_2' + a_3\mathbf{v}_3' = [a_1d_1, a_2d_2, a_3d_3]$, which can equal $[0, 0, 0]$ *only* if $a_1 = a_2 = a_3 = 0$.

EXAMPLE 5.10. Is the set
$\mathbf{v}_1' = [1, 2, 3]$, $\mathbf{v}_2' = [1, -1, 1]$, $\mathbf{v}_3' = [4, 1, -2]$ linearly dependent or independent?

When a set of vectors is in fact linearly dependent, it suffices to exhibit a set of numbers, not all zero, that makes $\sum_{i=1}^{k} a_i\mathbf{v}_i = \mathbf{0}$. But when the set is linearly independent, how can we prove this fact? How can we be sure, for instance, that absolutely *no* set of numbers other than $a_1 = a_2 = a_3 = 0$ will satisfy

$$a_1[1, 2, 3] + a_2[1, -1, 1] + a_3[4, 1, -2] = [0, 0, 0] \qquad \text{(a)}$$

in the last example above? The answer lies in rearranging the required condition (a) into the form of a set of homogeneous equations, thus:

$$\begin{bmatrix} 1 & 1 & 4 \\ 2 & -1 & 1 \\ 3 & 1 & -2 \end{bmatrix} \begin{bmatrix} a_1 \\ a_2 \\ a_3 \end{bmatrix} = \begin{bmatrix} 0 \\ 0 \\ 0 \end{bmatrix} \qquad \text{(b)}$$

As we already know, a set of homogeneous linear equations has a nontrivial solution if and only if its coefficient matrix is singular—that is (from Theorem 2), when the matrix has one or more zero eigenvalues. On the other hand, if the matrix is nonsingular (that is, has no zero eigenvalue), the only solution is the null vector, and we may conclude that the set of vectors under investigation is linearly independent.

The foregoing gives us a systematic method for testing whether a set of vectors is linearly dependent or independent, in the special case when the number of vectors in the set is equal to their dimensionality p. We have merely to put the vectors together as a matrix like that in Eq. b above, and determine the rank r (the number of nonzero eigenvalues) of this $p \times p$ matrix. *If* $r < p$, *the set of vectors is linearly dependent; if* $r = p$, *it is linearly independent*. We now show that an extension of this method is applicable even when the number of vectors in the set is not equal to their dimensionality. Furthermore, when the set is linearly dependent, the extended method will tell us the number of vectors that can form a linearly independent subset of the original set. It was in anticipation of this fact that we stated the criterion for linear dependence in the form $r < p$ rather than mere singularity of the matrix comprising the vectors.

At this point it is convenient to adopt a new, more general definition of the rank of a matrix, applicable to square and rectangular matrices alike. This change should not lead to any confusion, as it will presently be shown that the two definitions actually are equivalent.

> ***Definition.*** The rank of a $p \times q$ matrix is the largest number r such that there exists at least one set of r columns or r rows which , treated as a set of vectors, is linearly independent.

The following theorem from matrix algebra, which we state without proof, forms the basis on which to establish the said equivalence (which, in the case of rectangular matrices, will necessarily be indirect, since only square matrices have eigenvalues). In the sequel, we shall understand "rank" in the sense of maximal number of linearly independent rows or columns—until its equivalence with the earlier concept has been shown.

> ***Theorem 6*(a).** The rank of a product matrix $\mathbf{C} = \mathbf{AB}$ is *less than or equal to* the rank of $\mathbf{A}$ or $\mathbf{B}$, whichever is smaller.

(**b**) In the special case when **A** and **B** are mutual transposes, the ranks of **A** and **B** are, of course, equal, and the rank of **C** is precisely equal to (rather than "less than or equal to") the common rank of **A** and **B**. In other words, the ranks of both $\mathbf{AA}'$ and $\mathbf{A}'\mathbf{A}$ are equal to the rank of **A**.

Let **A** be a $p \times q$ matrix, where, for specificity, we assume $q \leq p$. Suppose that the $q \times q$ matrix $\mathbf{A}'\mathbf{A}$ has r nonzero eigenvalues. (Of course, $r \leq q$.) Then, in accordance with Eq. 5.34, if we take **V** as the $q \times r$ matrix whose columns are the r eigenvectors of $\mathbf{A}'\mathbf{A}$ associated with its nonzero eigenvalues, we have $\mathbf{V}'(\mathbf{A}'\mathbf{A})\mathbf{V} = \mathbf{\Lambda}$, the $r \times r$ diagonal matrix with the nonzero eigenvalues of $\mathbf{A}'\mathbf{A}$ as its diagonal elements. The rank of $\mathbf{\Lambda}$ is r, since the columns of a diagonal matrix with all diagonal elements nonzero form a linearly independent set, as shown in Example 5.9. Consequently, by Theorem 6a, the rank of $\mathbf{A}'\mathbf{A}$ cannot be *less* than r (since $\mathbf{A}'\mathbf{A}$ is one of three factors whose product has rank r). On the other hand, suppose that the rank of $\mathbf{A}'\mathbf{A}$ were *greater* than r, say s. Then, by definition, there would exist at least one set of s linearly independent rows and columns in $\mathbf{A}'\mathbf{A}$. From these, we could construct an $s \times s$ submatrix of $\mathbf{A}'\mathbf{A}$ that is nonsingular, and hence has $s > r$ nonzero eigenvalues. This means that a *submatrix* of $\mathbf{A}'\mathbf{A}$ can be diagonalized into a larger ($s \times s$) diagonal matrix than can the entire matrix $\mathbf{A}'\mathbf{A}$, which is absurd. (A rigorous demonstration of the absurdity of this conclusion is left as an exercise at the end of this section, with hints for the approach to be taken.) Thus, the rank of $\mathbf{A}'\mathbf{A}$ can be neither less than nor greater than r; so it must be equal to r. Hence, by Theorem 6b the rank of **A** is also equal to r.

We have thus demonstrated the equivalence of the concept of the rank of a matrix **A**, as the maximal number of linearly independent rows or columns, and the number of nonzero eigenvalues possessed by either **A** itself (when it is square) or by $\mathbf{A}'\mathbf{A}$ (when **A** is rectangular). In doing so, we have also extended the method, stated earlier, for testing whether a set of vectors is linearly dependent or independent, into the following.

Given a set of q p-dimensional vectors to be tested for linear independence; and, if the set is found to be dependent, to determine the maximal number of linearly independent vectors in the set.

1. Arrange the vectors to form the rows (or columns) of a matrix **A**.
2. Determine the eigenvalues of **A** (if $p = q$), or of $\mathbf{A}'\mathbf{A}$ or $\mathbf{AA}'$ (whichever has the smaller order) if $p \neq q$. Let the number of nonzero eigenvalues be r.
3. If $r = p$ or q and $q \leq p$, then the set is linearly independent.
4. If $r = p$ but $q > p$, then the set is linearly dependent, and p is the largest number of linearly independent vectors in the set.
5. If r is less than the smaller of p and q, the set is linearly dependent, and r is the largest number of linearly independent vectors in the set.

EXAMPLE 5.11. How many linearly independent vectors are there in the set

$$\mathbf{v}_1 = \begin{bmatrix} 2 \\ 0 \\ -1 \end{bmatrix} \qquad \mathbf{v}_2 = \begin{bmatrix} 0 \\ -2 \\ 3 \end{bmatrix}$$

$$\mathbf{v}_3 = \begin{bmatrix} 1 \\ -1 \\ 1 \end{bmatrix} \qquad \mathbf{v}_4 = \begin{bmatrix} 1 \\ 1 \\ -2 \end{bmatrix}$$

Arranging the given vectors as the columns of a matrix, we obtain

$$\mathbf{A} = \begin{bmatrix} 2 & 0 & 1 & 1 \\ 0 & -2 & -1 & 1 \\ -1 & 3 & 1 & -2 \end{bmatrix}$$

Since $\mathbf{A}$ is not square, and $\mathbf{AA}'$ has a smaller order than $\mathbf{A}'\mathbf{A}$, we compute

$$\mathbf{AA}' = \begin{bmatrix} 6 & 0 & -3 \\ 0 & 6 & -9 \\ -3 & -9 & 15 \end{bmatrix}$$

and determine its eigenvalues. The characteristic equation of $\mathbf{AA}'$ is:

$$\begin{vmatrix} 6-\lambda & 0 & -3 \\ 0 & 6-\lambda & -9 \\ -3 & -9 & 15-\lambda \end{vmatrix} = 0$$

which, on expansion and factorization, yields

$$\lambda(\lambda - 6)(\lambda - 21) = 0$$

Therefore, the eigenvalues of $\mathbf{AA}'$ are $\lambda_1 = 21$, $\lambda_2 = 6$, and $\lambda_3 = 0$, so that $r = 2$. We thus conclude that there are two linearly independent vectors in the given set of four vectors. (Verify that $\mathbf{v}_3$ and $\mathbf{v}_4$ can be expressed as linear combinations of $\mathbf{v}_1$ and $\mathbf{v}_2$.)

EXERCISE

Redo Examples 5.7 to 5.10 on p. 132 by the method just described.

Gramian Matrices. Throughout the foregoing development, a symmetric matrix of the form $\mathbf{AA}'$ (or $\mathbf{A}'\mathbf{A}$) has played a prominent role in ascertaining the linear dependence or independence of a set of vectors. In fact, the SSCP matrix, which was even earlier found to occupy a central position in multivariate statistical analysis, is also of this form: $\mathbf{S} = \mathbf{xx}'$. Such matrices are called *Gramian matrices*, and they have many special properties, some of which were tacitly assumed in this chapter and the preceding one. We now state several of these properties explicitly, but without proof.

> ***Property 1.*** Every *principal minor determinant* (defined below) of a Gramian matrix **G** is nonnegative (that is, it is positive or zero in value). In particular, if **G** is nonsingular, all of its principal minor determinants are positive.

A principal minor determinant of a matrix **A** is the determinant of any submatrix of **A** obtained by deleting any number of its *corresponding* row-and-column pairs (for example, deleting just its first row and first column; or those as well as the fourth row and fourth column). The diagonal elements of **A** (which are the "determinants" of submatrices remaining after deletion of all but one corresponding row-and-column pair), as well as $|\mathbf{A}|$ itself, are included among the principal minor determinants of **A**.

EXAMPLE 5.12. If

$$\mathbf{X} = \begin{bmatrix} 2 & 3 & 3 & 4 \\ 3 & 2 & 4 & 3 \\ 2 & 4 & 4 & 3 \end{bmatrix}$$

$$\mathbf{G} = \mathbf{XX}' = \begin{bmatrix} 38 & 36 & 40 \\ 36 & 38 & 39 \\ 40 & 39 & 45 \end{bmatrix}$$

is a Gramian matrix.
Its principal minor determinants are:

$$38,\ 38,\ 45 \qquad \begin{vmatrix} 38 & 36 \\ 36 & 38 \end{vmatrix} = 148 \qquad \begin{vmatrix} 38 & 40 \\ 40 & 45 \end{vmatrix} = 110$$

$$\begin{vmatrix} 38 & 39 \\ 39 & 45 \end{vmatrix} = 189 \qquad \text{and} \qquad |\mathbf{G}| = 382$$

all of which are positive.

EXAMPLE 5.13.

$$\mathbf{H} = \begin{bmatrix} 3 & 1 & 1 & 2 \\ 1 & 3 & 1 & 0 \\ 2 & 2 & 1 & 1 \end{bmatrix} \begin{bmatrix} 3 & 1 & 2 \\ 1 & 3 & 2 \\ 1 & 1 & 1 \\ 2 & 0 & 1 \end{bmatrix} = \begin{bmatrix} 15 & 7 & 11 \\ 7 & 11 & 9 \\ 11 & 9 & 10 \end{bmatrix}$$

is a Gramian matrix. Verify that all of its principal minor determinants are positive except for $|\mathbf{H}|$ itself, which is zero.

EXAMPLE 5.14.

$$\mathbf{K} = \begin{bmatrix} 9 & 6 & 3 \\ 6 & 4 & 2 \\ 3 & 2 & 1 \end{bmatrix}$$

has all of its principal minor determinants equal to zero except for its diagonal elements, which are positive. $\mathbf{K}$ is a Gramian matrix. The reader should try to determine a matrix $\mathbf{M}$ such that $\mathbf{MM}' = \mathbf{K}$.

Property 2. If $\mathbf{G}$ is a nonsingular Gramian matrix, $\mathbf{G}^{-1}$ is also a nonsingular Gramian matrix.

Property 3. If $\mathbf{G}$ is a Gramian matrix, then for all vectors $\mathbf{x}$, the quadratic form $Q = \mathbf{x}'\mathbf{Gx}$ has nonnegative values; such a quadratic form is said to be *positive semidefinite*. In particular, if $\mathbf{G}$ is nonsingular, then $Q > 0$ for all $\mathbf{x} \neq \mathbf{0}$; Q is then said to be *positive definite*.

This property (in combination with Property 2) was implicit in the assertion repeatedly made earlier, that $\mathbf{x}'\mathbf{\Sigma}^{-1}\mathbf{x} = C$, with an arbitrary positive constant C, represents an isodensity ellipsoid of a multivariate normal distribution. Indeed, unless this property were true, the multivariate normal distribution would not have the characteristic, possessed by the univariate normal distribution, that the maximum density occurs at the centroid. For, unless $\mathbf{x}'\mathbf{\Sigma}^{-1}\mathbf{x}$ is positive definite, there would be some point $\mathbf{x}$ at which $\mathbf{x}'\mathbf{\Sigma}^{-1}\mathbf{x} < 0$, and hence

$$[(2\pi)^{-p/2}|\mathbf{\Sigma}|^{-1/2}] \exp(-\tfrac{1}{2}\mathbf{x}'\mathbf{\Sigma}^{-1}\mathbf{x}) > (2\pi)^{-p/2}|\mathbf{\Sigma}|^{-1/2}$$

that is, the density at $\mathbf{x}$ is greater than the density at the centroid $(0, 0, \ldots, 0)$.

EXAMPLE 5.15. The quadratic form

$$Q = [x, y] \begin{bmatrix} 3 & -1 \\ -1 & 2 \end{bmatrix} \begin{bmatrix} x \\ y \end{bmatrix}$$

whose matrix is Gramian (see Property 5) and nonsingular, is positive definite. [This fact may be verified by expanding Q as $3x^2 - 2xy + 2y^2$, and noting that its discriminant (treating Q as a quadratic in x for fixed y) is $D = (2y)^2 - 4(3)(2y^2) = -20y^2 < 0$ for all $y \neq 0$. Therefore, for each fixed $y \neq 0$, the graph of Q is a parabola open upward and not intersecting the x axis. Hence $Q > 0$ for all values of x and y except $x = y = 0$.]

EXAMPLE 5.16.

$$Q = [x, y, z]\begin{bmatrix} 15 & 7 & 11 \\ 7 & 11 & 9 \\ 11 & 9 & 10 \end{bmatrix}\begin{bmatrix} x \\ y \\ z \end{bmatrix}$$

whose matrix is Gramian but singular (see Example 5.13, above), is positive semidefinite. See if you can determine values of x, y, z (not all zero) that make $Q = 0$.

Terminology. Although the terms "positive definite" and "positive semidefinite" properly apply to quadratic forms, it is customary to transfer them to the matrices of the quadratic forms as well. Thus, a Gramian matrix itself is said to be positive semidefinite or (if it is nonsingular) positive definite.

Property 4. All the eigenvalues of Gramian matrix are nonnegative; hence all the eigenvalues of a nonsingular Gramian matrix are positive.

This fact was implicitly proved when it was shown that $\text{Var}(Y_i) = \lambda_i$ (which is Eq. 5.26) for each principal component Y_i, where λ_i is the ith eigenvalue of the Gramian matrix $\mathbf{\Sigma}$.

Property 5. Any *symmetric* matrix whose principal minor determinants are *all* nonnegative is a Gramian matrix. (Note that symmetry is a prerequisite condition here. A nonsymmetric matrix obviously cannot be Gramian, even if all its principal minor determinants are nonnegative.) In other words, given that $\mathbf{A}$ is a symmetric, positive semidefinite (or, of course, positive definite) matrix, there always exists a matrix $\mathbf{B}$ such that $\mathbf{A} = \mathbf{B}'\mathbf{B}$.

This property, which is evidently the converse of Property 1, is of paramount importance as an "existence theorem" for the well-known multivariate technique of *factor analysis*. It assures us that, given a correlation matrix (or a "reduced" correlation matrix—see p.145—which still has the positive semidefiniteness property), we can always factor it as $\mathbf{R} = \mathbf{F}'\mathbf{F}$.

A slightly more restrictive version of this theorem is easy to prove: If $\mathbf{A}$ is a symmetric matrix, all of whose eigenvalues λ_i are positive and distinct, it is Gramian. For we can construct an orthogonal transformation matrix $\mathbf{V}$ (with the associated eigenvectors as its columns) such that (see Eq. 5.32)

$$\mathbf{V}'\mathbf{A}\mathbf{V} = \mathbf{\Lambda}$$

where $\mathbf{\Lambda}$ is the diagonal matrix with the λ_i's as its diagonal elements. Then, by pre- and postmultiplying both members by $\mathbf{V}$ and $\mathbf{V}'$, respectively, we obtain (since $\mathbf{V}$ is an orthogonal matrix):

$$\mathbf{A} = \mathbf{V}\mathbf{\Lambda}\mathbf{V}' \tag{5.35}$$

We have now only to define

$$\mathbf{B} = \mathbf{\Lambda}^{1/2}\mathbf{V}' \tag{5.36}$$

where $\mathbf{\Lambda}^{1/2}$ is the diagonal matrix whose ith diagonal element is $\sqrt{\lambda_i}$ (which is real, since $\lambda_i > 0$ for each i). It may readily be verified that

$$\mathbf{B}'\mathbf{B} = (\mathbf{\Lambda}^{1/2}\mathbf{V}')'(\mathbf{\Lambda}^{1/2}\mathbf{V}') = \mathbf{A}$$

because of Eq. 5.35.

EXAMPLE 5.17. The following matrix is symmetric and (as the reader should verify) positive definite:

$$\mathbf{A} = \begin{bmatrix} 10 & 3 & -2 \\ 3 & 9 & -3 \\ -2 & -3 & 10 \end{bmatrix}$$

Determine a matrix $\mathbf{B}$ such that $\mathbf{A} = \mathbf{B}'\mathbf{B}$.

The eigenvalues and vectors of $\mathbf{A}$ are found to be: $\lambda_1 = 15$, $\lambda_2 = 8$, $\lambda_3 = 6$, and $\mathbf{v}_1' = [1/\sqrt{3}, 1/\sqrt{3}, -1/\sqrt{3}]$, $\mathbf{v}_2' = [1/\sqrt{2}, 0, 1/\sqrt{2}]$, $\mathbf{v}_3' = [-1/\sqrt{6}, 2/\sqrt{6}, 1/\sqrt{6}]$.

Therefore, in accordance with Eq. 5.36, we take

$$\mathbf{B} = \begin{bmatrix} \sqrt{15} & & 0 \\ & \sqrt{8} & \\ 0 & & \sqrt{6} \end{bmatrix} \begin{bmatrix} 1/\sqrt{3} & 1/\sqrt{3} & -1/\sqrt{3} \\ 1/\sqrt{2} & 0 & 1/\sqrt{2} \\ -1/\sqrt{6} & 2/\sqrt{6} & 1/\sqrt{6} \end{bmatrix}$$

$$= \begin{bmatrix} \sqrt{5} & \sqrt{5} & -\sqrt{5} \\ 2 & 0 & 2 \\ -1 & 2 & 1 \end{bmatrix}$$

The reader should verify that

$$\mathbf{B}'\mathbf{B} = \begin{bmatrix} 10 & 3 & -2 \\ 3 & 9 & -3 \\ -2 & -3 & 10 \end{bmatrix} = \mathbf{A}$$

Theorems on Eigenvalues of Related Matrices. We have covered most of the important theorems on eigenvalues, eigenvectors, and related matters. All of these have dealt with a single given matrix. We conclude this section with a sequence of theorems concerning the relationship between the eigenvalues and eigenvectors of a given matrix and those of certain related matrices. Some of these theorems play an important role in the sequel; others are useful for certain computational purposes. All of them have the feature in common that they justify the following loose but intuitively appealing characterization of an eigenvector:

An eigenvector of a matrix is such a vector that, "in its presence" (that is, when multiplied by it), the matrix acts as though it were a scalar. (Needless to say, the scalar that the matrix simulates is the associated eigenvalue.)

All the theorems we state have a common premise, so we state it once and for all to avoid having to repeat it each time.

Given: Matrix **A** has an eigenvalue λ and an associated eigenvector **v**; that is,

$$\mathbf{Av} = \lambda\mathbf{v}.$$

Then, the following conclusions can be drawn:

Theorem 7. The matrix $b\mathbf{A}$ (where b is an arbitrary scalar) has $b\lambda$ as an eigenvalue, with **v** as the associated eigenvector.

Proof. Multiplying both members of $\mathbf{Av} = \lambda\mathbf{v}$ by b, we have

$$(b\mathbf{A})\mathbf{v} = b(\lambda\mathbf{v}) = (b\lambda)\mathbf{v}$$

Theorem 8. The matrix $\mathbf{A} + c\mathbf{I}$ (where c is an arbitrary scalar) has $(\lambda + c)$ as an eigenvalue, with **v** as the associated eigenvector.

Proof. $(\mathbf{A} + c\mathbf{I})\mathbf{v} = \mathbf{Av} + c\mathbf{v} = \lambda\mathbf{v} + c\mathbf{v} = (\lambda + c)\mathbf{v}$

Theorem 9. The matrix $\mathbf{A}^m$ (where m is any positive integer) has λ^m as an eigenvalue, with **v** as the associated eigenvector.

Proof. $\mathbf{A}^2\mathbf{v} = \mathbf{A}(\mathbf{Av}) = \mathbf{A}(\lambda\mathbf{v}) = \lambda(\mathbf{Av}) = \lambda(\lambda\mathbf{v}) = \lambda^2\mathbf{v}$. The result may be extended to any positive integral power $\mathbf{A}^m$ by induction.

Theorem 10. The matrix $\mathbf{A}^{-1}$ (if it exists) has $1/\lambda$ as an eigenvalue, with **v** as the associated eigenvector.

Proof. Premultiplying both members of $\mathbf{Av} = \lambda\mathbf{v}$, by $\mathbf{A}^{-1}$ we get

$$\mathbf{v} = \mathbf{A}^{-1}(\lambda\mathbf{v}) = \lambda(\mathbf{A}^{-1}\mathbf{v}) \qquad \text{whence} \qquad \mathbf{A}^{-1}\mathbf{v} = (1/\lambda)\mathbf{v}$$

As a final, gala demonstration of how like its eigenvalues a matrix behaves, we state the celebrated *Cayley-Hamilton theorem.*

Theorem 11. A matrix satisfies its own characteristic equation.

Proof (under restricted conditions). Although a general proof of this theorem requires a much deeper delving into matrix algebra than we have undertaken, a proof for the special case when the matrix is symmetric and nonsingular, and has all distinct eigenvalues, is relatively simple.

Let **A** be such a matrix, and let its characteristic equation

$$|\mathbf{A} - \lambda\mathbf{I}| = 0$$

be represented by

$$c_0\lambda^p + c_1\lambda^{p-1} + \cdots + c_n = 0$$

Then, for each i (= 1, 2, . . . , p), it follows from Theorems 7 and 9 that

$$\begin{aligned}(c_0\mathbf{A}^p + c_1\mathbf{A}^{p-1} + \cdots + c_n\mathbf{I})\mathbf{v}_i &= (c_0\lambda_i{}^p + c_1\lambda^{p-1} + \cdots + c_n)\mathbf{v}_i \\ &= 0\mathbf{v}_i = \mathbf{0}\end{aligned}$$

Consequently, by collecting $\mathbf{v}_1, \mathbf{v}_2, \ldots, \mathbf{v}_p$ into a matrix **V** (which, from Theorem 3 we know to be orthogonal), we have

$$(c_0\mathbf{A}^p + c_1\mathbf{A}^{p-1} + \cdots + c_n\mathbf{I})\mathbf{V} = \mathbf{0}_p$$

where $\mathbf{0}_p$ is the $p \times p$ null matrix. Postmultiplying both sides of this equation by $\mathbf{V}'$ $(= \mathbf{V}^{-1})$ yields

$$(c_0\mathbf{A}^p + c_1\mathbf{A}^{p-1} + \cdots + c_n\mathbf{I})\mathbf{I} = \mathbf{0}_p$$

or simply

$$c_0\mathbf{A}^p + c_1\mathbf{A}^{p-1} + \cdots + c_n\mathbf{I} = \mathbf{0}_p \tag{5.37}$$

as asserted.

An important implication of this theorem is that every positive integral power of a $p \times p$ matrix can be expressed as a linear combination of its powers no higher than the $(p - 1)$th, and the identity matrix.

EXAMPLE 5.18. The characteristic equation $|\mathbf{A} - \lambda\mathbf{I}| = 0$ of the matrix

$$\mathbf{A} = \begin{vmatrix} 7 & 0 & 1 \\ 0 & 7 & 2 \\ 1 & 2 & 3 \end{vmatrix}$$

is

$$-\lambda^3 + 17\lambda^2 - 86\lambda + 112 = 0.$$

The square and cube of **A** are:

$$\mathbf{A}^2 = \begin{bmatrix} 50 & 2 & 10 \\ 2 & 53 & 20 \\ 10 & 20 & 14 \end{bmatrix} \quad \text{and} \quad \mathbf{A}^3 = \begin{bmatrix} 360 & 34 & 84 \\ 34 & 411 & 168 \\ 84 & 168 & 92 \end{bmatrix}$$

Therefore,

$$-\mathbf{A}^3 + 17\mathbf{A}^2 - 86\mathbf{A} + 112\mathbf{I}$$

$$= \begin{bmatrix} -360 & -34 & -84 \\ -34 & -411 & -168 \\ -84 & -168 & -92 \end{bmatrix} + \begin{bmatrix} 850 & 34 & 170 \\ 34 & 901 & 340 \\ 170 & 340 & 238 \end{bmatrix}$$

$$+ \begin{bmatrix} -602 & 0 & -86 \\ 0 & -602 & -172 \\ -86 & -172 & -258 \end{bmatrix} + \begin{bmatrix} 112 & 0 & 0 \\ 0 & 112 & 0 \\ 0 & 0 & 112 \end{bmatrix}$$

$$= \begin{bmatrix} 0 & 0 & 0 \\ 0 & 0 & 0 \\ 0 & 0 & 0 \end{bmatrix}$$

which shows that **A** satisfies its own characteristic equation.

EXERCISES

1. Suppose that the following is a variance-covariance matrix:

$$\mathbf{\Sigma} = \begin{bmatrix} 8 & 0 & 1 \\ 0 & 8 & 3 \\ 1 & 3 & 5 \end{bmatrix}$$

Determine the eigenvalues $\lambda_1, \lambda_2, \lambda_3$ and vectors $\mathbf{v}_1, \mathbf{v}_2, \mathbf{v}_3$, of $\mathbf{\Sigma}$, and verify Theorems 1 to 5 of this section. That is, show that:

(a) $\lambda_1 + \lambda_2 + \lambda_3 = \text{tr}(\mathbf{\Sigma})$

(b) $\lambda_1\lambda_2\lambda_3 = |\mathbf{\Sigma}|$

(c) $\mathbf{v}_1'\mathbf{v}_2 = \mathbf{v}_1'\mathbf{v}_3 = \mathbf{v}_2'\mathbf{v}_3 = 0$

(d) $\mathbf{V}'\mathbf{\Sigma}\mathbf{V} = \mathbf{\Lambda}$

(e) the quadratic form $\mathbf{x}'\mathbf{\Sigma}^{-1}\mathbf{x}$ is an invariant of the rotation.

2. Assuming that you had not been told that it was a variance-covariance matrix, prove that the matrix $\mathbf{\Sigma}$ given in Problem 1 above is Gramian (see Property 5). Then determine a matrix **B** such that $\mathbf{B}'\mathbf{B} = \mathbf{\Sigma}$.

3. The matrix

$$\mathbf{A} = \begin{bmatrix} 6.8 & 2.4 \\ 2.4 & 8.2 \end{bmatrix}$$

has eigenvalues $\lambda_1 = 10$ and $\lambda_2 = 5$, with associated eigenvectors

$$\mathbf{v}_1 = \begin{bmatrix} .6 \\ .8 \end{bmatrix} \quad \text{and} \quad \mathbf{v}_2 = \begin{bmatrix} -.8 \\ .6 \end{bmatrix}$$

respectively. Verify this first, then show that:

(a)
$$5\mathbf{A} = \begin{bmatrix} 34 & 12 \\ 12 & 41 \end{bmatrix}$$

has eigenvalues $5\lambda_1 = 50$ and $5\lambda_2 = 25$, with associated eigenvectors $\mathbf{v}_1$ and $\mathbf{v}_2$ (the same as those of $\mathbf{A}$, given above), thus verifying Theorem 7.

(b)
$$\mathbf{A} + 2\mathbf{I} = \begin{bmatrix} 8.8 & 2.4 \\ 2.4 & 10.2 \end{bmatrix}$$

has eigenvalues $\lambda_1 + 2 = 12$ and $\lambda_2 + 2 = 7$, with associated eigenvectors $\mathbf{v}_1$ and $\mathbf{v}_2$ (as before), thus verifying Theorem 8.

(c)
$$\mathbf{A}^2 = \begin{bmatrix} 52 & 36 \\ 36 & 73 \end{bmatrix}$$

has eigenvalues $\lambda_1{}^2 = 100$ and $\lambda_2{}^2 = 25$, with associated eigenvectors $\mathbf{v}_1$ and $\mathbf{v}_2$ (as before), thus verifying Theorem 9.

(d)
$$\mathbf{A}^{-1} = \begin{bmatrix} .164 & -.048 \\ -.048 & .136 \end{bmatrix}$$

has eigenvalues $1/\lambda_1 = .10$ and $1/\lambda_2 = .20$, with associated eigenvectors $\mathbf{v}_1$ and $\mathbf{v}_2$ (as before), thus verifying Theorem 10.

4. In the course of showing that the rank of a matrix $\mathbf{A}$ is equal to the number of nonzero eigenvalues possessed by $\mathbf{A}'\mathbf{A}$, it was asserted that the rank of a matrix can never exceed the number of nonzero eigenvalues it has. Complete the proof of this assertion, started below.

   Suppose that a $p \times p$ matrix $\mathbf{B}$ had r nonzero eigenvalues and $s > r$ linearly independent rows and columns (which we may assume, without loss of generality, to be the first s rows and columns). Then, the upper left-hand $s \times s$ submatrix $\mathbf{B}_{ss}$ of $\mathbf{B}$ would be nonsingular, and hence have s nonzero eigenvalues. Denote by $\mathbf{V}_{ss}$ the matrix whose columns are the s eigenvectors of $\mathbf{B}_{ss}$ (all corresponding to nonzero eigenvalues, by the supposition above). We now define a $p \times s$ matrix $\mathbf{V}$ formed by augmenting $\mathbf{V}_{ss}$ with $p - s$ rows of zeros (that is, tacking on a null matrix of order $(p - s) \times s$, $\mathbf{0}_{p-s,\,s}$), thus:

$$\mathbf{V} = \begin{bmatrix} \mathbf{V}_{ss} \\ \hdashline \mathbf{O}_{p-s,\,s} \end{bmatrix}$$

By partitioning the original $p \times p$ matrix $\mathbf{B}$ as

$$\mathbf{B} = \begin{bmatrix} \mathbf{B}_{ss} & \mathbf{B}_{s,\,p-s} \\ \hdashline \mathbf{B}_{p-s,\,s} & \mathbf{B}_{p-s,\,p-s} \end{bmatrix}$$

it can be shown that, under the foregoing supposition, we would have

$$\mathbf{V}'\mathbf{B}\mathbf{V} = \mathbf{\Lambda} \tag{5.38}$$

the $s \times s$ diagonal matrix with the s nonzero eigenvalues of $\mathbf{B}_{ss}$ as its diagonal elements. But this means that $\mathbf{B}$ has $s > r$ nonzero eigenvalues, which is contrary to our hypothesis.

Prove Eq. 5.38. This also illustrates the process of forming the product of two partitioned matrices, which is often useful in multivariate analysis.

The general rule for this process is exactly the same as for ordinary matrix multiplication, with elements replaced by submatrices. Thus, the partitioning must be done in such a way that the various submatrix products are possible, as exemplified by the following schema:

$$\left[\begin{array}{c|c} \mathbf{A}_{mp} & \mathbf{A}_{mq} \\ \hline \mathbf{A}_{np} & \mathbf{A}_{nq} \end{array}\right] \left[\begin{array}{c|c} \mathbf{B}_{pr} & \mathbf{B}_{ps} \\ \hline \mathbf{B}_{qr} & \mathbf{B}_{qs} \end{array}\right] = \left[\begin{array}{c|c} \mathbf{C}_{mr} & \mathbf{C}_{ms} \\ \hline \mathbf{C}_{nr} & \mathbf{C}_{ns} \end{array}\right]$$

The four submatrices of the product matrix $\mathbf{AB} = \mathbf{C}$ are as follows:

$$\mathbf{C}_{mr} = \mathbf{A}_{mp}\mathbf{B}_{pr} + \mathbf{A}_{mq}\mathbf{B}_{qr} \qquad \mathbf{C}_{ms} = \mathbf{A}_{mp}\mathbf{B}_{ps} + \mathbf{A}_{mq}\mathbf{B}_{qs}$$
$$\mathbf{C}_{nr} = \mathbf{A}_{np}\mathbf{B}_{pr} + \mathbf{A}_{nq}\mathbf{B}_{qr} \qquad \mathbf{C}_{ns} = \mathbf{A}_{np}\mathbf{B}_{ps} + \mathbf{A}_{nq}\mathbf{B}_{qs}$$

5. If two matrices have the same eigenvalues and eigenvectors, the matrices are identical. Prove this theorem for the special case of symmetric matrices. (Hint: Use Eq. 5.31 to express a symmetric matrix in terms of its eigenvalues and vectors.)

5.6 APPLICATIONS OF PRINCIPAL COMPONENTS ANALYSIS

In the foregoing discussions, principal components analysis was described as a purely mathematical technique of principal-axes rotation, without regard to practical applications. We now briefly indicate two such applications.

Principal Components Analysis as a First-Stage Solution in Factor Analysis. The earliest use of principal components analysis for this purpose was made by Hotelling (1933), who developed an iterative computational procedure that is feasible without the aid of electronic computers for moderately large numbers of variables (up to about 10, say).

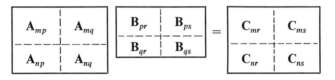

Actually, Hotelling himself, as well as T. L. Kelley (1935), did not regard principal components analysis as a "first-stage solution" for factor analysis, but as an end in itself—that is, as one approach to factor analysis in its own right. However, most modern factor analysts prefer to take the "first-stage solution" viewpoint, and do not consider a factor analysis complete until further rotations (usually *non*orthogonal, or oblique) are performed, subsequent to the principal-axes rotation, in order to achieve what is known as *simple structure* in the factor matrix. The concept of simple structure, introduced by Thurstone (1935) as a criterion for determining the final rotations is far from simple to describe. Very loosely speaking, it requires that each factor in the final solution have large correlations with a few of the original variables and essentially zero correlations with the other variables. Such a structure generally facilitates "meaningful" interpretation of the factors in psychological terms (or in terms of whatever domain is being investigated), and has hence come to be a major desideratum for the final factor solution. (See Thurstone, 1947 or Harman, 1967 for a detailed description of the simple-structure principle.)

Another difference between Hotelling's original work and the practice of most modern factor analysts is this: Whereas the former used the complete correlation matrix **R** (whose diagonal elements are 1s) as the starting point of analysis, the latter customarily starts with a reduced correlation matrix, whose diagonal elements are estimated *communalities*. The communality of a variable is the proportion of its variance that is accounted for by the *common factors*, that is, factors simultaneously involved in more than one of the set of variables being analyzed (see Harman, 1967, p. 17 for details). This difference has led many authors to distinguish between component analysis, in which the complete correlation matrix is used as the starting point, and factor analysis, in which the reduced correlation matrix is used.

The reader may well ask what is to be gained—apart from obtaining a set of uncorrelated variables—by a principal components analysis (or a principal-axes rotation) that is not followed by rotations to achieve simple structure for obtaining "psychologically meaningful" factors. A clue to the answer lies in the observation made earlier (p. 130) that *any degeneracy present in a multivariate distribution is made explicit by a principal-axes rotation.*

Certainly, we do not expect any real data to involve such degeneracy *exactly*—except when artificial constraints are imposed on the variables, as in the case of *ipsative measures*, described in the next subsection. However, a *near*-degeneracy may often be present (but concealed) in the original variance-covariance (or correlation) matrix. That is, the scatter of observation points in the p-variate space may *almost* be confined to a subspace of smaller dimensionality embedded in the total space. Such a state of affairs would be revealed by principal components analysis in the following manner: A certain number, say r', of the eigenvalues of the variance-covariance (or correlation) matrix would be substantially positive, but the remaining

$p - r'$ eigenvalues would be nearly equal to zero. We may then conclude that the rank of the variance-covariance matrix is "essentially equal to r'," and hence that the scatter of observation points is "practically confined" to an r'-dimensional subspace. Thus, replacing the original p variables by the first r' principal components and discarding the rest will result in very little loss of information. Principal components analysis, therefore, effects a parsimony of description, which is useful in its own right—provided we do not insist on "psychological meaningfulness" of the "factors."

The near-degeneracy described above may be expected to hold for the total correlation matrix whenever the original set of variables (battery of tests) consists of several "natural" subsets, the variables in each set measuring more or less the same trait (or "psychological dimension")—that is, whenever the battery of tests is of the sort that one ordinarily uses in factor analytic studies. However, the approximation to degeneracy is usually enhanced in the reduced correlation matrix with communalities instead of unities in the main diagonal because there we are concerned only with the common factor space. But, in replacing the unities by estimated communalities, we have to be careful to preserve the Gramian properties (in particular, the positive definiteness) of the matrix. Otherwise, we may end up with one or more negative eigenvalues, and (if these are large in absolute value) the solution would become meaningless—not only psychologically, but logically.

The question remains as to how the "essential dimensionality" r' should be determined. In other words, given the p eigenvalues $\lambda_1 > \lambda_2 > \cdots > \lambda_p > 0$ of a correlation matrix (reduced or complete), how do we decide on the smallest of these that will be treated as "truly" nonzero? This is a big question, indeed, for which there is no general consensus on how to go about answering, let alone what constitutes the "correct" answer. We do not propose to discuss this problem here in any detail, but will merely indicate, with brief comment, some of the approaches that have been advocated for its solution.

A statistical significance test for the number of factors required for adequately accounting for the observed correlation matrix, based on the likelihood-ratio principle, was developed by Lawley (1940). This method, and subsequent developments along similar lines by a number of authors are described in Lawley and Maxwell (1963). Although this approach is completely objective (to the extent that one can decide on the suitable significance level), strict adherence to it will often lead to the retention of more factors than are practically—as against statistically—significant, especially when the sample size is very large. Thus, one is led to temper one's decision based on the significance test with a consideration of the proportion of total variance (or total common variance) accounted for by the first r' principal components (or factors). The latter may be assessed by computing the ratio

$$\left(\sum_{i=1}^{r'} \lambda_i\right) \Big/ p \quad \text{or} \quad \left(\sum_{i=1}^{r'} \lambda_i\right) \Big/ \left(\sum_{i=1}^{p} h_i^2\right)$$

depending on whether the complete correlation matrix (unit diagonals) or the reduced correlation matrix with estimated communalities h_i^2 is being analyzed. If the relevant ratio exceeds .85 or .90, then the remaining $p - r'$ components (factors) taken together account for only 10% to 15% of the total relevant variance. Hence the $(r' + 1)$th dimension, even if statistically significant, will probably contribute such a miniscule proportion (perhaps 2% to 5%) to the explained variance that it would be of no practical consequence.

Perhaps the most widely used rule today is that proposed by Kaiser (1960): to retain only those components corresponding to eigenvalues of the (complete) correlation matrix that are greater than unity. This rule accords well with considerations of statistical significance and psychometric reliability among other things, but it, too, is apparently not immune to criticism. Cattell (1966), for example, argues that the 1.0 cut-off value properly applies to eigenvalues of the *population* correlation matrix, not its sample estimate. Also, subsequent rotation may well assign variances greater than 1 to factors beyond those qualifying under the 1.0 cutoff.

A graphical method known as the scree test was proposed by Cattell (1966). Essentially a refinement of the earliest test in this field, due to Tucker (1938), the scree test consists in examining the plot of the entire set of p eigenvalues against their ordinal numbers. Such a plot typically shows a steep initial descent (like a negative exponential curve), followed by a terminal arm that is very nearly a straight line with a gradual downward slope. The recommendation is to retain all the components down to and including the one associated with the largest eigenvalue whose plot lies on the "scree line." Although a tentative theoretical explanation is given for this method, it is (as Cattell says) primarily based on empirical generalization. If emphasis is placed on the gradual slope rather than near rectilinearity of the scree, it would seem that this method amounts to an "eyeballing" counterpart of a test suggested by Bartlett (1951), wherein the hypothesis that the $p - r'$ smallest eigenvalues in the population are equal is tested.

In order not to leave the reader in a state of utter despair at the plethora of conflicting criteria for deciding on the "essential dimensionality" r', it should be mentioned that in practice one usually employs a judicious combination of two or three such criteria. In the hands of experienced factor analysts, this leads to a modicum of consensus so that controversy is generally limited to the last one or two dimensions among the 10 to 15 usually identified in most large-scale studies.

Finally, we note that most of the tests mentioned above (in fact, all but the scree test) can be conducted step by step, without necessarily having the entire set of eigenvalues and vectors computed. It is therefore evident that if a computational procedure were available that yielded the solutions one at a time, it would be more economical than the method described earlier in this chapter, which requires finding all the eigenvalues at once. Hotelling's iterative procedure alluded to at the beginning of this section is, in fact, such a method. It first yields the eigenvector corresponding to the largest eigenvalue λ_1 (as well as λ_1 itself). A residual correlation matrix with the effects of the first component (factor) removed is then constructed, and the procedure is repeated on this matrix to yield the second largest eigenvalue and its associated vector; and so on. Hence, this method is advantageous when one anticipates that the first few components will suffice essentially to account for the observed correlation matrix. One pauses at that stage and uses one or more of the above-mentioned tests to check whether he needs to determine the next component. Details of this method, with a worked-out example, are given in Appendix D.

Principal Components Analysis as a Preliminary to Other Multivariate Analyses. It was mentioned in Chapter Two that when there is an artificial linear constraint on the set of predictor variables (or any subset thereof) in a multiple regression problem, the SSCP matrix becomes singular and hence cannot be inverted to obtain the regression weights in accordance with Eq. 2.23. Such constraints occur most typically with *ipsative measures*—that is, when the scoring system for a set of variables is such that the total score is the same for every individual. Let us first see why the SSCP matrix is singular in such cases.

Letting $X_{\alpha i}$ be the score of the αth individual on the ith variable, we have, by definition,

$$X_{\alpha 1} + X_{\alpha 2} + \cdots + X_{\alpha p} = C \tag{5.39}$$

for each α. Hence, the means of the p variables are also subject to the constraint

$$\bar{X}_1 + \bar{X}_2 + \cdots + \bar{X}_p = C$$

and it follows that the deviation scores satisfy the relation

$$x_{\alpha 1} + x_{\alpha 2} + \cdots + x_{\alpha p} = 0 \qquad [\alpha = 1, 2, \ldots, N]$$

Consequently, the columns $\mathbf{x}_i$ of the deviation-score matrix $\mathbf{x}$ are linearly dependent, since

$$\mathbf{x}_1 + \mathbf{x}_2 + \cdots + \mathbf{x}_p = \mathbf{0}$$

Therefore, the rank of $\mathbf{x}$ is less than p. (Most likely it is equal to $p - 1$.) Hence, the rank of the $p \times p$ SSCP matrix $\mathbf{S} = \mathbf{xx}'$ is, by Theorem 6b, also less than p, so that $\mathbf{S}$ is singular. It follows, of course, that the variance-covariance matrix and the correlation matrix are also singular.

When the variables are subject to the constraint (5.39) we could, of course, omit one of them and use only $p - 1$ variables as predictors without any loss of predictive efficiency. However, this procedure would make it impossible to compare the relative importance of all p variables for the prediction. An alternative method, which does not suffer from this defect, is to perform a principal components analysis of the p variables and use the $p - 1$ principal components $Y_1, Y_2, \ldots, Y_{p-1}$ (corresponding to the nonzero eigenvalues of $\mathbf{\Sigma}_x$) as the predictors in lieu of the original set. This approach may be used not only as a preliminary to multiple regression analysis, but also for discriminant and canonical analyses, described in Chapter Six. Here we outline the procedure with respect to multiple regression.

Let $\mathbf{V}$ be the $p \times (p - 1)$ transformation matrix defining the principal components:

$$\mathbf{Y}' = \mathbf{X}'\mathbf{V} \tag{a}$$

On carrying out a multiple regression analysis with $Y_1, Y_2, \ldots, Y_{p-1}$ (and, usually, other variables as well) as the predictors, and (say) Z as the criterion, we shall have a set of $p - 1$ regression weights to be applied to the Y_i's. Denoting by $\mathbf{d}$ the vector comprising these weights as elements, the regression equation will be of the form

$$\tilde{Z} = a + \mathbf{Y}'\mathbf{d} \tag{b}$$

(plus further terms if there are other predictors). But the Y_i's may be eliminated from Eq. b by substituting from (a)

$$\tilde{Z} = a + \mathbf{X}'(\mathbf{Vd}) \tag{5.40}$$

Thus, the relative importance of the X's in predicting Z may be assessed from the elements of the p-dimensional vector $\mathbf{Vd}$ after converting them to standard-score form. It also is evident that, in using the regression equation for predictive purposes, it is not necessary to convert a person's X scores into Y's, since Eq. 5.40 is already expressed in terms of the X's.

5.7 THEORETICAL SUPPLEMENT: GENERALIZED INVERSES

When the concept of the inverse of a square, nonsingular matrix was first introduced in Chapter Two, it was mentioned that, under certain conditions, a matrix $\mathbf{X}$ such that $\mathbf{XA} = \mathbf{I}$ can be found even when $\mathbf{A}$ is not square. Discussion of this matter was deferred because it required a knowledge of eigenvalues and eigenvectors. It is now time to describe how such a matrix, called a *left general inverse* of **A**, can be determined when it exists, and to specify the conditions for its existence. Although this concept will not be used in this book, it is an important one in the more advanced phases of multivariate analysis, especially in connection with the general linear hypothesis approach to multivariate analysis of variance, alluded to in Section 7.7.

Suppose that **A** is an $n \times m$ matrix with rank r, where $n > m \geq r$. Then **AA**′ is an $n \times n$ matrix and, by Theorem 6 (p. 133) and the ensuing discussions, it will have r nonzero eigenvalues, which may be denoted as $\lambda_1, \lambda_2, \ldots, \lambda_r$. Let **V** be the $n \times r$ matrix whose columns are the r normalized eigenvectors of **AA**′ associated with the r nonzero eigenvalues. Then, by definition, we have the relation

$$(\mathbf{AA}')\mathbf{V} = \mathbf{V}\boldsymbol{\Lambda} \tag{5.41}$$

where $\boldsymbol{\Lambda}$ is the diagonal matrix with diagonal elements λ_i $(i = 1, 2, \ldots, r)$. Similarly, let **U** be the $m \times r$ matrix whose columns are the normalized eigenvectors of the $m \times m$ matrix **A′A** corresponding to its r nonzero eigenvalues. It can be proved, although we shall not do so here, that these eigenvalues are the same as those of **AA**′, namely, $\lambda_1, \lambda_2, \ldots, \lambda_r$. It therefore follows that

$$(\mathbf{A}'\mathbf{A})\mathbf{U} = \mathbf{U}\boldsymbol{\Lambda} \tag{5.42}$$

Property 4 (p. 138) of Gramian matrices—which both **AA**′ and **A′A** are—assures us that $\lambda_1, \lambda_2, \ldots, \lambda_r$ are all positive. Hence we may form the square root of the diagonal matrix $\boldsymbol{\Lambda}$, which we shall denote by **D**:

$$\mathbf{D} = \boldsymbol{\Lambda}^{1/2}$$

whose ith diagonal element is $\sqrt{\lambda_i}$ $(i = 1, 2, \ldots, r)$.

An important theorem in matrix algebra, which is an extension of Eq. 5.35 proved in connection with the existence theorem of factor analysis, permits us to express **A** in terms of **D**, **U**, and **V** as follows:

$$\mathbf{A} = \mathbf{VDU}' \tag{5.43}$$

(If the reader wonders why the square root of $\boldsymbol{\Lambda}$ occurs in this expression whereas Eq. 5.35 involves $\boldsymbol{\Lambda}$ itself, he should recall that the $\boldsymbol{\Lambda}$ here has the nonzero eigenvalues of **AA**′ as its elements, and hence corresponds to the *square* of the $\boldsymbol{\Lambda}$ in Eq. 5.35. Thus, in the special case when **A** is a square, symmetric matrix, Eq. 5.43 reduces to Eq. 5.35, as it should.) The proof of this theorem is rather involved, and we shall not give it here. The interested reader is referred to Horst (1963, Chapter 18), who calls **VDU**′ the *basic structure* of matrix **A**.

Now let us define an $m \times n$ matrix $\mathbf{X}$ as follows

$$\mathbf{X} = \mathbf{UD}^{-1}\mathbf{V}' \tag{5.44}$$

and investigate the product $\mathbf{XA}$. Using the right-hand side expressions in Eqs. 5.44 and 5.43 for $\mathbf{X}$ and $\mathbf{A}$, respectively, and successively transforming, we find

$$\begin{aligned}
\mathbf{XA} &= (\mathbf{UD}^{-1}\mathbf{V}')(\mathbf{VDU}') \\
&= \mathbf{UD}^{-1}(\mathbf{V}'\mathbf{V})\mathbf{DU}' \\
&= \mathbf{UD}^{-1}\,(\mathbf{I}_r)\mathbf{DU}' && \text{(since the columns of } \mathbf{V} \text{ have} \\
&= \mathbf{U}(\mathbf{D}^{-1}\mathbf{D})\mathbf{U}' && \text{unit norms and are} \\
&= \mathbf{UU}' && \text{mutually orthogonal)}
\end{aligned} \tag{5.45}$$

Note that, although $\mathbf{U}'\mathbf{U} = \mathbf{I}_r$ (since the columns of $\mathbf{U}$, like those of $\mathbf{V}$, have unit norms and are mutually orthogonal), it is in general *not* true that $\mathbf{UU}' = \mathbf{I}_m$. However, consider the special case when $r = m$, that is, when the columns of $\mathbf{A}$ are linearly independent, and hence the rank of $\mathbf{A}$ is equal to its number of columns (which, it will be recalled, is here assumed to be smaller than its number of rows). In this case, $\mathbf{U}$ will be a square matrix of order m, and hence $\mathbf{U}'\mathbf{U} = \mathbf{I}_m$ will also imply $\mathbf{UU}' = \mathbf{I}_m$; that is, $\mathbf{U}$ is an orthogonal matrix whose transpose is its inverse. Thus, the right-hand side of Eq. 5.45 reduces to $\mathbf{I}_m$ in this case, and we arrive at the following conclusion: if a rectangular matrix $\mathbf{A}$ of order $n \times m$ (with $n > m$) has rank m, it possesses a left general inverse $\mathbf{X}$, satisfying the relation

$$\mathbf{XA} = \mathbf{I}_m \tag{5.46}$$

where $\mathbf{X}$ is as defined in Eq. 5.44.

Similarly, if $\mathbf{A}$ is an $n \times m$ matrix with $m > n$ (that is, with more columns than rows), and if its rows are linearly independent so that the rank of $\mathbf{A}$ is n, then $\mathbf{A}$ possesses a *right general inverse* $\mathbf{Y}$ (say) whose order is $m \times n$ and which satisfies

$$\mathbf{AY} = \mathbf{I}_n \tag{5.47}$$

This follows from the fact that, in this case, $\mathbf{A}'$ is a matrix with more rows than columns and whose rank is equal to its number of columns. Hence $\mathbf{A}'$ has a left general inverse as in Eq. 5.46. If we denote this by $\mathbf{Y}'$ (an $n \times m$ matrix), we have

$$\mathbf{Y}'\mathbf{A}' = \mathbf{I}_n$$

from which Eq. 5.47 results on taking the transposes of both sides.

In summary, we have shown that a rectangular matrix with maximal rank (that is, rank equal to its number of rows or its number of columns, whichever is smaller) possesses a left or right general inverse such that pre- or postmultiplication by the latter produces the identity matrix of order equal to the rank of the original matrix. What about a rectangular matrix of nonmaximal rank? It can be shown that such a matrix does not have either a left or right general inverse in the above sense. However, we note that even in this case the matrix **X** defined by Eq. 5.44 satisfies an interesting relation, namely

$$\mathbf{AXA} = \mathbf{A} \tag{5.48}$$

This can easily be proved by premultiplying the first and last members of Eq. 5.45 by **A**, substituting expression 5.43 in the right-hand side of the resulting equation, and utilizing the fact that $\mathbf{U}'\mathbf{U} = \mathbf{I}_r$. Details are left to the reader as an exercise.

In many applications to multivariate analysis, we do not need the full property of an inverse matrix $\mathbf{A}^{-1}$ even when it exists (that is, when **A** is a square nonsingular matrix). In other words, we often do not utilize the fact that $\mathbf{A}^{-1}\mathbf{A} = \mathbf{A}\mathbf{A}^{-1} = \mathbf{I}$ in its full force, but merely require a "weaker" consequence of this fact, namely that $\mathbf{A}\mathbf{A}^{-1}\mathbf{A} = \mathbf{A}$. Thus, the **X** defined in Eq. 5.44 suffices for this more modest role demanded of the matrix inverse, since it satisfies Eq. 5.48 even when Eq. 5.46 does not hold. The same is true, of course, of a matrix **Y** that would have been a right general inverse of an $n \times m$ matrix **A** (with $m > n$) had the latter been of maximal rank. For this reason, Rao (1965, p. 24) has given the name *generalized inverse* (or "*g*-inverse" for short) to any matrix **X** that satisfies Eq. 5.48, and denotes such a matrix by the symbol $\mathbf{A}^-$. The *g*-inverse is a very general concept, since an $\mathbf{A}^-$ exists for any given **A**, regardless of its order and rank—including the case when **A** is a square, singular matrix. In fact, any matrix *except* a square, nonsingular one possesses an infinite number of *g*-inverses. The regular inverse $\mathbf{A}^{-1}$ of a nonsingular matrix **A** is its only *g*-inverse as well.

Equation 5.44 (or its counterpart for a matrix with more columns than rows) thus represents just one of many possible ways for constructing a *g*-inverse, and hence a left (or right) general inverse when the matrix is of maximal rank. We have focussed on this method because of its theoretical elegance, deriving from its being based on the important theorem stated in Eq. 5.43—a generalized version of the existence theorem for factor analysis, as was mentioned earlier. Other methods are described by Rao (1965), each method producing a different $\mathbf{A}^-$ for a given **A**. One particularly elementary method, applicable only to rectangular matrices of maximal rank, is given in Exercise 5 below. Although it may be of limited practical utility, it has the advantage of clearly showing the infinitude of *g*-inverses.

We may now summarize all our discussions pertaining to matrix inverses, including those in Chapter Two, in the following integrated manner:

1. Any matrix **A** possesses one or more (more precisely, either *just* one or *infinitely* many) generalized inverses (*g*-inverse) $\mathbf{A}^-$ satisfying Eq. 5.48:

$$\mathbf{A}\mathbf{A}^-\mathbf{A} = \mathbf{A}$$

2. If **A** has maximal rank (that is, rank equal to its number of rows n or number of columns m, whichever is smaller; or rank equal to its order when **A** is square), then its *g*-inverse is also a left or right general inverse, satisfying Eq. 5.46 or 5.47, as the case may be:

$$\mathbf{A}^-\mathbf{A} = \mathbf{I}_m \text{ (if } m \leq n) \qquad \text{or} \qquad \mathbf{A}\mathbf{A}^- = \mathbf{I}_n \text{ (if } n \leq m)$$

3. In particular, if **A** is square and nonsingular (that is, if $|\mathbf{A}| \neq 0$) its *g*-inverse is unique, and is denoted as $\mathbf{A}^{-1}$. Both Eqs. 5.46 and 5.47 are satisfied in this case (since $m \leq n$ and $n \leq m$ are both true in this case). Thus,

$$\mathbf{A}^{-1}\mathbf{A} = \mathbf{A}\mathbf{A}^{-1} = \mathbf{I}$$

This $\mathbf{A}^{-1}$ is, of course, the inverse of **A** as defined in Chapter Two.

EXAMPLE 5.20. Given, the 3×2 matrix

$$\mathbf{A} = \begin{bmatrix} 8.2 & -2.4 \\ 3.0 & 4.0 \\ 2.4 & -6.8 \end{bmatrix}$$

whose columns are obviously linearly independent. Determine its left general inverse **X** in accordance with Eq. 5.44. We first compute the products $\mathbf{AA}'$ and $\mathbf{A}'\mathbf{A}$, obtaining

$$\mathbf{AA}' = \begin{bmatrix} 73 & 15 & 36 \\ 15 & 25 & -20 \\ 36 & -20 & 52 \end{bmatrix}$$

and

$$\mathbf{A}'\mathbf{A} = \begin{bmatrix} 82 & -24 \\ -24 & 68 \end{bmatrix}$$

The two nonzero eigenvalues of $\mathbf{AA}'$ (which are also the only two eigenvalues of $\mathbf{A}'\mathbf{A}$) are found to be $\lambda_1 = 100$ and $\lambda_2 = 50$. The normalized eigenvectors of $\mathbf{AA}'$ associated with these eigenvalues are

$$\mathbf{v}_1 = \begin{bmatrix} .8 \\ 0 \\ .6 \end{bmatrix} \qquad \text{and} \qquad \mathbf{v}_2 = \begin{bmatrix} 3/\sqrt{50} \\ 5/\sqrt{50} \\ -4/\sqrt{50} \end{bmatrix}$$

The normalized eigenvectors of **A′A** are

$$\mathbf{u}_1 = \begin{bmatrix} .8 \\ -.6 \end{bmatrix} \quad \text{and} \quad \mathbf{u}_2 = \begin{bmatrix} .6 \\ .8 \end{bmatrix}$$

(The reader should verify the foregoing results, as well as Eq. 5.43, according to which the triple product

$$\mathbf{VDU}' = \begin{bmatrix} .8 & 3/\sqrt{50} \\ 0 & 5/\sqrt{50} \\ .6 & -4/\sqrt{50} \end{bmatrix} \begin{bmatrix} 10 & 0 \\ 0 & \sqrt{50} \end{bmatrix} \begin{bmatrix} .8 & -.6 \\ .6 & .8 \end{bmatrix}$$

should equal the original matrix **A**.)

We then compute **X** in accordance with Eq. 5.44; that is,

$$\mathbf{UD}^{-1}\mathbf{V}' = \begin{bmatrix} .8 & .6 \\ -.6 & .8 \end{bmatrix} \begin{bmatrix} 1/10 & 0 \\ 0 & 1/\sqrt{50} \end{bmatrix} \times \begin{bmatrix} .8 & 0 & .6 \\ 3/\sqrt{50} & 5/\sqrt{50} & -4/\sqrt{50} \end{bmatrix}$$

$$= \begin{bmatrix} .8 & .6 \\ -.6 & .8 \end{bmatrix} \begin{bmatrix} .08 & 0 & .06 \\ .06 & .10 & -.08 \end{bmatrix}$$

$$= \begin{bmatrix} .10 & .06 & 0 \\ 0 & .08 & -.10 \end{bmatrix} = \mathbf{X}$$

We leave it to the reader to verify that $\mathbf{XA} = \mathbf{I}_2$.

In order to emphasize the nonuniqueness of left (or right) general inverses, we point out that the following is another left general inverse of the matrix **A** in this example:

$$\begin{bmatrix} -.650 & 1.310 & 1 \\ -.075 & .205 & 0 \end{bmatrix}$$

as the reader should verify. This was obtained by the method described in Exercise 5 below.

EXERCISES

1. Using Hotelling's iterative method (Appendix D), determine the eigenvalues and eigenvectors of the matrix

$$\mathbf{A} = \begin{bmatrix} .7 & 0 & .1 \\ 0 & .7 & .2 \\ .1 & .2 & .3 \end{bmatrix}$$

2. If **A** is a symmetric matrix with eigenvalues $\lambda_1, \lambda_2, \ldots, \lambda_r$ and eigenvectors $\mathbf{v}_1, \mathbf{v}_2, \ldots, \mathbf{v}_r$ (normalized to unity), then

$$\mathbf{A} = \lambda_1\mathbf{v}_1\mathbf{v}_1' + \lambda_2\mathbf{v}_2\mathbf{v}_2' + \cdots + \lambda_r\mathbf{v}_r\mathbf{v}_r'$$

(Hint: Denote the matrix indicated on the right-hand side by **B**. Then, by successively postmultiplying by $\mathbf{v}_1$, $\mathbf{v}_2$, and so on, show that

$$\mathbf{B}\mathbf{v}_1 = \lambda_1\mathbf{v}_1, \qquad \mathbf{B}\mathbf{v}_2 = \lambda_2\mathbf{v}_2, \qquad \text{etc.}$$

In other words, show that **B** has the same eigenvalues and eigenvectors as **A**. Hence, by Exercise 5 on p. 144, it follows that $\mathbf{B} = \mathbf{A}$.)

3. If the p variables subject to the linear constraint are the *only* predictors being used, the regression weights $d_1, d_2, \ldots, d_{p-1}$ for the principal components do not have to be computed from scratch, but may be computed on the basis of the SSCP matrix for the original p variables and the criterion. Namely,

$$\mathbf{d} = (\mathbf{V}'\mathbf{S}_{pp}\mathbf{V})^{-1}\mathbf{V}'\mathbf{S}_{pc} = \mathbf{\Lambda}^{-1}\mathbf{V}'\mathbf{S}_{pc}$$

where **V** is the $p \times (p-1)$ transformation matrix, and $\mathbf{S}_{pp}$, $\mathbf{S}_{pc}$ are the appropriate sections of the original SSCP matrix, as defined in Chapter Two.

4. Compute the left general inverse of the following matrix in accordance with Eq. 5.44:

$$\mathbf{A} = \begin{bmatrix} 8 & 3 \\ 0 & 5 \\ 6 & -4 \end{bmatrix}$$

5. *Alternative method for computing a left general inverse.* Let **A** be an $n \times m$ matrix with rank m ($< n$), partitioned as follows:

$$\mathbf{A} = \begin{bmatrix} \mathbf{A}_{mm} \\ \hline \mathbf{A}_{qm} \end{bmatrix}$$

where the subscripts indicate the orders of the submatrices, and $q = n - m$. Since the rank of **A** is m, the submatrix $\mathbf{A}_{mm}$ must be nonsingular. So we can meaningfully define a matrix

$$\mathbf{X}_{mm} = \mathbf{A}_{mm}^{-1} - \mathbf{X}_{mq}\mathbf{A}_{qm}\mathbf{A}_{mm}^{-1}$$

where $\mathbf{X}_{mq}$ is an arbitrary matrix of order $m \times q$. Now form an $m \times n$ matrix **X** by adjoining $\mathbf{X}_{mq}$ to the right of $\mathbf{X}_{mm}$, thus:

$$\mathbf{X} = \left[\mathbf{X}_{mm} \,\middle|\, \mathbf{X}_{mq}\right]$$

Prove that this **X** is a left general inverse of **A**—that is, that **X** satisfies Eq. 5.46.
(Hint: Review the rule of multiplying two partitioned matrices, given in Exercise 4 on p. 144.)

6. Verify that the matrix displayed at the end of Example 5.20 was obtained by the method described in Exercise 5 above, using

$$\mathbf{X}_{21} = \begin{bmatrix} 1 \\ 0 \end{bmatrix}$$

as the arbitrary matrix $\mathbf{X}_{mq}$ occurring in the expressions for $\mathbf{X}_{mm}$ and **X** above. Also, try using a different $\mathbf{X}_{21}$ to obtain yet another left general inverse of matrix **A** of that example.

Chapter Six

Discriminant Analysis and Canonical Correlation

In Chapter Four we discussed multivariate procedures for testing the significance of the overall difference among several group centroids. If a significant difference is found, we may be further interested in studying the directions or dimensions along which the major differences occur. In Chapter Five we noted, in connection with principal components analysis, that the geometric concept of directions or dimensions is closely related to the algebraic notion of linear combinations. Hence, the problem of studying the direction of group differences is, equivalently, a problem of finding a linear combination of the original predictor variables that shows large differences in group means. Discriminant analysis is a method for determining such linear combinations. For a more detailed mathematical treatment than given in this chapter, the reader is referred to a monograph by McKeon (1964). A more elementary and less mathematical treatment may be found in Tatsuoka (1970).

6.1 THE DISCRIMINANT CRITERION

The first step toward determining a linear combination of a set of variables such that several group means on this linear combination will differ widely among themselves is to decide on a criterion for measuring such group-mean differences. Now, once a linear combination has been constructed, we are dealing with a single transformed variable. Hence, the familiar F-ratio for testing the significance of the overall difference among several group means on a single variable suggests an appropriate criterion.

When we have K groups with a total of N individuals, the F-ratio is given by

$$F = \frac{SS_b/(K-1)}{SS_w/(N-K)} = \frac{SS_b}{SS_w}\,\frac{N-K}{K-1}$$

Since the second factor, $(N - K)/(K - 1)$, in the last expression is a constant for any given problem (where N and K are fixed), the first factor, SS_b/SS_w, is the only essential quantity for measuring how widely a set of group means differ among themselves, relative to the amount of variability present within the groups. Let us, therefore, write this ratio for a linear combination of a set of variables, and investigate what sort of function it is of the combining weights used.

If there are p predictor variables, $X_1, X_2, \ldots, X_p$, and we form a linear combination,

$$Y = v_1X_1 + v_2X_2 + \cdots + v_pX_p \tag{6.1}$$

of these variables, the within-groups and between-groups sums-of-squares of Y both turn out to be expressible as quadratic forms analogous to that in Eq. 5.18. The formula for $SS_w(Y)$ is a direct consequence of this equation, obtained by applying it to each of the K groups separately and then adding the results. Thus, if we denote the sum-of-squares of Y for the kth group by $SS_k(Y)$, and let $\mathbf{v}' = [v_1, v_2, \ldots, v_p]$, we have

$$\begin{aligned} SS_w(Y) &= SS_1(Y) + SS_2(Y) + \cdots + SS_K(Y) \\ &= \mathbf{v}'\mathbf{S}_1\mathbf{v} + \mathbf{v}'\mathbf{S}_2\mathbf{v} + \cdots + \mathbf{v}'\mathbf{S}_K\mathbf{v} \\ &= \mathbf{v}'(\mathbf{S}_1 + \mathbf{S}_2 + \cdots + \mathbf{S}_K)\mathbf{v} \end{aligned}$$

or

$$SS_w(Y) = \mathbf{v}'\mathbf{W}\mathbf{v} \tag{6.2}$$

since

$$\sum_{k=1}^{K} \mathbf{S}_k = \mathbf{W}$$

is the within-groups SSCP matrix as defined in Eq. 3.20.

To obtain the corresponding formula for the between-groups sum-of-squares $SS_b(Y)$ involves a little more work. We first define the between-groups SSCP matrix $\mathbf{B}$ for the original p variables as follows. The diagonal elements of $\mathbf{B}$ are the usual between-groups sums-of-squares for the variables taken one at a time; that is,

$$b_{ii} = \sum_{k=1}^{K} n_k(\bar{X}_{ik} - \bar{X}_i)^2 \tag{6.3}$$

where n_k is the size of the kth group, $\bar{X}_{ik}$ is the kth group mean of X_i, and $\bar{X}_i$ is the grand mean of X_i. The off-diagonal elements of $\mathbf{B}$ are the between-groups sums-of-products for pairs of variables. Thus, the (i, j)-element is

$$b_{ij} = \sum_{k=1}^{K} n_k(\bar{X}_{ik} - \bar{X}_i)(\bar{X}_{jk} - \bar{X}_j) \tag{6.4}$$

Equations 6.3 and 6.4 provide an adequate definition of the between-groups SSCP matrix for computational purposes, but it is convenient to express **B** in terms of more basic matrices in order to relate it to $SS_b(Y)$. This may be done as follows. Define an $N \times p$ matrix $\bar{\mathbf{X}}$ in which the first n_1 rows are each equal to $[\bar{X}_{11}, \bar{X}_{21}, \ldots, \bar{X}_{p1}]$, the next n_2 rows are each equal to $[\bar{X}_{12}, \bar{X}_{22}, \ldots, \bar{X}_{p2}]$, and so on until the last n_K rows are each equal to $[\bar{X}_{1K}, \bar{X}_{2K}, \ldots, \bar{X}_{pK}]$; that is, $\bar{\mathbf{X}}$ consists of n_1 repeated listings of the group 1 means of the p variables, n_2 repeated listings of the group 2 means of the p variables, and so forth. Next, define another $N \times p$ matrix $\bar{\bar{\mathbf{X}}}$, all of whose rows are listings of the grand means of the p variables, $[\bar{X}_1, \bar{X}_2, \ldots, \bar{X}_p]$. (Note that this is just the matrix we earlier denoted by $\bar{\mathbf{X}}$, first in Chapter Two and then in Chapter Five. But we now have two classes of means—the group means and the grand means; hence the double bar in the symbol for the matrix of grand means.) It can then be shown, as the reader should verify, that the matrix **B** can be expressed as

$$\mathbf{B} = (\bar{\mathbf{X}} - \bar{\bar{\mathbf{X}}})'(\bar{\mathbf{X}} - \bar{\bar{\mathbf{X}}}) \tag{6.5}$$

Now if we pre- and postmultiply both sides of Eq. 6.5 by $\mathbf{v}'$ and $\mathbf{v}$, respectively, we get

$$\begin{aligned} \mathbf{v}'\mathbf{B}\mathbf{v} &= \mathbf{v}'(\bar{\mathbf{X}} - \bar{\bar{\mathbf{X}}})'(\bar{\mathbf{X}} - \bar{\bar{\mathbf{X}}})\mathbf{v} \\ &= (\bar{\mathbf{X}}\mathbf{v} - \bar{\bar{\mathbf{X}}}\mathbf{v})'(\bar{\mathbf{X}}\mathbf{v} - \bar{\bar{\mathbf{X}}}\mathbf{v}) \end{aligned}$$

But $\bar{\mathbf{X}}\mathbf{v}$ and $\bar{\bar{\mathbf{X}}}\mathbf{v}$ are, in accordance with Eq. 5.13, vectors whose elements consist of the group means of Y and the grand mean of Y, respectively. More specifically, $\bar{\mathbf{X}}\mathbf{v}$ is an N-dimensional column vector whose first n_1 elements are equal to $\bar{Y}_1$, the next n_2 elements are equal to $\bar{Y}_2$, and so on until the last n_K elements are equal to $\bar{Y}_K$. $\bar{\bar{\mathbf{X}}}\mathbf{v}$ is an N-vector with all elements equal to $\bar{Y}$. Therefore, it is not difficult to see that the product $(\bar{\mathbf{X}}\mathbf{v} - \bar{\bar{\mathbf{X}}}\mathbf{v})'(\bar{\mathbf{X}}\mathbf{v} - \bar{\bar{\mathbf{X}}})\mathbf{v}$ is equal to

$$\sum_{k=1}^{K} n_k(\bar{Y}_k - \bar{Y})^2$$

which is none other than the between-groups sum-of-squares of the transformed variable Y. We have thus shown that

$$SS_b(Y) = \mathbf{v}'\mathbf{B}\mathbf{v} \tag{6.6}$$

Using Eq. 6.2 and 6.6, we are now able to write the ratio of the between-groups to within-groups sums-of-squares of Y as a function of the vector of combining weights $\mathbf{v}$, thus:

$$\frac{SS_b(Y)}{SS_w(Y)} = \frac{\mathbf{v}'\mathbf{B}\mathbf{v}}{\mathbf{v}'\mathbf{W}\mathbf{v}} \equiv \lambda \tag{6.7}$$

Following the argument presented earlier, we take the ratio λ, defined by Eq. 6.7, as a criterion for measuring the group differentiation along the dimension specified by the vector **v**. Fisher (1936) was the first to propose this criterion in connection with his two-group discriminant function. We shall refer to λ as the *discriminant criterion.*

6.2 MAXIMIZING THE DISCRIMINANT CRITERION

Once we have decided on a criterion for group differentiation, our task reduces to that of determining a set of weights, $[v_1, v_2, \ldots, v_p]$, which maximizes the discriminant criterion. This is accomplished by taking the partial derivative of λ with respect to each component v_i of **v** and setting the result equal to zero. Symbolically, we may find the derivative of λ with respect to the column vector **v** and equate the result to the $p \times 1$ null vector. (For details see Appendix C.) The vector equation thus obtained is as follows:

$$\frac{\partial \lambda}{\partial \mathbf{v}} = \frac{2[(\mathbf{Bv})(\mathbf{v}'\mathbf{Wv}) - (\mathbf{v}'\mathbf{Bv})(\mathbf{Wv})]}{(\mathbf{v}'\mathbf{Wv})^2} = \mathbf{0}$$

Dividing both numerator and denominator of the middle member by $\mathbf{v}'\mathbf{Wv}$ and using the definition (6.7) of λ, this equation reduces to

$$\frac{2[\mathbf{Bv} - \lambda\mathbf{Wv}]}{\mathbf{v}'\mathbf{Wv}} = \mathbf{0}$$

which is equivalent to

$$(\mathbf{B} - \lambda\mathbf{W})\mathbf{v} = \mathbf{0} \tag{6.8}$$

It may safely be assumed that **W** is nonsingular, and hence possesses an inverse $\mathbf{W}^{-1}$, except when a linear restriction of the form

$$a_i X_i + a_j X_j + \cdots + a_q X_q = 0$$

holds, by design, among some subset of the p predictor variables, as would be the case if ipsative measures are included among the predictors. In such a case, one of the variables in this subset may be eliminated or, alternatively, the principal components of this subset may be computed and used in place of these variables in the analysis, as described in connection with multiple regression analysis in Chapter Five.[1] Thus we may, without loss of generality,

[1]A third alternative is simply to solve Eq. 6.8 as it stands, without attempting to reduce it to the form of (6.9). It is then called a generalized eigenvalue problem, and the characteristic equation is

$$|\mathbf{B} - \lambda\mathbf{W}| = 0$$

which is more tedious to formulate with a desk calculator than is

$$|\mathbf{A} - \lambda\mathbf{I}| = 0$$

However, with electronic computers, it makes little difference, and most discriminant analysis programs use this approach.

assume that $\mathbf{W}^{-1}$ exists, and premultiply both sides of Eq. 6.8 by it to obtain

$$(\mathbf{W}^{-1}\mathbf{B} - \lambda\mathbf{I})\mathbf{v} = \mathbf{0} \tag{6.9}$$

This equation is of the form

$$(\mathbf{A} - \lambda\mathbf{I})\mathbf{v} = \mathbf{0}$$

procedures for solving which were fully described in Chapter Five. Its solution, yielding the eigenvalues λ_m and associated eigenvectors $\mathbf{v}_m$ of the matrix $\mathbf{A}$, is therefore well known by now, and we have thus solved the problem of maximizing the discriminant criterion. (Strictly speaking, Eq. 6.9 represents only a necessary condition for maximizing the discriminant criterion, and we should examine the second-order derivative of λ with respect to $\mathbf{v}$ in order to verify that the solutions of this equation yield maxima rather than minima or points of inflexion. We shall not carry out this extra step here, but take it as a proven fact that (6.9) is a sufficient as well as a necessary condition for maximizing λ.)

6.3 DISCRIMINANT FUNCTIONS

It was shown in the preceding chapter that the number of nonzero eigenvalues of a square matrix $\mathbf{A}$ is equal to the rank of $\mathbf{A}$. In the present context, with $\mathbf{W}^{-1}\mathbf{B}$ playing the role of $\mathbf{A}$, the number of nonzero eigenvalues depends on the rank of $\mathbf{B}$, since the rank of the product of two matrices cannot exceed the smaller of the two factor matrices' ranks, and $\mathbf{W}^{-1}$ (being nonsingular) must be of full rank p, while the rank of $\mathbf{B}$ is usually smaller than p.

The last-mentioned fact may be seen by referring to Eq. 6.5, which shows that the rank of $\mathbf{B}$ is equal to that of $\bar{\mathbf{X}} - \bar{\bar{\mathbf{X}}}$. Now $\bar{\mathbf{X}} - \bar{\bar{\mathbf{X}}}$ is an $N \times p$ matrix with only K different rows. (Recall that the first n_1 rows of $\bar{\mathbf{X}}$ are all equal, the next n_2 rows are different from the first n_1 but equal among themselves, and so on until the last n_K rows are different from all preceding rows but equal among themselves, and that $\bar{\bar{\mathbf{X}}}$ has all N rows identical.) Moreover, these K different rows, which may be denoted $\mathbf{b}_1', \mathbf{b}_2', \ldots, \mathbf{b}_K'$ in their order of occurrence, are themselves subject to a linear restriction; namely,

$$n_1\mathbf{b}_1' + n_2\mathbf{b}_2' + \cdots + n_K\mathbf{b}_K' = \mathbf{0}'$$

because, for each variable X_i, the well-known identity

$$\sum_{k=1}^{K} n_k(\bar{X}_{ik} - \bar{X}_i) = 0$$

holds. Therefore, there are only $K - 1$ linearly independent rows in $\bar{\mathbf{X}} - \bar{\bar{\mathbf{X}}}$.

Thus the rank of $\bar{\mathbf{X}} - \bar{\bar{\mathbf{X}}}$, and hence that of **B**, must be equal to $K - 1$ or p, whichever is smaller. Usually $K - 1$ will be smaller than p, because we would rarely perform a discriminant analysis using fewer variables than the number of groups being studied. For generality, however, let us denote the rank of **B** by $r = \min(K - 1, p)$.

Thus, when Eq. 6.9 is solved, we get r nonzero eigenvalues, which will be denoted as $\lambda_1, \lambda_2, \ldots, \lambda_r$ in descending order of magnitude, and r associated eigenvectors $\mathbf{v}_1, \mathbf{v}_2, \ldots, \mathbf{v}_r$. As usual, the eigenvectors are determined only up to an arbitrary multiplier, because if $\mathbf{v}$ satisfies (6.9) for some λ, it is clear that $c\mathbf{v}$ also satisfies the equation for the same λ (where c is an arbitrary constant). It is customary to choose the multiplier for each eigenvector in one of two ways: (1) so that its norm will be unity (that is, $\mathbf{v}_m'\mathbf{v}_m = 1$, for each m); (2) so that its largest element will be unity.

Now, from the fact that the eigenvalues λ_m are, by definition, the values assumed by the discriminant criterion for linear combinations using the elements of the corresponding eigenvectors $\mathbf{v}_m$ as combining weights, it is clear that the eigenvector $\mathbf{v}_1' = [v_{11}, v_{12}, \ldots, v_{1p}]$ provides a set of weights such that the transformed variable

$$Y_1 = v_{11}X_1 + v_{12}X_2 + \cdots + v_{1p}X_p$$

has the largest discriminant-criterion, λ_1, achievable by any linear combination of the p predictor variables. What are the properties of the remaining eigenvectors, $\mathbf{v}_2, \mathbf{v}_3, \ldots, \mathbf{v}_r$?

It can be shown (see Exercise 4 at the end of this chapter) that if the elements of $\mathbf{v}_2$ are used as combining weights to form a second linear combination,

$$Y_2 = v_{21}X_1 + v_{22}X_2 + \cdots + v_{2p}X_p$$

then Y_2 has this property: its discriminant-criterion value, λ_2, is the largest achievable by any linear combination of the X's that is uncorrelated (in the total sample) with Y_1. Similarly,

$$Y_3 = v_{31}X_1 + v_{32}X_2 + \cdots + v_{3p}X_p$$

has the largest possible discriminant-criterion value (λ_3) among all linear combinations of the X's that are uncorrelated with both Y_1 and Y_2; and so on until Y_r, using the elements of $\mathbf{v}_r$ as weights, has the largest possible discriminant-criterion value among linear combinations that are uncorrelated with all the preceding linear combinations $Y_1, Y_2, \ldots, Y_{r-1}$. The linear combinations $Y_1, Y_2, \ldots, Y_r$ are called the first, second, ..., rth (linear) *discriminant functions* for optimally differentiating among the K given groups.

We thus see that, although we started out by seeking to maximize the discriminant criterion, we obtain several discriminant functions, the first of which has the largest possible discriminant-criterion value, and each of the others has a conditionally maximal discriminant-criterion value. It is in this sense, then, that discriminant analysis reveals the "dimensions" of group differences. By this technique we find, simultaneously, the dimension along which maximum group differentiation occurs; the dimension along which is observed the largest group differences not accounted for by the first dimension; and so forth.

The situation here is reminiscent of principal components analysis. There, the dimension corresponding to the first component had maximum variance; the second-component dimension had maximum variance among those uncorrelated with the first; and so on. In discriminant analysis, the ratio of between- to within-groups sums-of-squares merely takes the place of variance as the criterion in determining the successive dimensions. However, an important difference between the dimensions identified in discriminant analysis and those in component analysis is that the former are generally not mutually orthogonal in test space, even though they are uncorrelated. That is, the axes representing the discriminant functions are not a subset of axes obtainable by rigid rotation of the original system of p axes; the discriminant rotation is an oblique rotation.

Just as in component analysis, the dimensions represented by the discriminant functions *may* be susceptible to meaningful interpretations. Even if they are not, we shall still have achieved parsimony by having reduced the dimensionality of the space in which to describe group differences. In seeking to interpret the discriminant functions, we would want to know which of the original p variables contribute most to each function. For this purpose, comparison of the relative magnitudes of the combining weights as given by the elements of each eigenvector of $\mathbf{W}^{-1}\mathbf{B}$ is inappropriate because these are weights to be applied to the predictors in raw-score scales, and are hence affected by the particular unit used for each variable. To eliminate the spurious effects of units on the magnitudes of combining weights, we must compare the weights that would be applied to the predictors in standardized form. The relative magnitudes of these standardized weights may be assessed by multiplying each raw-score weight by the standard deviation of the corresponding variable as computed from the within-groups SSCP matrix. This amounts to multiplying each element of a given eigenvector $\mathbf{v}_m$ by the square root of the corresponding diagonal element of $\mathbf{W}$. Thus, for each m, we define

$$v_{mi}^* = \sqrt{w_{ii}}\, v_{mi} \qquad i = 1, 2, \ldots, p \tag{6.10}$$

as the standardized discriminant weights. The relative contribution of the ith predictor to the mth discriminant function may then be gauged by the magnitude of v_{mi}^* in comparison with the other weights v_{mj}^*.

Up to this point, we have taken the dimensionality of the discriminant space to be equal to the number of nonzero eigenvalues of $\mathbf{W}^{-1}\mathbf{B}$, which is the smaller of the two numbers, $K - 1$ and p. It may often happen, however, that the number of *significant* discriminant dimensions may be even smaller. That is, not all of the discriminant functions may represent dimensions along which statistically significant group differences occur. We now turn to a test for determining the number of discriminant functions that are significant at a prescribed level.

6.4 SIGNIFICANCE TESTS IN DISCRIMINANT ANALYSIS

It was seen in Chapter Four that a basic quantity in testing the significance of the overall difference among several group centroids was the ratio of the determinants of the within-groups and the total SSCP matrices, known as Wilks' Λ criterion. We shall now see that this criterion figures prominently not only in the overall test, but also in the fine-grained test to determine how many of the discriminant dimensions contribute significantly to group differentiation. The relevance of Λ for this purpose hinges on an interesting algebraic relation existing between it and the discriminant-criterion values for the successive discriminant functions.

By equating the reciprocals of the two members of Eq. 4.28 and successively transforming, it is seen that

$$\begin{aligned} 1/\Lambda &= |\mathbf{T}|/|\mathbf{W}| = |\mathbf{W}^{-1}\mathbf{T}| && (\text{because } |\mathbf{AB}| = |\mathbf{A}||\mathbf{B}|) \\ &= |\mathbf{W}^{-1}(\mathbf{W} + \mathbf{B})| && (\text{because } \mathbf{T} = \mathbf{W} + \mathbf{B}) \\ &= |\mathbf{I} + \mathbf{W}^{-1}\mathbf{B}| \end{aligned}$$

Hence, by use of Theorems 2 and 8 of Chapter Five, it follows that

$$1/\Lambda = (1 + \lambda_1)(1 + \lambda_2)\cdots(1 + \lambda_r) \tag{6.11}$$

where $\lambda_1, \lambda_2, \ldots, \lambda_r$ are the nonzero eigenvalues of $\mathbf{W}^{-1}\mathbf{B}$. Consequently, Bartlett's V statistic for testing the significance of an observed Λ value, given in Eq. 4.29, can be expressed as

$$\begin{aligned} V &= -[N - 1 - (p + K)/2] \ln \Lambda \\ &= [N - 1 - (p + K)/2] \ln[(1 + \lambda_1)(1 + \lambda_2)\cdots(1 + \lambda_r)] \\ &= [N - 1 - (p + K)/2] \sum_{m=1}^{r} \ln(1 + \lambda_m) \end{aligned} \tag{6.12}$$

This statistic, it will be recalled, is distributed approximately as a chi-square with $p(K - 1)$ degrees of freedom.

Now, because of the uncorrelatedness of the successive discriminant functions, the successive terms $\ln(1 + \lambda_m)$ in the last expression above are statistically independent (assuming multivariate normality of the original p variables). As a result, the additive components of V are each approximately distributed as a chi-square variate. More specifically, the mth component,

$$V_m = [N - 1 - (p + K)/2] \ln(1 + \lambda_m) \tag{6.13}$$

is approximately a chi-square with $p + K - 2m$ degrees of freedom. [It may be readily verified that the sum of the n.d.f.'s of the r components, that is, $(p + K - 2) + (p + K - 4) + \cdots + (p + K - 2r)$, is equal to $p(K - 1)$ regardless of whether $r = K - 1$ or p.] Consequently, when we cumulatively subtract V_1, V_2, and so on from V, the remainder each time is also a chi-square variate; and these successive remainders become appropriate statistics for testing whether the *residual discrimination* after removing (or "partialling out") the first discriminant function, the first and second discriminant functions, and so forth, is statistically significant. The successive test statistics and their n.d.f.'s may be summarized as follows:

Residual After Removing	Approximate χ^2-Statistic	n.d.f.
First discriminant function	$V - V_1$	$p(K - 1) - (p + K - 2) = (p - 1)(K - 2)$
First 2 discriminant functions	$V - V_1 - V_2$	$(p - 1)(K - 2) - (p + K - 4) = (p - 2)(K - 3)$
First 3 discriminant functions	$V - V_1 - V_2 - V_3$	$(p - 2)(K - 3) - (p + K - 6) = (p - 3)(K - 4)$
⋮	⋮	⋮

As soon as the residual after removing the first s discriminant functions becomes smaller than the prescribed centile point (that is, the $100(1 - \alpha)$th centile) of the appropriate chi-square distribution, we may conclude that only the first s discriminant functions are significant at that α level.

If the number of significant discriminant functions thus found is smaller than r (as will often be the case), we will have effected a further reduction in the dimensionality of the space required to describe the differences among the K populations from which our sample groups were drawn. The remaining $r - s$ dimensions may be regarded as immaterial for population differentiation, since our sample differences along these dimensions can be attributed to sampling error.

6.5 NUMERICAL EXAMPLE

Let us follow up the example given in Chapter Four (pp. 90–91) in connection with the use of Wilks' Λ criterion in multivariate significance testing, and compute the discriminant functions for those data. The within-groups SSCP matrix, already computed there, was as follows:

$$\mathbf{W} = \begin{bmatrix} 3967.8301 & 351.6142 & 76.6342 \\ 351.6142 & 4406.2517 & 235.4365 \\ 76.6342 & 235.4365 & 2683.3164 \end{bmatrix}$$

To illustrate the computations for the between-groups SSCP matrix **B**, we reproduce the means on the three variables in each group and in the total sample, below.

			Means On		
(k)	Group	n_k	X_1	X_2	X_3
1.	Passenger agents	85	12.59	24.22	9.02
2.	Mechanics	93	18.54	21.14	10.14
3.	Operations men	66	15.58	15.45	13.24
	Total sample	$N = 244$	15.66	20.68	10.59

Equations 6.3 and 6.4 may now be used to compute the diagonal and off-diagonal elements of **B**, respectively. For instance,

$$\begin{aligned} b_{11} &= 85(12.59 - 15.66)^2 + 93(18.54 - 15.66)^2 + 66(15.58 - 15.66)^2 \\ &= 1572.92 \end{aligned}$$

and

$$\begin{aligned} b_{12} = {} & 85(12.59 - 15.66)(24.22 - 20.68) \\ & + 93(18.54 - 15.66)(21.14 - 20.68) \\ & + 66(15.58 - 15.66)(15.45 - 20.68) = -772.94 \end{aligned}$$

Proceeding in this manner, the between-groups SSCP matrix is found to be

$$\mathbf{B} = \begin{bmatrix} 1572.7441 & -773.0506 & 273.6214 \\ -773.0506 & 2889.3193 & -1405.9955 \\ 273.6214 & -1405.9955 & 691.6068 \end{bmatrix}$$

(The discrepancies between the values of b_{11} and b_{12} shown in the matrix and those computed above are due to the fact that the raw-score forms of

Eq. 6.3 and 6.4 were used for greater accuracy in the actual calculations yielding the matrix; that is,

$$b_{ij} = \sum_{k=1}^{K} (T_{ik}T_{jk})/n_k - T_iT_j/N$$

where T_{ik} and T_{jk} are the totals, in group k, of variables X_i and X_j, and T_i and T_j are their grand totals. The reader may verify the more accurate results by reconstructing the group totals for each variable by multiplying the group means given above by the group sizes and rounding to the nearest integer.)

Before continuing it is advisable, as a computational check, to verify that $\mathbf{W} + \mathbf{B} = \mathbf{T}$ (where $\mathbf{T}$ is the total SSCP matrix, given on p. 90). We leave this verification for the reader to carry out. Next, we compute the matrix $\mathbf{W}^{-1}\mathbf{B}$. The result is

$$\mathbf{W}^{-1}\mathbf{B} = \begin{bmatrix} .413255 & -.246239 & .093744 \\ -.214244 & .706315 & -.341803 \\ .108967 & -.578917 & .285056 \end{bmatrix}$$

which is the matrix whose eigenvalues and eigenvectors we need to find.

The characteristic equation, $|\mathbf{W}^{-1}\mathbf{B} - \lambda\mathbf{I}| = 0$, is found to be

$$\lambda^3 - 1.404626\lambda^2 + .350182\lambda = 0$$

(If the reader carries out the computations, he will find a constant term of the order of .00002. But we know that the rank of $\mathbf{B}$, and hence that of $\mathbf{W}^{-1}\mathbf{B}$, must be 2, which is the smaller of the two numbers $K - 1$ and p for this example. Therefore, the nonzero constant term is due to rounding error, and we simply omit it from the characteristic equation.) The roots of this equation are, in descending order of magnitude,

$$\lambda_1 = 1.080548 \qquad \lambda_2 = .324078 \qquad \lambda_3 = 0$$

Next, we compute the eigenvectors $\mathbf{v}_1$ and $\mathbf{v}_2$ corresponding to the two nonzero eigenvalues λ_1 and λ_2, following the steps described on pp. 120–121. The results (after normalizing each vector) are as follows:

$$\mathbf{v}_1 = \begin{bmatrix} .3524 \\ -.7331 \\ .5818 \end{bmatrix} \quad \text{and} \quad \mathbf{v}_2 = \begin{bmatrix} .9145 \\ .1960 \\ -.3540 \end{bmatrix}$$

The elements of these vectors may be interpreted geometrically as discussed in Chapter Five; namely, as the cosines of the angles between the original axes (representing the variables X_1, X_2, and X_3) and the new axes representing the two discriminant functions.

Our next task is to ascertain whether both or only one of these discriminant functions is statistically significant. (In general, the possibility exists, of course, that none of the functions is significant; but this event is already excluded in the present example, because the F-ratio based on Wilks' Λ criterion was found to be significant in the example at the end of Chapter Four.) We first compute the approximate chi-square statistic V for the overall group centroid differences in accordance with Eq. 6.12:

$$\begin{aligned} V &= [N - 1 - (p + K)/2][\ln(1 + \lambda_1) + \ln(1 + \lambda_2)] \\ &= [244 - 1 - (3 + 3)/2](2.3026)(\log 2.0805 + \log 1.3241) \\ &= (240)(2.3026)(.31817 + .12192) \\ &= 175.82 + 67.37 = 243.19 \end{aligned}$$

which agrees with the value found in the example of Chapter Four using Λ, thus verifying the fact that $1/\Lambda = (1 + \lambda_1)(1 + \lambda_2)$, stated in Eq. 6.11. The two numbers given just before the final result, 243.19, for V are the terms V_1 (= 175.82) and V_2 (= 67.37) as computed from Eq. 6.13. Thus, the quantity $V - V_1$ to be used for testing the significance of the "residual discrimination" after removing the first discriminant function is equal simply to V_2. [In general, $V - V_1 - V_2 - \cdots - V_s = V_r + V_{r-1} + \cdots + V_{s+1}$; that is, the statistic for testing the residual after the first s discriminants may be obtained as the cumulative sum from V_r (corresponding to the last, or smallest, nonzero eigenvalue λ_r) up through V_{s+1}, based on the next smaller eigenvalue after the sth.] Now the quantity $V - V_1 = V_2 = 67.37$ is an approximate chi-square variate with $(p - 1)(K - 2) = (2)(1) = 2$ degrees of freedom, as stated in the tabular display on p. 165. This value far exceeds the 99.9th centile of the chi-square distribution with 2 d.f.'s, which is 13.82. We therefore reject the hypothesis that the group differences after removal of the first discriminant function are due to sampling error, and conclude that both discriminant functions are statistically significant beyond the .001 level.

Having decided to retain both of the discriminant functions as significant, we may next try to see if some intuitively meaningful interpretations can be given to the two dimensions along which our three groups were found to differ. In this attempt it is useful, besides determining the standardized discriminant weights in accordance with Eq. 6.10, to examine the three group means on each of the discriminant functions. These means may be computed by forming the linear combinations of the group means on X_1, X_2, and X_3, using in turn, the elements of $\mathbf{v}_1$ and those of $\mathbf{v}_2$ as the weights in Eq. 6.1—or, what amounts to the same thing, by using the matrix Eq. 5.13:

$$\begin{aligned} \bar{\mathbf{Y}} = \bar{\mathbf{X}}\mathbf{V} &= \begin{bmatrix} 12.59 & 24.22 & 9.02 \\ 18.54 & 21.14 & 10.14 \\ 15.58 & 15.45 & 13.24 \end{bmatrix} \begin{bmatrix} .3524 & .9145 \\ -.7331 & .1960 \\ .5818 & -.3540 \end{bmatrix} \\ &= \begin{bmatrix} -8.07 & 13.07 \\ -3.06 & 17.51 \\ 1.87 & 12.59 \end{bmatrix} \end{aligned}$$

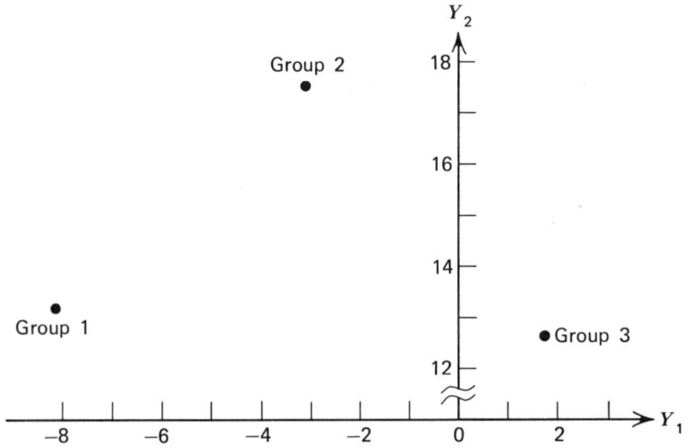

FIGURE 6.1 ***Discriminant function centroids of the three groups.***

The group centroids ($\bar{Y}_{1k}$, $\bar{Y}_{2k}$) may be represented graphically as in Fig. 6.1. It should be noted, however, that our use of a rectangular coordinate system is merely a matter of convenience, and does not imply that the axes representing the discriminant functions in the original three-dimensional test space are mutually orthogonal. (The actual angle between the discriminant axes is such that its cosine is equal to

$$\mathbf{v}_1'\mathbf{v}_2 = (.3524)(.9145) + (-.7331)(.1960) + (.5818)(-.3540) = -.0274$$

Consulting a table of trigonometric functions, the angle is found to be 91°34′.) The plot of discriminant function centroids in Fig. 6.1 shows at a glance that the first discriminant dimension (Y_1) separates the three groups in roughly equal steps: the Group 2 mean is about as far from the Group 1 mean as the Group 3 mean is from Group 2 along this dimension. In terms of the second dimension, Groups 1 and 3 are practically indistinguishable, but Group 2 is separated from them by a considerable distance.

Let us now determine the relative contributions of the three original variables to group differentiation along each of the two dimensions, by computing the standardized discriminant weights in accordance with Eq. 6.10. The results are,

$$\mathbf{v}_1^* = \begin{bmatrix} 22.2 \\ -48.7 \\ 30.1 \end{bmatrix} \quad \text{and} \quad \mathbf{v}_2^* = \begin{bmatrix} 57.6 \\ 13.0 \\ -18.3 \end{bmatrix}$$

Actually, we are interested only in rough indications of the relative importance of each variable for each discriminant dimension. So we may take small integers roughly proportional to the elements of $\mathbf{v}_1^*$ and $\mathbf{v}_2^*$ to assess the relative contributions of X_1, X_2, and X_3. We thus find these to be in the proportion 2:−5:3 for the first dimension, and 6:1:−2 in the second dimension. (In this example, the rank order of the standardized weights and that of the raw-score weights are identical, because the diagonal elements of **W** are of the same order of magnitude; but this will not always be the case.) Recalling that X_1, X_2, and X_3 stand for the outdoor, convivial, and conservative scales, respectively, of an Activity Preference Questionnaire, let us try to attach some meaningful labels to the two discriminant dimensions in the light of the relative weights of the three variables.

The Y_1 dimension is most highly weighted, in the negative direction, with the convivial scale; that is, persons with few convivial activity preferences (or, equivalently, with many solitary activity preferences) will score high on Y_1. But the conservative scale also contributes fairly heavily to the first discriminant, with a weight of 3 compared to the −5 for X_2. Therefore, we might say that Y_1 represents a "solitary-conservative" syndrome. Examination of Fig. 6.1, which shows the group means on Y_1 to be ranked in the order (from high to low) $\bar{Y}_{13}$, $\bar{Y}_{12}$, $\bar{Y}_{11}$, corroborates this interpretation. We would expect operations control personnel, as a group, to be less gregarious (either by nature or by occupational habit) and perhaps more conservative in outlook than the other two groups. Passenger agents (Group 1) would certainly be expected to be the most convivial and sociable lot, while mechanics would fall in between the other two groups in this respect.

In the Y_2 dimension, X_1 (outdoor activity preference) far outweighs the other two variables in importance. Without doubt, it is the person with many outdoor interests who will score high on the second discriminant. Figure 6.1 shows the mechanics (Group 2) to be a class by itself, scoring much higher than the other two groups on Y_2, which is what we would expect.

6.6 SPECIAL CASE OF TWO GROUPS

When there are only two groups under study, the computations for determining the single ($K - 1 = 1$) discriminant function can be simplified a great deal. To see this simplification, we first show that discriminant functions (not only for the two-group case, but in general) can be determined just as well from Eq. 6.14

$$(\mathbf{T}^{-1}\mathbf{B} - \mu\mathbf{I})\mathbf{v} = \mathbf{0} \tag{6.14}$$

as from Eq. 6.9. We rewrite this latter equation in the form

$$\mathbf{Bv} = \lambda\mathbf{Wv}$$

and add $\lambda\mathbf{Bv}$ to both members, obtaining

$$(1 + \lambda)\mathbf{Bv} = \lambda(\mathbf{W} + \mathbf{B})\mathbf{v}$$

or,

$$\left(\mathbf{T}^{-1}\mathbf{B} - \frac{\lambda}{1 + \lambda}\mathbf{I}\right)\mathbf{v} = \mathbf{0}$$

Hence, if $\mathbf{v}_0$ is the eigenvector of $\mathbf{W}^{-1}\mathbf{B}$ associated with the eigenvalue λ_0, it is also an eigenvector of $\mathbf{T}^{-1}\mathbf{B}$, and its associated eigenvalue is $\lambda_0/(1 + \lambda_0)$. In other words, the vectors $\mathbf{v}$ satisfying Eq. 6.14 are identical to those satisfying Eq. 6.9, and the eigenvalues stand in the relation

$$\mu = \frac{\lambda}{1 + \lambda}$$

Now, in the two-group case, the between-groups SSCP matrix $\mathbf{B}$ assumes a particularly simple form, as follows. First, the expression (6.4) for its general element b_{ij} reduces to

$$b_{ij} = n_1(\bar{X}_{i1} - \bar{X}_i)(\bar{X}_{j1} - \bar{X}_j) + n_2(\bar{X}_{i2} - \bar{X}_i)(\bar{X}_{j2} - \bar{X}_j)$$

But the grand mean $\bar{X}_i$ of the ith variable is, in this case, equal to

$$\frac{n_1\bar{X}_{i1} + n_2\bar{X}_{i2}}{n_1 + n_2}$$

Consequently,

$$\bar{X}_{i1} - \bar{X}_i = \bar{X}_{i1} - \frac{n_1\bar{X}_{i1} + n_2\bar{X}_{i2}}{n_1 + n_2} = \frac{n_2}{n_1 + n_2}(\bar{X}_{i1} - \bar{X}_{i2})$$

and, similarly,

$$\bar{X}_{i2} - \bar{X}_i = \frac{n_1}{n_1 + n_2}(\bar{X}_{i2} - \bar{X}_{i1}) = \frac{-n_1}{n_1 + n_2}(\bar{X}_{i1} - \bar{X}_{i2})$$

Substituting these expressions (and corresponding ones for $\bar{X}_{j1} - \bar{X}_j$ and $\bar{X}_{j2} - \bar{X}_j$) in the formula for b_{ij}, we obtain:

$$\begin{aligned}
b_{ij} &= n_1\left[\frac{n_2}{n_1 + n_2}(\bar{X}_{i1} - \bar{X}_{i2})\right]\left[\frac{n_2}{n_1 + n_2}(\bar{X}_{j1} - \bar{X}_{j2})\right] \\
&\quad + n_2\left[\frac{-n_1}{n_1 + n_2}(\bar{X}_{i1} - \bar{X}_{i2})\right]\left[\frac{-n_1}{n_1 + n_2}(\bar{X}_{j1} - \bar{X}_{j2})\right] \\
&= \left[\frac{n_1 n_2^2}{(n_1 + n_2)^2} + \frac{n_2 n_1^2}{(n_1 + n_2)^2}\right](\bar{X}_{i1} - \bar{X}_{i2})(\bar{X}_{j1} - \bar{X}_{j2}) \\
&= \frac{n_1 n_2}{n_1 + n_2}(\bar{X}_{i1} - \bar{X}_{i2})(\bar{X}_{j1} - \bar{X}_{j2})
\end{aligned}$$

Therefore, if we define a p-dimensional row vector

$$\mathbf{d}' = [\bar{X}_{11} - \bar{X}_{12}, \bar{X}_{21} - \bar{X}_{22}, \ldots, \bar{X}_{p1} - \bar{X}_{p2}] \tag{6.15}$$

of the differences between the two group means on the p variables, we see that

$$\mathbf{B} = \frac{n_1 n_2}{n_1 + n_2} \mathbf{dd}' \tag{6.16}$$

When this special form of **B** for the two-group case is substituted in Eq. 6.14, we get

$$(c\mathbf{T}^{-1}(\mathbf{dd}') - \mu\mathbf{I})\mathbf{v} = \mathbf{0}$$

where $c = n_1 n_2/(n_1 + n_2)$. Distributing the postmultiplication by $\mathbf{v}$ and transposing, we obtain

$$\begin{aligned}\mu\mathbf{v} &= c\mathbf{T}^{-1}(\mathbf{dd}')\mathbf{v} \\ &= c(\mathbf{T}^{-1}\mathbf{d})(\mathbf{d}'\mathbf{v})\end{aligned}$$

In the last expression $\mathbf{d}'\mathbf{v}$, although unknown, is a scalar quantity, since d′ is $1 \times p$ and $\mathbf{v}$ is $p \times 1$. We may collect all the scalar quantities into a single multiplier and write

$$\begin{aligned}\mathbf{v} &= [c(\mathbf{d}'\mathbf{v})/\mu]\mathbf{T}^{-1}\mathbf{d} \\ &= m\mathbf{T}^{-1}\mathbf{d}\end{aligned} \tag{6.17}$$

where m is an unknown scalar.

Although Eq. 6.17 involves an unknown multiplier, it actually gives a solution for the eigenvector $\mathbf{v}$ of $\mathbf{T}^{-1}\mathbf{B}$ in the two-group case, because $\mathbf{v}$ is in any case determined only up to an arbitrary proportionality constant. (See discussion on p. 162; the constant may be determined by normalizing $\mathbf{v}$ to unity, or by setting its leading element equal to unity.) We thus arrive at the interesting result that, *in the two-group case*, the single discriminant function can be obtained without solving an eigenvalue problem. We need only postmultiply the inverse of the total SSCP matrix by the column of mean differences on the p predictors in order to get the vector of discriminant-function coefficients. Let us now examine the nature of the coefficients thus obtained.

The total SSCP matrix **T** is simply the SSCP matrix of the p predictors, computed for the two groups thrown together as a single sample; it is identical to what we denoted by $\mathbf{S}_{pp}$ in Section 2.5. Comparing Eq. 6.17 with Eq. 2.29, we see a striking resemblance, except for the arbitrary multiplier m. We may, in fact, rewrite Eq. 6.17 as

$$\mathbf{v} = \mathbf{S}_{pp}^{-1}(m\mathbf{d})$$

which suggests that the elements of $\mathbf{v}$ might be regarded as some sort of multiple regression weights—provided the elements of $m\mathbf{d}$ are interpretable as sums-of-products between predictors and criterion. But what criterion variable do we have in a two-group discriminant analysis? The answer is, a variable indicating each subject's membership in one or the other of the two groups. This would be a dichotomous variable taking the values 1 and 0 (for example) for members of group 1 and group 2, respectively.

Now, an appropriate coefficient of correlation between a continuous predictor X and a dichotomous criterion Y is the point-biserial coefficient, given in most statistics texts as

$$r_{pb} = \frac{\overline{X}_1 - \overline{X}_2}{s_x} \frac{\sqrt{n_1 n_2}}{N}$$

where $\overline{X}_1$ and $\overline{X}_2$ are the two group means on the predictor, s_x is the predictor standard deviation in the total sample of N $(= n_1 + n_2)$ cases, and n_1 and n_2 are the numbers of cases with $Y = 1$ and $Y = 0$ (that is, the two group sizes), respectively. From this coefficient, we may work backward to infer that the sum-of-products is given by

$$\sum xy = \frac{n_1 n_2}{n_1 + n_2} (\overline{X}_1 - \overline{X}_2)$$

We thus see that the elements of

$$m\mathbf{d}' = m[\overline{X}_{11} - \overline{X}_{12}, \overline{X}_{21} - \overline{X}_{22}, \ldots, \overline{X}_{p1} - \overline{X}_{p2}]$$

are indeed proportional to the predictor-criterion sum-of-products, where the criterion is a dichotomous variable indicating group-membership.

Our conclusion, then, is that *in the two-group case* the discriminant weights are proportional to the weights for a multiple regression equation of a dichotomous group-membership variable on the p predictors. That is to say, discriminant analysis reduces, in this case, to a multiple regression analysis in which all Group 1 members are assigned the score 1, and all Group 2 members the score 0, on a "dummy" criterion variable Y. This fact led many early writers (for example, Garrett, 1943 and Wherry, 1947) to state that discriminant analysis *in general* was nothing more than a special case of multiple regression analysis. It cannot be overemphasized that this reduction holds only in the two-group case. When there are more than two groups under study, discriminant analysis reduces, not to multiple regression, but to canonical correlation analysis, as we shall see in the next section.

EXAMPLE 6.1. In this miniature numerical example we illustrate the computation of a two-group discriminant function as a multiple regression equation, and compare it with the result obtained by the general method of discriminant analysis. Table 6.1 shows the scores of members of two groups on two predictor variables X_1 and X_2, and on a "dummy" criterion variable Y. All members of Group 1 are assigned $Y = 1$, and all members of Group 2 are given $Y = 0$.

TABLE 6.1 Roster for Two Groups on Two Predictors and a "Dummy" Criterion Variable Y.

	Group 1 ($n_1 = 8$)			Group 2 ($n_2 = 10$)		
	X_1	X_2	Y	X_1	X_2	Y
	20	6	1	17	12	0
	21	10	1	11	11	0
	15	12	1	15	14	0
	15	8	1	20	16	0
	11	11	1	14	16	0
	24	17	1	13	18	0
	18	13	1	16	13	0
	14	4	1	12	6	0
				17	19	0
				10	16	0
$\sum X_i$	138	81		145	141	
$\sum X_i^2$	2508	939		2189	2119	
$\sum X_1 X_2$	1449			2076		

The matrix $\mathbf{S}_{pp}$, whose inverse occurs in Eq. 2.29, is the upper left-hand 2×2 submatrix of an augmented SSCP matrix of the form shown on p. 35. To compute this matrix, we merge the two groups into a single sample of $n_1 + n_2$ (= 18) cases. In other words, the values of $\sum X_i$, $\sum X_i^2$, and $\sum X_1 X_2$, listed separately for each group at the foot of Table 6.1, are to be added across groups before computing the deviation-score sums-of-squares and sums-of-products. Further, it should be clear that the various sums involving Y for the total sample are given by

$$\sum Y = \sum Y^2 = n_1 (= 8) \qquad \text{and} \qquad \sum X_i Y = \sum_{\text{Gr.1}} X_i$$

Using these intermediate results, the augmented SSCP matrix is found to be

$$\left[\begin{array}{cc|c} 247.61 & 34.67 & 12.22 \\ 34.67 & 320.00 & -17.67 \\ \hline 12.22 & -17.67 & 4.44 \end{array}\right]$$

Therefore, according to Eqs. 2.23 and 6.17, the multiple regression weights, to which the two-group discriminant function weights are proportional, are given by the elements of the vector

$$\mathbf{b} = \begin{bmatrix} 247.61 & 34.67 \\ 34.67 & 320.00 \end{bmatrix}^{-1} \begin{bmatrix} 12.22 \\ -17.67 \end{bmatrix}$$

$$= (10^{-2}) \begin{bmatrix} .4101 & -.0444 \\ -.0444 & .3173 \end{bmatrix} \begin{bmatrix} 12.22 \\ -17.67 \end{bmatrix} = \begin{bmatrix} .0580 \\ -.0615 \end{bmatrix}$$

Let us now compute the discriminant function by the general method of this chapter. The within-groups and between-groups SSCP matrices of the predictor variables are:

$$\mathbf{W} = \begin{bmatrix} 214.00 & 83.25 \\ 83.25 & 249.78 \end{bmatrix}$$

and

$$\mathbf{B} = \begin{bmatrix} 33.61 & -48.58 \\ -48.58 & 70.22 \end{bmatrix}$$

We then obtain

$$\mathbf{W}^{-1}\mathbf{B} = \begin{bmatrix} .2674 & -.3864 \\ -.2836 & .4099 \end{bmatrix}$$

as the matrix to be substituted in Eq. 6.9. The characteristic equation, $|\mathbf{W}^{-1}\mathbf{B} - \lambda\mathbf{I}| = 0$, is found to be

$$\lambda^2 - .6773\lambda = 0$$

whose single nonzero root is $\lambda_1 = .6773$. The adjoint of $\mathbf{W}^{-1}\mathbf{B} - \lambda_1\mathbf{I}$ is

$$\begin{bmatrix} -.2674 & .3864 \\ .2836 & -.4099 \end{bmatrix}$$

and hence the eigenvector $\mathbf{v}_1$ of $\mathbf{W}^{-1}\mathbf{B}$, with the larger element set equal to unity, is

$$\mathbf{v}_1 = \begin{bmatrix} -.2674/.2836 \\ 1 \end{bmatrix} = \begin{bmatrix} -.9429 \\ 1 \end{bmatrix}$$

When the vector of regression weights obtained earlier is similarly rescaled, we find

$$\mathbf{b}^* = \begin{bmatrix} .0580/(-.0615) \\ 1 \end{bmatrix} = \begin{bmatrix} -.9431 \\ 1 \end{bmatrix}$$

which agrees, within rounding error, with the $\mathbf{v}_1$ just obtained by the general method of discriminant analysis.

Theoretical Supplement. As another matter pertaining to the special case of two groups, we recall the relationship between Wilks' Λ and Hotelling's T^2, mentioned without proof at the end of Chapter Four. We are now in a position to prove this relationship:

$$\frac{1 - \Lambda}{\Lambda} = \frac{T^2}{N - 2} \qquad \text{(where } N = n_1 + n_2\text{)}$$

We saw, in Eq. 6.11, that

$$1/\Lambda = (1 + \lambda_1)(1 + \lambda_2)\cdots(1 + \lambda_r)$$

where $\lambda_1, \lambda_2, \ldots, \lambda_r$ are the nonzero eigenvalues of $\mathbf{W}^{-1}\mathbf{B}$. However, in the two-group case $\mathbf{W}^{-1}\mathbf{B}$ has only one nonzero eigenvalue λ_1, since $\mathbf{B}$ is of rank one. It therefore follows, in this case, that

$$1/\Lambda = 1 + \lambda_1$$

or

$$\frac{1 - \Lambda}{\Lambda} = \lambda_1 \qquad \text{(the only nonzero eigenvalue of } \mathbf{W}^{-1}\mathbf{B}\text{)}$$

On the other hand, from Eq. 4.26 it follows that

$$T^2/(N - 2) = c\mathbf{d}'\mathbf{W}^{-1}\mathbf{d}$$

where $c = n_1 n_2/(n_1 + n_2)$ and $\mathbf{d}$ is as defined in Eq. 6.15. Premultiplying both sides of this equation by $\mathbf{d}$, we get

$$\mathbf{d}[T^2/(N - 2)] = \mathbf{d}[c\mathbf{d}'\mathbf{W}^{-1}\mathbf{d}]$$

or, on regrouping

$$[T^2/(N - 2)]\mathbf{d} = [(c\mathbf{d}\mathbf{d}')\mathbf{W}^{-1}]\mathbf{d}$$

which becomes, by virtue of eq. 6.16,

$$[T^2/(N - 2)]\mathbf{d} = (\mathbf{B}\mathbf{W}^{-1})\mathbf{d}$$

This means that $T^2/(N - 2)$ is the sole nonzero eigenvalue of $\mathbf{BW}^{-1}$ (with $\mathbf{d}$ as the associated eigenvector). But, since both $\mathbf{B}$ and $\mathbf{W}^{-1}$ are symmetric matrices, $\mathbf{BW}^{-1} = (\mathbf{W}^{-1}\mathbf{B})'$. Furthermore, any matrix and its transpose have the same eigenvalue(s). (Prove this.) Therefore, $T^2/(N - 2)$ is also the only nonzero eigenvalue of $\mathbf{W}^{-1}\mathbf{B}$. The relation to be proved follows immediately.

6.7 A CANONICAL CORRELATIONS APPROACH TO DISCRIMINANT ANALYSIS

A natural extension of the multiple-regression formulation of two-group discriminant analysis to discriminant analysis for any number of groups was made by Brown (1947) prior to the developments described in Sections 6.1 to 6.3. In this formulation, a set of "dummy criterion variables" one fewer than the number of groups is used, and the predictor and criterion sets are treated by the method of canonical correlation analysis. A mathematical proof that the discriminant criterion and canonical correlation approaches yield identical results was given by Tatsuoka (1953).

Suppose that $K = 5$ groups are under study. Then, for carrying out a discriminant analysis via the canonical correlation approach, $K - 1 = 4$ dummy criterion variables Y_1, Y_2, Y_3, Y_4 are used, and "scores" on these are assigned to members of the five groups in the following manner:

	Y_1	Y_2	Y_3	Y_4
All Group 1 members get	1	0	0	0
All Group 2 members get	0	1	0	0
All Group 3 members get	0	0	1	0
All Group 4 members get	0	0	0	1
All Group 5 members get	0	0	0	0

The generalization to the case of K groups should be obvious: for each k between 1 and $K - 1$ inclusive, all members of the kth group get a 1 on Y_k and 0's on the other Ys; while members of the last (Kth) group get 0's on all $K - 1$ dummy criterion variables $Y_1, Y_2, \ldots, Y_{K-1}$.

Upon thus assigning "scores" on the $K - 1$ dummy variables to each individual in the entire sample, we shall have $p + (K - 1)$ "observations" on $N (= n_1 + n_2 + \cdots + n_k)$ individuals; namely, p predictor-variable scores and $K - 1$ "criterion" scores. Whereas in the two-group case we had only $2 - 1 = 1$ "criterion" variable, and hence multiple regression analysis enabled us to determine a linear combination of the p predictors having maximum correlation with the criterion, we now have $q = K - 1$ criterion variables. Something more elaborate than multiple regression analysis must now be invoked.

Canonical correlation analysis is a technique by which we determine a linear combination of p predictors on the one hand, and a linear combination of q ($= K - 1$ in the present context) criterion variables on the other, such that the correlation between these linear combinations in the total sample is as large as possible. Formulated mathematically, the problem is as follows. We want to determine one set of weights

$$\mathbf{u}' = [u_1, u_2. \dots, u_p]$$

for the predictor variables, and another set of weights

$$\mathbf{v}' = [v_1, v_2, \dots, v_q]$$

for the criterion variables, in such a way that the correlation r_{zw} between

$$Z = u_1 X_1 + u_2 X_2 + \cdots + u_p X_p$$

and

$$W = v_1 Y_1 + v_2 Y_2 + \cdots + v_q Y_q$$

is the largest achievable for our particular sample.

Postponing the discussion of the rationale until the next section, we here describe the mechanics of the solution to this problem. The first step is to compute the SSCP matrix of all $p + q$ variables (in the present context, $p + K - 1$ variables) for the total sample of N individuals. This may be done in the usual manner regardless of whether the criterion variables are genuine variables such as test scores, or are "dummy" variables such as the group-membership variables in the discriminant-analysis context. In the latter case, however, some obvious simplifications of the formulas for the raw-score sums, sums-of-squares, and sums of products involving the Y's may be noted. Namely,

$$\Sigma Y_k = \Sigma Y_k^2 = n_k \qquad (k = 1, 2, \dots, K - 1)$$
$$\Sigma Y_h Y_k = 0 \qquad (h, k = 1, 2, \dots, K - 1;\ h \neq k)$$
$$\Sigma X_i Y_k = \sum_{\text{Gr.}k} X_i \qquad \text{(that is, the sum of } X_i \text{ in Group } k\text{)}$$

The SSCP matrix is then partitioned into four parts, similar to what was done for the multiple regression problem (see p. 35), except that the part referring to the "criterion" variables is now a $q \times q$ matrix instead of a single scalar, and the parts interrelating predictor and "criterion" variables are $p \times q$ and $q \times p$ matrices instead of column and row vectors, respectively. The partitioning is indicated as follows:

$$\mathbf{S} = \begin{bmatrix} \mathbf{S}_{pp} & \mathbf{S}_{pc} \\ \mathbf{S}_{cp} & \mathbf{S}_{cc} \end{bmatrix} \begin{matrix} p \text{ rows} \\ q \text{ rows} \end{matrix}$$
$$\quad p \text{ columns} \quad q \text{ columns}$$

Next, the quadruple matrix product

$$\mathbf{A} = \mathbf{S}_{pp}^{-1}\mathbf{S}_{pc}\mathbf{S}_{cc}^{-1}\mathbf{S}_{cp} \tag{6.18}$$

is computed.

Finally, the eigenvalues μ_i^2 and the eigenvectors $\mathbf{u}_i$ of the matrix $\mathbf{A}$ are computed. The largest eigenvalue μ_1^2 is the *square* of the maximum r_{zw}, called the first *canonical correlation* between the predictor and criterion sets. The elements of the corresponding eigenvector $\mathbf{u}_1$ are the weights to be used in combining the predictor variables to obtain the optimal linear combination Z_1. (The weights $\mathbf{v}_1$ to be applied to the criterion set to get W_1 are of no interest in the present context, where the "criteria" are simply dummy variables indicating group membership. An equation for computing $\mathbf{v}_i$ is given in the next section, where canonical correlation analysis is considered in greater generality.)

As the reader may have anticipated, the elements of the eigenvector $\mathbf{u}_2$, associated with the second largest eigenvalue μ_2^2, give the weights to be applied to the predictor set to construct a linear combination Z_2 which, among all predictor-set linear combinations that are uncorrelated with Z_1, has the largest possible correlation with any criterion-set linear combination that is uncorrelated with W_1. Similar interpretations hold for the other eigenvectors $\mathbf{u}_3, \mathbf{u}_4, \ldots, \mathbf{u}_{K-1}$.

The set of linear combinations $Z_1, Z_2, \ldots, Z_{K-1}$ thus obtained are identical (within proportionality) to the discriminant functions obtainable by the approach, described earlier, of maximizing the discriminant criterion. Furthermore, each discriminant-criterion value λ_i is related to the corresponding squared canonical-correlation value μ_i^2 by the equation (proved in Tatsuoka, 1953)

$$\lambda_i = \mu_i^2/(1 - \mu_i^2) \tag{6.19}$$

The following numerical example illustrates these facts.

EXAMPLE 6.2. For this example we add a third group of $n_3 = 12$ individuals to the two-group example of the preceding section. Table 6.2 shows the rosters of the three groups,

TABLE 6.2 Roster for Three Groups on Two Predictors and Two "Dummy" Criterion Variables Y_1 and Y_2.

	Group 1 ($n_1 = 8$)				Group 2 ($n_2 = 10$)				Group 3 ($n_3 = 12$)			
	X_1	X_2	Y_1	Y_2	X_1	X_2	Y_1	Y_2	X_1	X_2	Y_1	Y_2
	20	6	1	0	17	12	0	1	6	11	0	0
	21	10	1	0	11	11	0	1	10	8	0	0
	15	12	1	0	15	14	0	1	4	6	0	0
	15	8	1	0	20	16	0	1	14	7	0	0
	11	11	1	0	14	16	0	1	13	11	0	0
	24	17	1	0	13	18	0	1	18	11	0	0
	18	13	1	0	16	13	0	1	12	9	0	0
	14	4	1	0	12	6	0	1	12	9	0	0
					17	19	0	1	5	7	0	0
					10	16	0	1	10	5	0	0
									3	14	0	0
									6	12	0	0
$\sum X_i$	138	81			145	141			113	110		
$\sum X_i^2$	2508	939			2189	2119			1299	1088		
$\sum X_1 X_2$	1449				2076				1024			

including "scores" on the two "dummy" criterion variables Y_1 and Y_2. The SSCP matrix of the two predictors and two "criterion" variables for the total sample of $N = 8 + 10 + 12 = 30$ individuals, partitioned into four submatrices as indicated above, is found to be as follows:

$$\mathbf{S} = \begin{bmatrix} 768.80 & 166.60 & 32.40 & 13.00 \\ 166.60 & 471.87 & -7.53 & 30.33 \\ \\ 32.40 & -7.53 & 5.87 & -2.67 \\ 13.00 & 30.33 & -2.67 & 6.67 \end{bmatrix}$$

The computation of the upper left-hand submatrix $\mathbf{S}_{pp}$ should require no comment. The other submatrices $\mathbf{S}_{pc}$ and $\mathbf{S}_{cc}$, too, really involve nothing new, since their elements are of the forms

$$\sum x_i y_j = \sum X_i Y_j - (\sum X_i)(\sum Y_j)/N$$

and

$$\sum y_i y_j = \sum Y_i Y_j - (\sum Y_i)(\sum Y_j)/N$$

However, to illustrate use of the simplified formulas displayed earlier for the various sums involving Y_1 and Y_2, we show the calculations for a few of these elements below. (The reader should compare the figures below with the various entries of Table 6.2 and trace the substitutions made in the formulas.)

$$\begin{aligned}(\mathbf{S}_{pc})_{11} &= \sum X_1 Y_1 - (\sum X_1)(\sum Y_1)/N \\ &= 138 - (138 + 145 + 113)(8)/30 = 32.40\end{aligned}$$

$$\begin{aligned}(\mathbf{S}_{pc})_{12} &= \sum X_1 Y_2 - (\sum X_1)(\sum Y_2)/N \\ &= 145 - (396)(10)/30 = 13.00\end{aligned}$$

$$\begin{aligned}(\mathbf{S}_{cc})_{11} &= \sum Y_1^2 - (\sum Y_1)^2/N \\ &= 8 - (8)^2/30 = 5.87\end{aligned}$$

$$\begin{aligned}(\mathbf{S}_{cc})_{12} &= \sum Y_1 Y_2 - (\sum Y_1)(\sum Y_2)/N \\ &= 0 - (8)(10)/30 = -2.67\end{aligned}$$

After constructing the SCCP matrix $\mathbf{S}$ in this manner, we compute the quadruple matrix product $\mathbf{S}_{pp}^{-1}\mathbf{S}_{pc}\mathbf{S}_{cc}^{-1}\mathbf{S}_{cp}$, whose eigenvalues and eigenvectors we need to determine.

The result is

$$\mathbf{A} = \mathbf{S}_{pp}^{-1}\mathbf{S}_{pc}\mathbf{S}_{cc}^{-1}\mathbf{S}_{cp}$$

$$= \begin{bmatrix} 768.80 & 166.60 \\ 166.60 & 471.87 \end{bmatrix}^{-1} \begin{bmatrix} 32.40 & 13.00 \\ -7.53 & 30.33 \end{bmatrix}$$

$$\times \begin{bmatrix} 5.87 & -2.67 \\ -2.67 & 6.67 \end{bmatrix}^{-1} \begin{bmatrix} 32.40 & -7.53 \\ 13.00 & 30.33 \end{bmatrix}$$

$$= \begin{bmatrix} .403219 & .063240 \\ .059355 & .279504 \end{bmatrix}$$

The characteristic equation, $|\mathbf{A} - \mu^2\mathbf{I}| = 0$, is

$$\mu^4 - .6827\mu^2 + .1089 = 0$$

whose roots are

$$\mu_1{}^2 = .4286 \qquad \text{and} \qquad \mu_2{}^2 = .2541$$

The associated eigenvectors, each with the largest element set equal to unity, are:

$$\mathbf{u}_1' = [1, .3986] \qquad \text{and} \qquad \mathbf{u}_2' = [-.4246, 1]$$

Hence, the two discriminant functions may be written as

$$Z_1 = X_1 + .3986X_2$$

and

$$Z_2 = -.4246X_1 + X_2$$

We now carry out the discriminant analysis by the discriminant-criterion approach. The within-groups and between-groups SSCP matrices are found to be

$$\mathbf{W} = \begin{bmatrix} 448.9167 & 71.4167 \\ 71.4167 & 329.4417 \end{bmatrix}$$

and

$$\mathbf{B} = \begin{bmatrix} 319.8833 & 95.1833 \\ 95.1833 & 142.4250 \end{bmatrix}$$

From these, we compute

$$\mathbf{W}^{-1}\mathbf{B} = \begin{bmatrix} .690413 & .148368 \\ .139253 & .400158 \end{bmatrix}$$

The characteristic equation, $|\mathbf{W}^{-1}\mathbf{B} - \lambda\mathbf{I}| = 0$, is

$$\lambda^2 - 1.0906\lambda + .2556 = 0$$

with roots

$$\lambda_1 = .7496 \qquad \text{and} \qquad \lambda_2 = .3410$$

The associated eigenvectors are

$$\mathbf{v}_1' = [1, .3986] \qquad \text{and} \qquad \mathbf{v}_2' = [-.4247, 1]$$

which agree, within rounding error, with the results just obtained by the canonical correlation approach. Note also that

$$\mu_1^2/(1 - \mu_1^2) = .4286/.5714 = .7501 \simeq \lambda_1$$

and

$$\mu_2^2/(1 - \mu_2^2) = .2541/.7459 = .3407 \simeq \lambda_2$$

as asserted in Eq. 6.19.

6.8 CANONICAL CORRELATION ANALYSIS

In the preceding section we introduced canonical correlation analysis in a backhanded sort of way, as an alternative computational technique for multi-group discriminant functions. We now turn to the method of canonical analysis in its more usual and general application: that of seeking relationships between two sets of variables. The method was developed by Hotelling (1935).

The type of problem for which canonical analysis is useful may be illustrated by the following. An investigator wants to explore the possible relationship between personality variables on one hand, and academic achievement on the other, among high school students. He may administer a personality inventory such as Cattell and Eber's (1967–1968) 16 P. F. Questionnaire (which yields scores on 16 factor-analytically determined personality scales), and also give a battery of achievement tests in various subject-matter areas, to a group of high school students. The essential question is, "What sort of personality profile tends to be associated with what sort of pattern of academic achievement?" Canonical analysis helps answer this question by determining linear combinations of the personality scales that are most highly correlated with linear combinations of the achievement tests. The technique may therefore be loosely characterized as a sort of "double-barrelled principal components analysis." It identifies the "components" of one set of variables that are most highly related (linearly) to the "components" of the other set of variables.

Let the variables in the first set be denoted $X_1, X_2, \ldots, X_p$, and those in the second set, $Y_1, Y_2, \ldots, Y_q$. We construct a linear combination

$$Z = u_1X_1 + u_2X_2 + \cdots + u_pX_p$$

of the first set of variables, and a linear combination

$$W = v_1 Y_1 + v_2 Y_2 + \cdots + v_q Y_q$$

of the second set. Since we want to determine the two sets of coefficients, $\mathbf{u}' = [u_1, u_2, \ldots, u_p]$ and $\mathbf{v}' = [v_1, v_2, \ldots, v_q]$ so as to maximize the correlation between the two linear combinations, our first task is to express r_{zw} as a function of **u** and **v**.

The quantities $\sum z^2$ and $\sum w^2$, which occur in the denominator of the expression $\sum zw/\sqrt{(\sum z^2)(\sum w^2)}$ for r_{zw}, are expressible as quadratic forms, as we already saw in Chapter Five, Eq. 5.18. That is,

$$\sum z^2 = \mathbf{u}'\mathbf{S}_{xx}\mathbf{u} \tag{6.20}$$

and

$$\sum w^2 = \mathbf{v}'\mathbf{S}_{yy}\mathbf{v} \tag{6.21}$$

where $\mathbf{S}_{xx}$ and $\mathbf{S}_{yy}$ are the SSCP matrices of the two sets of variables, respectively. By a similar line of reasoning, it can be shown that

$$\sum zw = \mathbf{u}'\mathbf{S}_{xy}\mathbf{v} \tag{6.22}$$

where $\mathbf{S}_{xy}$ is the $p \times q$ matrix of sums-of-products between the X's and the Y's.

Bringing together Eqs. 6.20 to 6.22, the formula for r_{zw} may be written as

$$r_{zw} = \frac{\mathbf{u}'\mathbf{S}_{xy}\mathbf{v}}{\sqrt{(\mathbf{u}'\mathbf{S}_{xx}\mathbf{u})(\mathbf{v}'\mathbf{S}_{yy}\mathbf{v})}} \tag{6.23}$$

As was the case with discriminant analysis, the weights **u** and **v**, which maximize this expression, are determined only up to proportionality constants. For, if a and b are two aribtrary constants of the same sign, the value of r_{zw} obtained by using the elements of $a\mathbf{u}$ and $b\mathbf{v}$ as the combining weights is readily seen to be equal to the value resulting from the use of u_i and v_i as weights. It is most convenient, in the present situation, to choose the proportionality constants so that

$$\mathbf{u}'\mathbf{S}_{xx}\mathbf{u} = \mathbf{v}'\mathbf{S}_{yy}\mathbf{v} = 1 \tag{6.24}$$

and hence the entire denominator of expression (6.23) becomes unity. Then, introducing Lagrange multipliers $\lambda/2$ and $\mu/2$ (where the factors 1/2 are for numerical convenience), the function to be maximized is

$$F(\mathbf{u}, \mathbf{v}) = \mathbf{u}'\mathbf{S}_{xy}\mathbf{v} - (\lambda/2)(\mathbf{u}'\mathbf{S}_{xx}\mathbf{u} - 1) - (\mu/2)(\mathbf{v}'\mathbf{S}_{yy}\mathbf{v} - 1)$$

Taking the symbolic partial derivatives (see Appendix C) of $F(\mathbf{u}, \mathbf{v})$ with respect to $\mathbf{u}$ and $\mathbf{v}'$, and setting each of these equal to the null vector, we have

$$\frac{\partial F}{\partial \mathbf{u}} = \mathbf{S}_{xy}\mathbf{v} - \lambda \mathbf{S}_{xx}\mathbf{u} = \mathbf{0} \tag{6.25}$$

and

$$\frac{\partial F}{\partial \mathbf{v}'} = \mathbf{u}'\mathbf{S}_{xy} - \mu \mathbf{v}'\mathbf{S}_{yy} = \mathbf{0}' \tag{6.26}$$

as the equations to be satisfied by $\mathbf{u}$ and $\mathbf{v}$ in order to maximize r_{zw} under the side conditions (6.24). (It can also be shown that Eqs. 6.25 and 6.26 constitute sufficient conditions for the desired maximization.)

Next, we premultiply all members of Eq. 6.25 by $\mathbf{u}'$, and postmultiply all members of 6.26 by $\mathbf{v}$. These operations result in the relations

$$\mathbf{u}'\mathbf{S}_{xy}\mathbf{v} = \lambda(\mathbf{u}'\mathbf{S}_{xx}\mathbf{u}) = \mu(\mathbf{v}'\mathbf{S}_{yy}\mathbf{v})$$

which, by virtue of conditions (6.24), reduce to

$$\mathbf{u}'\mathbf{S}_{xy}\mathbf{v} = \lambda = \mu$$

In other words, both λ and μ are equal to the maximum value achievable by the correlation coefficient r_{zw}, which results from using as combining weights the elements of $\mathbf{u}$ and $\mathbf{v}$ that satisfy Eqs. 6.25 and 6.26. Such being the case, the λ in Eq. 6.25 may be replaced by μ. Making this replacement, taking the transpose of both members of Eq. 6.26, and writing $\mathbf{S}_{yx}$ for $\mathbf{S}'_{xy}$, the pair of Eqs. 6.25 and 6.26 may be rewritten as

$$\mathbf{S}_{xy}\mathbf{v} = \mu \mathbf{S}_{xx}\mathbf{u}$$
$$\mathbf{S}_{yx}\mathbf{u} = \mu \mathbf{S}_{yy}\mathbf{v}$$

Assuming $\mathbf{S}_{yy}$ to be nonsingular, we may solve the second of these equations to express $\mathbf{v}$ in terms of $\mathbf{u}$, as follows:

$$\mathbf{v} = (1/\mu)\mathbf{S}_{yy}^{-1}\mathbf{S}_{yx}\mathbf{u} \tag{6.27}$$

This expression may then be substituted for $\mathbf{v}$ in the first of the above pair of equations to yield

$$\mathbf{S}_{xy}\left(\frac{1}{\mu}\mathbf{S}_{yy}^{-1}\mathbf{S}_{yx}\mathbf{u}\right) = \mu \mathbf{S}_{xx}\mathbf{u}$$

or, premultiplying both members by $\mu\mathbf{S}_{xx}^{-1}$ and rearranging the terms,

$$(\mathbf{S}_{xx}^{-1}\mathbf{S}_{xy}\mathbf{S}_{yy}^{-1}\mathbf{S}_{yx} - \mu^2\mathbf{I})\mathbf{u} = \mathbf{0} \tag{6.28}$$

It is thus seen that the largest eigenvalue $\mu_1{}^2$ of the quadruple matrix product $\mathbf{S}_{xx}^{-1}\mathbf{S}_{xy}\mathbf{S}_{yy}^{-1}\mathbf{S}_{yx}$ gives the square of the maximum r_{zw}, and that the elements of the associated eigenvector $\mathbf{u}_1$ provide the weights by which one set of variables (the X's) should be linearly combined in order to achieve this maximum correlation. The vector of combining weights $\mathbf{v}_1$ for the other set of variables (the Y's) may be obtained by substituting μ_1 and $\mathbf{u}_1$ in Eq. 6.27; that is,

$$\mathbf{v}_1 = \frac{1}{\mu_1} \mathbf{S}_{yy}^{-1}\mathbf{S}_{yx}\mathbf{u}_1$$

Equation 6.28 will, in general, yield other eigenvalues and vectors besides $\mu_1^{\,2}$ and $\mathbf{u}_1$, and their meaning should be familiar to the reader by now. The situation is exactly parallel to those holding in principal components analysis and in multigroup discriminant analysis. In each case, we start out by seeking a set of combining weights (or two such sets, in canonical correlation analysis) that will maximize a specified criterion for the resulting linear combination (or pair of linear combinations). In each case, the elements of the vector associated with the largest eigenvalue of a certain matrix constitute the weights leading to the desired absolute maximum. In addition, the elements of the vectors associated with the second, third, and subsequent eigenvalues (in descending order of magnitude) yield conditional maxima in the following sense: the linear combination formed by the elements of the second vector has the largest value on the relevant criterion among those that are uncorrelated with the first linear combination; the third linear combination has the maximum criterion value among linear combinations uncorrelated with the first two linear combinations; and so forth. In the case of canonical correlation analysis, the linear combinations occur in two sequences, one for each of the two sets of variables. The uncorrelatedness holds both within each sequence and between unmatched pairs of linear combinations across the two sequences. Thus, not only is Z_2 uncorrelated with Z_1, and W_2 with W_1, but Z_2 is also uncorrelated with W_1, and W_2 with Z_1. In short, the only nonzero correlations are those between the corresponding members, Z_1 and W_1, Z_2 and W_2, and so on, of the two sequences. These pairs of linear combinations are called the *canonical variates*. The number of canonical-variate pairs will be equal to the number of variables in the smaller set, that is, the smaller of the two numbers, p and q. This is because the rank of the quadruple product matrix $\mathbf{S}_{xx}^{-1}\mathbf{S}_{xy}\mathbf{S}_{yy}^{-1}\mathbf{S}_{yx}$ whose eigenvalues and eigenvectors determine the canonical variates, is equal to p or q, whichever is smaller.

6.9 SIGNIFICANCE TESTS OF CANONICAL VARIATES

Two kinds of significance tests are of interest in canonical correlation analysis. The first is an overall test to decide whether there is *any* significant linear relationship between the two sets of variables. If overall significance is found, we would then want to know how many of the canonical-variate pairs are significant. As might be expected (since we have seen that discriminant analysis may be regarded as a special case of canonical analysis) the significance tests here are closely related to those described earlier for discriminant analysis.

To see this relationship, we recall that Eq. 6.19 holds between each eigenvalue λ_i from the basic equation,

$$(\mathbf{W}^{-1}\mathbf{B} - \lambda\mathbf{I})\mathbf{v} = \mathbf{0}$$

for discriminant analysis via the discriminant criterion approach and the corresponding eigenvalue μ_i^2 from the basic equation,

$$(\mathbf{S}_{xx}^{-1}\mathbf{S}_{xy}\mathbf{S}_{yy}^{-1}\mathbf{S}_{yx} - \mu^2\mathbf{I})\mathbf{u} = \mathbf{0}$$

in the canonical correlation approach. Namely, the corresponding eigenvalues resulting from these two basic equations are related by the equality

$$\lambda_i = \mu_i^2/(1 - \mu_i^2)$$

As a consequence, Wilks' Λ criterion, which was shown in Eq. 6.11 to be expressible as

$$\Lambda = 1/\prod_{i=1}^{r} (1 + \lambda_i)$$

may also be expressed in terms of the μ_i^2 as follows:

$$\begin{aligned}\Lambda &= 1/\prod[1 + \mu_i^2/(1 - \mu_i^2)] \\ &= 1/\prod[1/(1 - \mu_i^2)]\end{aligned}$$

or

$$\Lambda = \prod_{i=1}^{r} (1 - \mu_i^2) \tag{6.29}$$

The above demonstration, of course, shows only that this alternative expression for Λ holds when the μ_i^2 results from canonical analysis as applied to the problem of discriminant analysis. However, it is quite plausible that Eq. 6.29 will continue to hold for canonical analysis in general. For each μ_i^2, it will be recalled, is a conditionally maximal value of r_{zw}^2, the squared correlation between corresponding pairs of canonical variates constructed from the two sets of variables. Thus, each factor of the product

$$(1 - \mu_1^2)(1 - \mu_2^2)\cdots(1 - \mu_r^2)$$

is in fact the coefficient of alienation between a particular pair of canonical variates. This is consistent with the fact that Λ is a statistic that is *inversely* related to the magnitude of differences or strength of relationship: the smaller the value of Λ, the greater the difference or relationship in question.

There is a problem, however, in that the definition of Λ as the ratio $|\mathbf{W}|/|\mathbf{T}|$ (first given in Chapter Four and used again in connection with discriminant analysis in this chapter) does not make sense in the context of canonical analysis. The sample in this situation is not composed of several subgroups, and hence there is no such thing as a within-groups SSCP matrix $\mathbf{W}$. The resolution of this difficulty lies in introducing a more general concept of which the $\mathbf{W}$ matrix is a special instance applicable to multigroup significance tests and discriminant analysis. The general concept is the *error SSCP matrix*, which we will denote by $\mathbf{S}_e$. Thus, a definition of Λ more general than that given in Eq. 4.28 is as follows:

$$\Lambda = \frac{|\mathbf{S}_e|}{|\mathbf{T}|} \tag{6.30}$$

In the applications of Λ encountered up to now, the within-groups SSCP matrix was the appropriate error SSCP matrix. (This was referred to as "the simplest case" when we introduced Eq. 4.28). In the context of canonical analysis, the error SSCP matrix is the residual SSCP matrix after the effects of the correlations between the canonical-variate pairs have been removed. Of course, this matrix need not actually be computed in order to determine the value of Λ, since Λ may be obtained from Eq. 6.29 once the required eigenvalues are found. It is necessary only to realize that, in computing Λ from Eq. 6.29, we are indirectly calculating the determinantal ratio $|\mathbf{S}_e|/|\mathbf{T}|$ appearing in Eq. 6.30.

After Λ has been computed from Eq. 6.29, the overall significance test may be carried out by either the chi-square approximation or the F-ratio approximation described in Chapter Four, with K (the number of groups) replaced by $q + 1$. This is consistent with the fact that, in using the canonical correlation approach to discriminant analysis, $K - 1$ "dummy criterion variables" were employed; that is, the number of groups is *one more* than the number of variables in the second set. Thus, Bartlett's chi-square approximation becomes

$$\begin{aligned} V &= -[N - 3/2 - (p + q)/2] \ln \Lambda \\ &= -[N - 3/2 - (p + q)/2] \sum_{j=1}^{r} ln(1 - \mu_j^2) \end{aligned} \tag{6.31}$$

with pq degrees of freedom.

Similarly, Rao's F-ratio approximation is now written as

$$R = \frac{1 - \Lambda^{1/s}}{\Lambda^{1/s}} \cdot \frac{ms - pq/2 + 1}{pq} \tag{6.32}$$

where

$$m = N - 3/2 - (p + q)/2 \quad \text{and} \quad s = \sqrt{\frac{p^2q^2 - 4}{p^2 + q^2 - 5}}$$

and R is to be referred to the F-distribution with pq degrees of freedom in the numerator, and $[ms - pq/2 + 1]$ in the denominator. It should be noted that the special cases (shown in Table 4.1) in which the statistic R becomes an exact F-variate, hold just as well in the present context. In fact, the conditions characterizing these special cases are now simpler to state than before. Noting that $K = 2$ and $K = 3$ correspond, respectively, to $q = 1$ and $q = 2$, we see that the conditions stated in Table 4.1 now reduce to this simple rule:

If either of the two sets of variables consists of just *one* or *two* variables, then (no matter how many variables there are in the other set), the R-statistic defined by Eq. 6.32 becomes an exact F-variate.

(The case when at least one set consists of a single variable is trivial in the sense that the canonical correlation then reduces to, at most, a multiple correlation.)

Beyond the overall significance test described above there are the tests for deciding how many of the canonical correlations should be regarded as significant. The procedure here depends again on the fact that each term of the sum in Eq. 6.31 for V is itself an approximate chi-square variate. That is, for each j,

$$V_j = -[N - 3/2 - (p + q)/2] \ln(1 - \mu_j^2) \qquad (6.33)$$

is distributed approximately as a chi-square with $p + q - (2j - 1)$ degrees of freedom. Consequently, the cumulative differences between V and V_1, V_2, and so on, are also approximate chi-square variates, and they permit our testing whether a significant (linear) relationship exists between the two sets of variables after the effects of the first, second, and so forth, canonical-variate pairs have cumulatively been removed. The successive test statistics and their n.d.f.'s may be summarized as follows:

Residual After Removing	Approximate χ^2-statistic	n.d.f.
First canonical pair	$V - V_1$	$pq - (p + q - 1)$ $= (p - 1)(q - 1)$
First two canonical pairs	$V - V_1 - V_2$	$(p - 1)(q - 1) - (p + q - 3)$ $= (p - 2)(q - 2)$
First three canonical pairs	$V - V_1 - V_2 - V_3$	$(p - 2)(q - 2) - (p + q - 5)$ $= (p - 3)(q - 3)$
⋮	⋮	⋮

As soon as the residual after removing the effects of the first s canonical-variate pairs becomes smaller than the prescribed centile point of the appropriate chi-square distribution, we may conclude that only the first s canonical correlations are significant.

6.10 INTERPRETATION OF THE CANONICAL VARIATES

As was mentioned earlier, canonical correlation analysis may be regarded as a sort of "double-barrelled principal components analysis." Therefore, the rules (or perhaps the art) of interpreting the canonical variates are exactly the same as those for interpreting principal components as well as discriminant functions: one examines the relative magnitudes and the signs of the several combining weights defining each canonical variate, and sees if a meaningful psychological interpretation can be given.

For example, returning to the hypothetical study of relationships between personality attributes and academic achievement alluded to earlier, suppose that the variables with the largest weights for the first two pairs of canonical variates were as shown below. (The scales on the 16 P. F. Questionnaire are bipolar. A low score on each scale corresponds to the pole characterized by the first adjective; a high score, to the pole described by the second adjective. Hence, a large positive weight for a scale implies that a person close to the *second*-adjective pole of that personality dimension will tend to score high on the canonical variate in question; a large negative weight implies that a person close to the *first*-adjective pole will tend to score high on the said canonical variate.)

First Canonical Variate

Personality-Variable Set	Achievement-Variable Set
High Positive Weights	*High Positive Weights*
Practical versus Imaginative	Art
Submissive versus Dominant	Literature
Conservative versus Experimenting	
High Negative Weights	*High Negative Weight*
Emotional versus Calm	Algebra
Impulsive versus Controlled	

Second Canonical Variate

Personality-Variable Set	Achievement-Variable Set
High Positive Weights	*High Positive Weights*
Conservative versus Experimenting	Physics
Group-dependent versus Self-sufficient	Algebra

High Negative Weights	*High Negative Weight*
Submissive versus Dominant	Literature
Serious versus Gay	

Given these results, the interpretation is self-evident. (Of course, this is a hypothetical example.) It appears that a personality syndrome characterized as "highly imaginative, dominant (self-asserting), experimenting (nonconforming), emotional (easily upset), and impulsive (following own urges)" tends to go with high achievement in the artistic-literary areas, while the "experimenting, self-sufficient (not-very-sociable), submissive, and serious (or taciturn)" student tends to excel in the physical-science and mathematical areas.

With real data, one would seldom expect to find such clear-cut (and stereotype-confirming) results. But this does not detract from the potential value of canonical analysis. It would simply mean that the dimensions of one domain (such as personality) that are strongly associated with those of another domain (such as academic achievement) are not necessarily susceptible to "meaningful" verbal descriptions within the framework of our intuitive, everyday concepts. It may be that subsequent research will show that precisely these "nonintuitive" dimensions represented by the canonical variates are of greater scientific import.

Although we spoke of "large weights" without qualification in the foregoing discussion, the reader should be aware by now that the relative magnitudes of the weights must be compared in the standardized scale. Thus, it is not the relative magnitudes of the elements of vectors **u** and **v** as defined by Eqs. 6.28 and 6.27 that are relevant to the interpretation of canonical variates, but the relative magnitudes of the weights rescaled to be applicable to the respective variables in standardized form. These weights are obtained from the raw-score weights **u** and **v** by multiplying each element by the standard deviation of the corresponding variable. That is,

$$u^*_{mi} = s_{x_i} u_{mi} \qquad \text{and} \qquad v^*_{mi} = s_{y_i} v_{mi}$$

are the weights whose relative magnitudes we should compare.

EXERCISES

1. In a study attempting to determine the dimensions along which three curricular groups at the Blank Institute of Technology differ among one another, scores on three tests administered at time of admission were obtained for a random sample of 100 seniors in each of the three groups: A. General Engineering; B. Physical Science; and C. Industrial Management.

The three tests used were: (1) S.A.T. Mathematics Test; (2) S.A.T. Verbal Test; and (3) Persuasiveness Scale on a personality inventory. The mean scores (suitably scaled down to reduce the arithmetic) on the three tests for each group were as follows:

	Test			
Group	1	2	3	n_g
A. General Engineering	6.14	5.30	1.62	100
B. Physical Science	6.69	5.84	1.21	100
C. Industrial Management	6.05	5.61	1.70	100

The within-groups and between-groups SSCP matrices were as follows:

$$\mathbf{W} = \begin{bmatrix} 167.69 & 65.24 & -12.18 \\ 65.24 & 269.64 & -61.39 \\ -12.18 & -61.39 & 141.62 \end{bmatrix}$$

and

$$\mathbf{B} = \begin{bmatrix} 23.96 & 13.95 & -18.08 \\ 13.95 & 14.71 & -10.36 \\ -18.08 & -10.36 & 13.64 \end{bmatrix}$$

Carry out a discriminant analysis for the above data, including significance tests to determine whether one or both of the discriminant functions are significant at the .05 level. Also plot the discriminant-function means of the three groups (on a line or a plane, as the case may be), and try to interpret the dimension(s) of group differentiation. (Computational Hint: Carry five significant digits throughout your calculations. To facilitate this, it will be convenient to work with $\mathbf{W} \times 10^{-1}$ instead of $\mathbf{W}$ itself. Note the effect that this has on the eigenvalues.)

2. The means on three scales of a personality test for samples from three clinical groups were as shown below:

	Scale			
Group	1	2	3	Sample Size
Anxiety neurosis	2.92	1.16	1.72	50
Psychopathic reaction	3.80	1.90	1.80	30
Obsessive-compulsive	4.70	1.60	2.10	20

The within-groups and between-groups SSCP matrices were as follows:

$$\mathbf{W} = \begin{bmatrix} 223.18 & 24.40 & 45.99 \\ 24.40 & 58.92 & 3.47 \\ 45.99 & 3.47 & 57.72 \end{bmatrix}$$

and

$$\mathbf{B} = \begin{bmatrix} 48.16 & 15.98 & 9.44 \\ 15.98 & 10.69 & 2.02 \\ 9.44 & 2.02 & 2.08 \end{bmatrix}$$

Carry out a discriminant analysis for the above data, including all relevant significance tests. (See end of Exercise 1 for computational hint.)

3. Do Exercise 2 by the canonical correlation approach. Note that you will first have to add $\mathbf{W}$ and $\mathbf{B}$ to get the total SSCP matrix $\mathbf{T}$, which becomes $\mathbf{S}_{pp}$ in this context. (See numerical example in Section 6.7 for computation of $\mathbf{S}_{pc}$ and $\mathbf{S}_{cc}$.)
4. Prove that the second discriminant function (whose weights are the elements of the eigenvector $\mathbf{v}_2$ associated with the second largest eigenvalue λ_2 of $\mathbf{W}^{-1}\mathbf{B}$) has the largest discriminant-criterion value among those linear combinations of the X_i that are uncorrelated with the first discriminant function in the total sample.
[Hint: From Section 6.2, we know that each discriminant function has a relative (or conditional) maximum value for its discriminant criterion. We therefore need only show that Y_2 is uncorrelated with Y_1. Noting that this correlation is proportional to $\mathbf{v}_1'\mathbf{T}\mathbf{v}_2$ (where $\mathbf{T} = \mathbf{W} + \mathbf{B}$), we have to prove that $\mathbf{v}_1'\mathbf{T}\mathbf{v}_2 = 0$. This may be done by a slight modification of the proof of Theorem 3 in Chapter Five (Section 5.5), starting as follows: From Eq. 6.8, we have

$$(\mathbf{B} - \lambda_i \mathbf{W})\mathbf{v}_i = \mathbf{0}, \qquad \text{for each } i.$$

Hence,

$$\mathbf{B}\mathbf{v}_1 = \lambda_1 \mathbf{W}\mathbf{v}_1 \qquad \text{and} \qquad \mathbf{B}\mathbf{v}_2 = \lambda_2 \mathbf{W}\mathbf{v}_2$$

Premultiplying these equations by $\mathbf{v}_2'$ and $\mathbf{v}_1'$ respectively, taking the transposes of both sides of the first resulting equation (and remembering that both $\mathbf{B}$ and $\mathbf{W}$ are symmetric matrices), we arrive at a pair of equations whose left-hand sides are identical. The right-hand sides of these equations must, therefore, also be equal.]

Chapter Seven

Multivariate Analysis of Variance

The multivariate significance tests discussed up to this point, including discriminant analysis, may all be regarded as multivariate extensions of analysis of variance (ANOVA) as applied to one-factor experiments or one-way classification designs. Extensions to multivariate experiments involving more than one factor, that is, factorial experiments, may be made in a closely parallel manner. We describe in detail the multivariate analysis of variance (MANOVA) procedures for two-factor designs with equal cell frequencies. Extensions to other designs can be made in exactly the same way by anyone who is sufficiently well versed in the corresponding univariate ANOVA. Guidelines to effecting such extensions are given in Section 7.5.

In the discussions to follow, it is assumed that the reader is familiar with the basic concepts and terminology used in ANOVA for factorial designs in the univariate (one dependent variable) case, including the choice of correct error terms depending on whether a fixed-effects, random-effects, or mixed model is involved. Those who need to review these matters may refer to Glass and Stanley (1970, Ch. 18), Ferguson (1966, Ch. 19), or Hays (1963, Chs. 12 and 13).

7.1 TWO-FACTOR DESIGNS WITH MULTIPLE DEPENDENT VARIABLES: ADDITIVE COMPONENTS OF THE TOTAL SSCP MATRIX

Just as in the usual two-factor design with one dependent variable, we are here concerned with the analysis of data from experiments involving two treatment or classification variables whose levels are represented by the rows and columns of a two-way layout. For instance, in a classroom learning experiment, each row may represent a different method for teaching shorthand,

and each column, a different condition of massed or distributed practice. But, instead of having one dependent or criterion variable by which to measure the outcome of the experiment, we now have two or more criterion measures, such as a speed score and an accuracy score in the shorthand-teaching experiment. Thus, each experimental unit (or subject) will be associated with a set of p observations $X_1, X_2, \ldots, X_p$. Hence, for complete identification of a particular score, we need four subscripts, specifying, respectively, the variable number (α or $\beta = 1, 2, \ldots, p$), row number ($r = 1, 2, \ldots, R$), column number ($c = 1, 2, \ldots, C$), and subject number within each cell ($i = 1, 2, \ldots, n$). That is, $X_{\alpha rci}$ will denote the score on the αth dependent variable earned by the ith individual in the rth row, cth column cell (hereafter called the (r, c)-cell for short).

The several classes of totals and means are defined in the usual manner for each dependent variable X_α:

$$T_{\alpha rc.} = \sum_{i=1}^{n} X_{\alpha rci} \quad \text{is the total in the } (r, c) \text{ cell and}$$

$$\bar{X}_{\alpha rc.} = T_{\alpha rc.}/n \quad \text{is the } (r, c) \text{ cell mean;}$$

$$T_{\alpha r..} = \sum_{c=1}^{C} T_{\alpha rc.} \quad \text{is the } r\text{th row total and}$$

$$\bar{X}_{\alpha r..} = T_{\alpha r..}/nC \quad \text{is the } r\text{th row mean;}$$

$$T_{\alpha .c.} = \sum_{r=1}^{R} T_{\alpha rc.} \quad \text{is the } c\text{th column total and}$$

$$\bar{X}_{\alpha .c.} = T_{\alpha .c.}/nR \quad \text{is the } c\text{th column mean;}$$

$$T_{\alpha ...} = \sum_{r=1}^{R} T_{\alpha r..} = \sum_{c=1}^{C} T_{\alpha .c.} \quad \text{is the grand total and}$$

$$\bar{X}_{\alpha ...} = T_{\alpha ...}/nRC \quad \text{is the grand mean}$$

Corresponding to the total sum-of-squares in univariate ANOVA, we have the total SSCP matrix $\mathbf{S}_t$, which is the same thing as the matrix denoted by $\mathbf{T}$ in Chapter Three (Eq. 3.28). That is to say, the (α, β)-element of $\mathbf{S}_t$ is given by

$$(\mathbf{S}_t)_{\alpha\beta} = \sum_{r=1}^{R} \sum_{c=1}^{C} \sum_{i=1}^{n} (X_{\alpha rci} - \bar{X}_{\alpha ...})(X_{\beta rci} - \bar{X}_{\beta ...}) \tag{7.1}$$

or, for computational purposes,

$$(\mathbf{S}_t)_{\alpha\beta} = \sum_{r=1}^{R} \sum_{c=1}^{C} \sum_{i=1}^{n} X_{\alpha rci} X_{\beta rci} - T_{\alpha ...} T_{\beta ...}/nRC \tag{7.1a}$$

Just as the total sum-of-squares in two-way ANOVA can be broken down into four additive components,

$$SS_t = SS_r + SS_c + SS_{rc} + SS_w$$

the total SSCP matrix in two-way MANOVA may likewise be expressed as

$$\mathbf{S}_t = \mathbf{S}_r + \mathbf{S}_c + \mathbf{S}_{rc} + \mathbf{S}_w \tag{7.2}$$

where the SSCP matrices on the right relate, respectively, to the row effect, column effect, interaction effect, and within-cell variation and covariation. The general element of each of these component SSCP matrices is a natural extension of the corresponding sum-of-squares in univariate ANOVA. Their definitional and computational formulas are as follows:

$$(\mathbf{S}_r)_{\alpha\beta} = nC \sum_{r=1}^{R} (\overline{X}_{\alpha r..} - \overline{X}_{\alpha ...})(\overline{X}_{\beta r..} - \overline{X}_{\beta ...}) \tag{7.3}$$

$$(\mathbf{S}_r)_{\alpha\beta} = \sum_{r=1}^{R} T_{\alpha r..}T_{\beta r..}/nC - T_{\alpha ...}T_{\beta ...}/nRC \tag{7.3a}$$

$$(\mathbf{S}_c)_{\alpha\beta} = nR \sum_{c=1}^{C} (\overline{X}_{\alpha .c.} - \overline{X}_{\alpha ...})(\overline{X}_{\beta .c.} - \overline{X}_{\beta ...}) \tag{7.4}$$

$$(\mathbf{S}_c)_{\alpha\beta} = \sum_{c=1}^{C} T_{\alpha .c.}T_{\beta .c.}/nR - T_{\alpha ...}T_{\beta ...}/nRC \tag{7.4a}$$

$$(\mathbf{S}_{rc})_{\alpha\beta} = n \sum_{r=1}^{R} \sum_{c=1}^{C} \Big[(\overline{X}_{\alpha rc.} - \overline{X}_{\alpha r..} - \overline{X}_{\alpha .c.} + \overline{X}_{\alpha ...}) \times (\overline{X}_{\beta rc.} - \overline{X}_{\beta r..} - \overline{X}_{\beta .c.} + \overline{X}_{\beta ...})\Big] \tag{7.5}$$

$$(\mathbf{S}_{rc})_{\alpha\beta} = \sum_{r=1}^{R} \sum_{c=1}^{C} T_{\alpha rc.}T_{\beta rc.}/n - \sum_{r=1}^{R} T_{\alpha r..}T_{\beta r..}/nC - \sum_{c=1}^{C} T_{\alpha .c.}T_{\beta .c.}/nR + T_{\alpha ...}T_{\beta ...}/nRC \tag{7.5a}$$

$$(\mathbf{S}_w)_{\alpha\beta} = \sum_{r=1}^{R} \sum_{c=1}^{C} \left[\sum_{i=1}^{n} (X_{\alpha rci} - \overline{X}_{\alpha rc.})(X_{\beta rci} - \overline{X}_{\beta rc.})\right] \tag{7.6}$$

$$(\mathbf{S}_w)_{\alpha\beta} = \sum_{r=1}^{R} \sum_{c=1}^{C} \left[\sum_{i=1}^{n} X_{\alpha rci}X_{\beta rci} - T_{\alpha rc.}T_{\beta rc.}/n\right] \tag{7.6a}$$

The reader should verify, using the computational formulas 7.3a through 7.6a, that the sum of the (α, β)-elements of the four component SSCP matrices is equal to the (α, β)-element of the total SSCP matrix given in Eq. 7.1a.

These SSCP matrices may be collected into a MANOVA summary table, together with their respective degrees of freedom, which are the same as those for the corresponding sums-of-squares in univariate ANOVA. Table 7.1 represents such a summary table.

TABLE 7.1 MANOVA Summary Table.

Source	SSCP Matrix	Computed from Eq.	n.d.f.
Row effect	$\mathbf{S}_r$	7.3a	$\nu_r = R - 1$
Column effect	$\mathbf{S}_c$	7.4a	$\nu_c = C - 1$
Interaction	$\mathbf{S}_{rc}$	7.5a	$\nu_{rc} = (R - 1)(C - 1)$
Within-cells	$\mathbf{S}_w$	7.6a	$\nu_w = RC(n - 1)$
Total	$\mathbf{S}_t$	7.1a	$\nu_t = nRC - 1$

7.2 SIGNIFICANCE TESTS IN MANOVA

The Λ criterion may again be used in carrying out the significance tests for the main effects and interaction effect in MANOVA, but a further generalization of the definition of this statistic must first be made. Recall that Λ was originally introduced in Chapter Four as the ratio $|\mathbf{S}_w|/|\mathbf{S}_t|$, and was subsequently generalized to $|\mathbf{S}_e|/|\mathbf{S}_t|$ for use in connection with canonical correlation analysis in Chapter Six; that is, the earlier matrix $\mathbf{S}_w$ was there replaced by a more general matrix $\mathbf{S}_e$, of which the former was a special case. This time, it is the matrix whose determinant stands in the denominator of the Λ-ratio that has to be generalized. That is to say, the total SSCP matrix $\mathbf{S}_t$ must now be replaced by a more general matrix which, in effect, reduces to $\mathbf{S}_t$ in the situations previously considered. Let us therefore examine what $\mathbf{S}_t$ stood for in the problems treated so far.

In K-sample significance tests for group centroids (in which context Wilks' Λ was first introduced) as well as in discriminant analysis, we had the relation

$$\mathbf{S}_t = \mathbf{S}_w + \mathbf{S}_b \qquad (\text{or } \mathbf{T} = \mathbf{W} + \mathbf{B} \text{ in our earlier notation}).$$

In canonical analysis, we could have written

$$\mathbf{S}_t = \mathbf{S}_e + \mathbf{S}_{can}$$

where $\mathbf{S}_{can}$ stands for the SSCP due to canonical correlation. We did not do so, however, because our main concern at that point was to argue that the earlier within-groups SSCP matrix $\mathbf{S}_w$ could, conceptually at least, be replaced by the more general error SSCP matrix, $\mathbf{S}_e$. In each of these situations, therefore, the total SSCP matrix stood for the sum of the error SSCP matrix and the SSCP matrix attributable to the effect posited as an alternative to the *single* null hypothesis being tested—equality of the K population centroids, or zero canonical correlation in the population, as the case may be.

In higher-order designs, however, there are several null hypotheses to be tested. In the two-factor case, with which we are here concerned, the null hypotheses relate to the row effect, column effect, and interaction effect, represented by $\mathbf{S}_r$, $\mathbf{S}_c$, and $\mathbf{S}_{rc}$, respectively. It is clear from Eq. 7.2 that, in this case, the SSCP matrix for the effect being tested by any one of the null hypotheses plus the error SSCP matrix does not equal the total SSCP matrix, for $\mathbf{S}_t$ now has four additive components. Moreover, the error SSCP matrix itself will now depend on whether Model I (fixed effects), Model II (random effects), or Model III (mixed model) applies to the particular experiment; either $\mathbf{S}_w$ or $\mathbf{S}_{rc}$ will be the appropriate error SSCP matrix. (In three-factor designs or more complicated two-factor designs such as when one factor is nested under the other, there will be a larger number of matrices from among which the appropriate error SSCP matrix must be chosen.)

The foregoing considerations suggest that $\mathbf{S}_t$, whose determinant formed the denominator of the Λ-ratio in its previous applications, must now be replaced by a matrix having the general form $\mathbf{S}_h + \mathbf{S}_e$, where $\mathbf{S}_h$ is the SSCP matrix for the effect being tested by a particular null hypothesis, and $\mathbf{S}_e$ is the error SSCP matrix appropriate for the particular effect under the applicable model. This is, in fact, the conclusion that emerges from the mathematical theory of likelihood-ratio criteria, which (as mentioned in Chapter Four) forms the basis for Wilks' Λ-ratio. We thus arrive at the completely general definition of the Λ criterion, as follows:

$$\Lambda_h = \frac{|\mathbf{S}_e|}{|\mathbf{S}_h + \mathbf{S}_e|} \tag{7.7}$$

where the subscript h designates the particular null hypothesis being tested. Note that, on substituting $\mathbf{S}_b$ for $\mathbf{S}_h$ and $\mathbf{S}_w$ for $\mathbf{S}_e$ in this expression, we get the Λ-ratio appropriate for K-sample (one-way MANOVA) significance tests, first introduced in Eq. 4.28. The appropriate substitutions for $\mathbf{S}_h$ and $\mathbf{S}_e$ in the case of two-factor designs with equal cell frequencies (or, more generally,

TABLE 7.2 SSCP Matrices to Be Substituted for $\mathbf{S}_h$ and $\mathbf{S}_e$ in Expression 7.7 in Testing the Various Effects in Two-Way MANOVA with Equal (or Proportional) Cell Frequencies.

Effect	$\mathbf{S}_h$	$\mathbf{S}_e$		
		I	II	III
Rows	$\mathbf{S}_r$	$\mathbf{S}_w$	$\mathbf{S}_{rc}$	$\mathbf{S}_{rc}$
Columns	$\mathbf{S}_c$	$\mathbf{S}_w$	$\mathbf{S}_{rc}$	$\mathbf{S}_w$
Interaction	$\mathbf{S}_{rc}$	$\mathbf{S}_w$	$\mathbf{S}_w$	$\mathbf{S}_w$

cell frequencies proportional to the marginals), are summarized in Table 7.2. The substitution for $\mathbf{S}_h$ of course depends only on the hypothesis being tested, but that for $\mathbf{S}_e$ depends also on which model is applicable. In Model III (mixed), it is assumed that the rows represent a fixed effect and the columns, a random effect.

As before, the actual computation of Λ can be made routine by utilizing the relationship between the determinant and eigenvalues of a matrix, provided a computer program for solving eigenvalue problems is available. From Eq. 7.7, we get

$$1/\Lambda_h = |\mathbf{S}_e + \mathbf{S}_h|/|\mathbf{S}_e| = |\mathbf{I} + \mathbf{S}_e^{-1}\mathbf{S}_h|$$

Hence, if the nonzero eigenvalues of the matrix $\mathbf{S}_e^{-1}\mathbf{S}_h$ are denoted by $\lambda_1, \lambda_2, \ldots, \lambda_t$, we have

$$1/\Lambda_h = \prod_{i=1}^{t} (1 + \lambda_i)$$

or

$$\Lambda_h = 1/\prod_{i=1}^{t} (1 + \lambda_i) \tag{7.8}$$

which is formally identical with Eq. 6.11. The number t of nonzero eigenvalues is equal to ν_h (that is, $\nu_r = R - 1$, $\nu_c = C - 1$, or $\nu_{rc} = (R - 1)(C - 1)$, as the case may be) or to p (the number of dependent variables), whichever is smaller. Using the eigenvalue approach to compute Λ is especially preferred if one wishes to follow up the significance tests with a discriminant analysis as described in Section 7.4.

After each Λ_h has been computed from Eq. 7.7 or 7.8, the significance test proceeds as before, using either Bartlett's V or Rao's R as the approximate test statistic. Minor details of the formulas for these statistics have to be modified as follows: The K (number of groups), which appeared in Eq. 4.29 for V and 4.30 for R, is now replaced by $\nu_h + 1$. Similarly, N is replaced by $\nu_e + \nu_h + 1$, where ν_e is equal to $\nu_w = RC(n - 1)$ or to $\nu_{rc} = (R - 1)(C - 1)$, as the case may be, according to the appropriate error SSCP matrix indicated in Table 7.2. (The reader should satisfy himself that these replacements are consistent with the replacements of $\mathbf{S}_b$ and $\mathbf{S}_t$ by $\mathbf{S}_h$ and $\mathbf{S}_e + \mathbf{S}_h$, respectively.) Thus, the formulas for V and R applicable to two-factor designs assume the forms shown in Eqs. 7.9 and 7.10 below. (In fact, with the appropriate definitions of ν_e and ν_h, these formulas may be used in connection with a MANOVA for any type of design.)

For Bartlett's chi-square approximation, the test statistic is:

$$\begin{aligned} V &= -[\nu_e + \nu_h - (p + \nu_h + 1)/2] \ln \Lambda_h \\ &= [\nu_e + \nu_h - (p + \nu_h + 1)/2] \prod_{i=1}^{t} \ln(1 + \lambda_i) \end{aligned} \tag{7.9}$$

which is distributed approximately as a chi-square with $p\nu_h$ degrees of freedom. If necessary, Schatzoff's correction factor may be applied to the selected centile point of the relevant chi-square distribution in order to obtain the exact critical value for V, as described in Chapter Four.

The formula for Rao's R statistic now becomes as follows:

$$R = \frac{1 - \Lambda^{1/s}}{\Lambda^{1/s}} \frac{ms - p\nu_h/2 + 1}{p\nu_h} \tag{7.10}$$

with

$$m = \nu_e + \nu_h - (p + \nu_h + 1)/2$$

and

$$s = \sqrt{\frac{(p\nu_h)^2 - 4}{p^2 + \nu_h^2 - 5}}$$

R having an approximate F-distribution with $p\nu_h$ and $ms - p\nu_h/2 + 1$ degrees of freedom. As usual, when $p = 1$ (univariate ANOVA) or $p = 2$, or when $\nu_h = 1$ or 2 (corresponding to $K = 2$ or 3 in Table 4.2), R reduces to an exact F-variate with degrees of freedom as shown in that table, with the substitutions $K = \nu_h + 1$ and $N = \nu_e + \nu_h + 1$.

7.3 NUMERICAL EXAMPLE

An experiment was conducted for comparing two methods of teaching shorthand to female seniors in a vocational high school. Also of interest were the effects of distributed *versus* massed practice, represented by the following conditions:

C_1: 2 hours of instruction per day for 6 weeks
C_2: 3 hours of instruction per day for 4 weeks
C_3: 4 hours of instruction per day for 3 weeks

Ten subjects were assigned at random to each of the $2 \times 3 = 6$ treatment groups.

On completion of instruction, all 60 *S*s were given a standardized shorthand test with two subscores, X_1 = speed, X_2 = accuracy. Results are shown in Table 7.3.

Let us first compute the within-cells SSCP matrix, $\mathbf{S}_w$, because this is obtained most directly from the summary quantities listed at the foot of the columns of scores in each cell. Note that the expression in brackets in Eq. 7.6a represents the (α, β)-element of the SSCP matrix for each cell treated as an intact group. We may denote these cell-by-cell SSCP matrices by $\mathbf{S}_{11}, \mathbf{S}_{12}, \ldots, \mathbf{S}_{23}$, and compute them separately, and then add them to

TABLE 7.3 Speed (X_1) and Accuracy (X_2) Scores for 60 *S*s Taught Shorthand by Two Methods Under Three Conditions of Distributed Versus Massed Practice.

	C_1		C_2		C_3	
	X_1	X_2	X_1	X_2	X_1	X_2
	36	26	46	17	26	14
	34	22	34	21	31	14
	28	21	31	17	30	16
	34	23	31	18	34	16
Method *A*	34	21	36	23	30	13
	29	19	26	19	27	13
	48	25	35	16	21	12
	28	20	33	19	31	15
	34	21	23	15	37	14
	38	20	30	14	29	14
$\sum X_\alpha$	343	218	325	179	296	141
$\sum X_\alpha^2$	12,077	4,798	10,909	3,271	8,934	2.003
$\sum X_1X_2$	7,553		5,855		4,204	
	42	25	32	18	28	11
	47	24	39	19	28	10
	51	29	37	17	25	10
	35	25	31	17	22	12
Method *B*	37	26	36	19	27	11
	44	28	32	19	25	12
	44	25	31	17	33	14
	49	24	41	21	31	13
	43	24	36	18	28	12
	36	26	40	20	23	11
$\sum X_\alpha$	428	256	355	185	270	116
$\sum X_\alpha^2$	18,586	6,580	12,733	3,439	7,394	1,360
$\sum X_1X_2$	10,970		6,601		3,153	

obtain $\mathbf{S}_w$. The details for computing $\mathbf{S}_{11}$ are, in accordance with the expression in brackets with r and c both set equal to 1, as follows:

$$(\mathbf{S}_{11})_{11} = 12{,}077 - (343)^2/10 = 312.1$$
$$(\mathbf{S}_{11})_{12} = (\mathbf{S}_{11})_{21} = 7{,}553 - (343)(218)/10 = 75.6$$
$$(\mathbf{S}_{11})_{22} = 4{,}798 - (218)^2/10 = 45.6$$

Carrying out parallel computations for all six cells, we find the cell-by-cell SSCP matrices to be as follows:

$$\mathbf{S}_{11} = \begin{bmatrix} 312.1 & 75.6 \\ 75.6 & 45.6 \end{bmatrix} \quad \mathbf{S}_{12} = \begin{bmatrix} 346.5 & 37.5 \\ 37.5 & 66.9 \end{bmatrix} \quad \mathbf{S}_{13} = \begin{bmatrix} 172.4 & 30.4 \\ 30.4 & 14.9 \end{bmatrix}$$

$$\mathbf{S}_{21} = \begin{bmatrix} 267.6 & 13.2 \\ 13.2 & 26.4 \end{bmatrix} \quad \mathbf{S}_{22} = \begin{bmatrix} 130.5 & 33.5 \\ 33.5 & 16.5 \end{bmatrix} \quad \mathbf{S}_{23} = \begin{bmatrix} 104.0 & 21.0 \\ 21.0 & 14.4 \end{bmatrix}$$

Adding these six matrices, we get

$$\mathbf{S}_w = \sum_r \sum_c \mathbf{S}_{rc} = \begin{bmatrix} 1333.10 & 211.20 \\ 211.20 & 184.70 \end{bmatrix}$$

To compute the other SSCP matrices in accordance with Eqs. 7.3a to 7.5a, it is convenient first to arrange the vectors of cell totals, marginal totals, and grand totals in the following pattern:

[343 218]	[325 179]	[296 141]	[964 538]
[428 256]	[355 185]	[270 116]	[1053 557]
[771 474]	[680 364]	[566 257]	[2017 1095]

For computing $\mathbf{S}_r$ and $\mathbf{S}_c$, we need only the marginal and grand-total vectors in the above display. Thus, the elements of $\mathbf{S}_r$ are computed from Eq. 7.3a as:

$$(\mathbf{S}_r)_{11} = [(964)^2 + (1053)^2]/(10)(3) - (2017)^2/(10)(2)(3) = 132.02$$
$$(\mathbf{S}_r)_{12} = [(964)(538) + (1053)(557)]/30 - (2017)(1095)/60 = 28.18$$
$$(\mathbf{S}_r)_{22} = [(538)^2 + (557)^2]/30 - (1095)^2/60 = 6.02$$

It may be noted in passing that, if the row vectors of row totals

$$[T_{1r..}, T_{2r..}, \ldots, T_{pr..}]$$

are denoted by $\mathbf{T}'_{r.}$ ($r = 1, 2, \ldots, R$), and the row vector of grand totals, by $\mathbf{T}'$, then the element-by-element formula (7.3a) can be replaced by a matrix equation expression $\mathbf{S}_r$ in terms of these vectors, thus:

$$\mathbf{S}_r = \sum_{r=1}^{R} \mathbf{T}_{r.}\mathbf{T}'_{r.}/nC - \mathbf{TT}'/nRC \tag{7.3b}$$

which is formally even more similar to the corresponding univariate ANOVA formula,

$$SS_r = \sum_{r=1}^{R} T_{r.}^2/nC - T^2/nRC$$

than is Eq. 7.3a. Similar remarks hold with regard to Eqs. 7.4a and 7.5a.

Exactly parallel computations, following Eq. 7.4a or the analogue of Eq. 7.3b involving the vectors of column totals, lead to the matrix $\mathbf{S}_c$. The SSCP matrices for the two main effects are thus found to be:

$$\mathbf{S}_r = \begin{bmatrix} 132.02 & 28.18 \\ 28.18 & 6.02 \end{bmatrix} \quad \text{and} \quad \mathbf{S}_c = \begin{bmatrix} 1055.03 & 1111.55 \\ 1111.55 & 1177.30 \end{bmatrix}$$

Computation of the interaction SSCP matrix $\mathbf{S}_{rc}$ involves the vectors of cell totals as well as those of the marginal and grand totals, as is evident from Eq. 7.5a. We show only the calculations for the off-diagonal elements; each of the diagonal elements is, of course, calculated just as one would compute SS_{rc} in univariate ANOVA.

$$\begin{aligned}(\mathbf{S}_{rc})_{12} = (\mathbf{S}_{rc})_{21} &= [(343)(218) + (325)(179) + \cdots + (270)(116)]/10 \\ &\quad - [(964)(538) + (1053)(557)]/30 \\ &\quad - [(771)(474) + (680)(363) + (566)(257)]/20 + (2017)(1095)/60 \\ &= 174.82\end{aligned}$$

Calculation of the diagonal elements in the manner described above completes our computation of $\mathbf{S}_{rc}$. The final result is shown, together with the other SSCP matrices, in the MANOVA summary table below, following the pattern of Table 7.1. (Although $\mathbf{S}_t$ does not enter into the significance tests as such, it is advisable to compute this matrix independently from Eq. 7.1a, as an arithmetic check; the four SSCP matrices computed above should sum to $\mathbf{S}_t$, as stated in Eq. 7.2. The calculation of $\mathbf{S}_t$ requires the addition, over cells, of the five summary quantities shown in the bottom sectors of the six cells in Table 7.3. This is left as an exercise for the reader.)

The MANOVA summary table (Table 7.4) shows, in addition to the quantities indicated in Table 7.1, the determinants needed for calculating the three Λ_h values in accordance with Eq. 7.7, as well as the Λ-ratios themselves. Since the fixed-effects model (Model I) is the most reasonable one to be regarded as applying in the experiment of this example,[1] $\mathbf{S}_w$ is taken as the error SSCP matrix for testing all three effects (see Table 7.2).

[1]The row effect (methods) is obviously fixed; we could not possibly think of generalizing beyond the two methods used. The status of the column effect is somewhat ambiguous, but it seems more reasonable to regard the conditions stated at the outset of this section as constituting the entire population of interest, rather than as a random sample from the set of all possible schedules totalling 60 hours of instruction.

TABLE 7.4 MANOVA Summary Table

Source	SSCP Matrix	n.d.f.	$\lvert \mathbf{S}_h + \mathbf{S}_e \rvert$ or $\lvert \mathbf{S}_e \rvert$	Λ_h
Row effect (Methods *A* and *B*)	$\begin{bmatrix} 132.02 & 28.18 \\ 28.18 & 6.02 \end{bmatrix}$	1	$\begin{vmatrix} 1465.11 & 239.38 \\ 239.38 & 190.72 \end{vmatrix} = 222{,}123$	.9077
Column effect (distribution of practice)	$\begin{bmatrix} 1055.03 & 1111.55 \\ 1111.55 & 1177.30 \end{bmatrix}$	2	$\begin{vmatrix} 2388.13 & 1322.75 \\ 1322.75 & 1362.00 \end{vmatrix} = 1502{,}966$	.1341
Interaction	$\begin{bmatrix} 308.04 & 174.82 \\ 174.82 & 99.23 \end{bmatrix}$	2	$\begin{vmatrix} 1641.14 & 386.02 \\ 386.02 & 283.93 \end{vmatrix} = 316{,}957$	.6361
Within-cells (error)	$\begin{bmatrix} 1333.10 & 211.20 \\ 211.20 & 184.70 \end{bmatrix}$	54	$\begin{vmatrix} 1333.10 & 211.20 \\ 211.20 & 184.70 \end{vmatrix} = 201{,}618$	—
Total SSCP	$\begin{bmatrix} 2828.18 & 1525.75 \\ 1525.75 & 1467.25 \end{bmatrix}$	59	— —	—

Exact significance tests are available for this example because p (the number of dependent variables) is equal to 2. Thus, we need not use either of the approximate test statistics V and R, but may employ the relevant F-variate shown in Table 4.2, with the replacements for K and N noted earlier. That is, we use

$$\frac{1 - \Lambda_h^{1/2}}{\Lambda_h^{1/2}} \frac{\nu_e - 1}{\nu_h}$$

as an F-variate with $2\nu_h$ and $2(\nu_e - 1)$ degrees of freedom.

For testing the row effect (Methods), we have

$$F_r = \frac{1 - \sqrt{.9077}}{\sqrt{.9077}} \frac{54 - 1}{1} = 2.63$$

This value falls a little short of the 95th centile of the F-distribution with 2 d.f.'s in the numerator and 106 in the denominator, which is found to be about 3.10, by rough interpolation in Table E.4. Thus, if we are using the "conventional" α-value of .05, we would conclude that there is not enough evidence to warrant rejecting the null hypothesis of "zero row effect" in the population. The implication is that Methods A and B probably do not differ in their effectiveness when the shorthand acquisition is measured by the two subtests of our standardized test.

For the column effect (distributed versus massed practice), the test statistic takes the value

$$F_c = \frac{1 - \sqrt{.1341}}{\sqrt{.1341}} \frac{53}{2} = 45.85$$

which far exceeds even the 99.9th centile of the F-distribution with 4 and 106 degrees of freedom. We may therefore conclude that there is strong evidence of nonchance differences due to the extent of distribution of practice. Examination of the column-total vectors indicates that, among the three conditions studied, achievement as measured by both speed and accuracy scores steadily increases with increasing degree of distribution of practice. (The interpretation will often not be as clear-cut as in this example because the trends of increase may differ from one dependent variable to another. For this reason, it will usually be instructive to follow the significance tests up with discriminant analyses, as described in the next section.)

Attending to the interaction effect, the value of the test statistic is

$$F_{rc} = \frac{1 - \sqrt{.6361}}{\sqrt{.6361}} \frac{53}{2} = 6.73$$

which exceeds the 99.9th centile (approximately 5.03) of the F-distribution with 4 and 106 degrees of freedom. We may conclude with high confidence that the relative effectiveness of the two methods depends on which one of the three conditions of distributed practice is actually used. Examination of the cell-total vectors suggests that Method B is more effective than Method A under the high- and medium-distributed practice conditions (C_1 and C_2), but that the reverse is true when the highly massed schedule (C_3: 4 hours' daily instruction for 3 weeks) is used. Again, the interpretation is clear-cut in this example because both criterion measures show the same trend; but we cannot expect this to be the case in general.

7.4 DISCRIMINANT ANALYSIS IN FACTORIAL DESIGNS

When a particular effect has been found significant in MANOVA, the question still remains of just how the several groups representing the levels of that effect differ in terms of the criterion variables used. It would be an exception rather than the rule to find the same trend or pattern of differences to be repeated for all the variables involved, as we did in the miniature example in the preceding section. A modified version of discriminant analysis may help to answer this question by identifying the dimension or dimensions along which the relevant subgroups differ most conspicuously.

The modification of discriminant analysis that is required for this purpose is exactly parallel to the modification (or generalization) of the Λ-ratio which was made in the foregoing. That is, the coefficients of the discriminant function(s) associated with a particular effect are now obtained as the elements of the vector(s) satisfying Eq. 7.11

$$(\mathbf{S}_h - \lambda \mathbf{S}_e)\mathbf{v} = \mathbf{0} \tag{7.11}$$

instead of the earlier Eq. 6.8:

$$(\mathbf{B} - \lambda \mathbf{W})\mathbf{v} = \mathbf{0}$$

(Observe that the parallel with the modification of Λ lies in the fact that the denominator of this ratio has now become $|\mathbf{S}_h + \mathbf{S}_e|$ instead of the previous $|\mathbf{S}_t| = |\mathbf{B} + \mathbf{W}|$ for significance testing in the usual multigroup discriminant analysis.)

Thus, assuming (as we shall ordinarily be justified in doing) that $\mathbf{S}_e$ is nonsingular, our problem amounts to finding the eigenvector(s) of the matrix $\mathbf{S}_e^{-1}\mathbf{S}_h$, for Eq. 7.11 may be replaced by its equivalent,

$$(\mathbf{S}_e^{-1}\mathbf{S}_h - \lambda \mathbf{I})\mathbf{v} = \mathbf{0} \tag{7.12}$$

This, incidentally, is the basis for our earlier assertion that it would be preferable to compute Λ_h via the eigenvalue(s) of $\mathbf{S}_e^{-1}\mathbf{S}_h$, as indicated in

Eq. 7.8, rather than by the definitional formula, 7.7, especially when a follow-up discriminant analysis is to be carried out.

After the discriminant functions associated with each of the significant effects have been computed, the procedures for interpretation are exactly the same as for the case of ordinary multigroup discriminant analysis. In fact, since the number of "groups" representing the levels of each independent variable is generally fairly small, there is a good chance that only one or two discriminant functions will be significant for a given main effect. This means that the discriminant function means or centroids for the several levels can be plotted on graph paper, and the interpretation should be simpler than in situations where three or more significant functions are found—as is quite likely in multigroup problems with a large number of groups.

EXAMPLE 7.1. We illustrate the computation of discriminant functions in MANOVA in the context of the miniature example of the preceding section—even though, in practice, there would hardly be any point in carrying out this follow-up analysis when all criterion measures show the same trend.

Since the row effect (methods) was not significant, we do not compute the discriminant function between Methods *A* and *B*. Those for the column effect (extent of distributed practice) and interaction are the only ones of possible interest. Computations for these discriminant functions are shown below.

Since $\mathbf{S}_w$ is the error SSCP matrix for all effects in this problem, the matrices $\mathbf{S}_e^{-1}\mathbf{S}_h$ whose eigenvectors we need to find in order to get the desired discriminant functions are $\mathbf{S}_w^{-1}\mathbf{S}_c$ and $\mathbf{S}_w^{-1}\mathbf{S}_{rc}$. Referring to Table 7.4, where the several SSCP matrices have been collected, we first compute

$$\mathbf{S}_w^{-1} = \begin{bmatrix} 1333.10 & 211.20 \\ 211.20 & 184.70 \end{bmatrix}^{-1}$$

$$= \begin{bmatrix} .9161 & -1.0475 \\ -1.0475 & 6.6120 \end{bmatrix} \times 10^{-3}$$

Postmultiplying this by $\mathbf{S}_c$ as given in Table 7.4, we get

$$\mathbf{S}_w^{-1}\mathbf{S}_c = \begin{bmatrix} -.1978 & -.2149 \\ 6.2444 & 6.6200 \end{bmatrix}$$

as the matrix to be substituted in Eq. 7.12 for computing the discriminant functions associated with the column effect.

The characteristic equation $|\mathbf{S}_w^{-1}\mathbf{S}_c - \lambda\mathbf{I}| = 0$ is found to be

$$\lambda^2 - 6.4222\lambda + .0325 = 0$$

whose roots are

$$\lambda_1 = 6.4171 \quad \text{and} \quad \lambda_2 = .0051$$

At this point we may verify that computing Λ_c via the eigenvalues of $\mathbf{S}_w^{-1}\mathbf{S}_c$ in accordance with Eq. 7.8 yields the same result as that obtained from the definitional Eq. 7.7. We find:

$$\Lambda_c = 1/(7.4171)(1.0051) = .1341$$

which agrees exactly (to four decimal places) with the value shown in Table 7.4, which was computed from the definitional formula.

The next step is to ascertain whether both or only the first of these eigenvalues will lead to significant discriminant functions. This is done, as described in Chapter Six, by examining the partial sums (in this case simply the second term) in the expression for Bartlett's approximate chi-square statistic, V. In the present example, it is evident by mere inspection that only the first eigenvalue is significant (for it accounts for over 99.9% of the trace of $\mathbf{S}_w^{-1}\mathbf{S}_c$). We carry out the test solely for illustrative purposes. Substituting the eigenvalues in Eq. 7.9 with $\nu_e = \nu_w = 54$, $\nu_h = \nu_c = 2$, and $p = 2$, we get

$$\begin{aligned} V &= [54 + 2 - (2 + 2 + 1)/2](\ln 7.4171 + \ln 1.0051) \\ &= (53.5)(2.0038 + 0.0051) = 107.21 + 0.27 \end{aligned}$$

From the description following Eq. 6.13, with K replaced by $\nu_h + 1$, it is seen that the two terms in the indicated sum should be compared to chi-squares with $p + \nu_h - 1$ and $p + \nu_h - 3$ degrees of freedom, respectively. For this example, these n.d.f.'s become 3 and 1, respectively. (Note that their sum is $4 = p\nu_h$, which is the n.d.f. for V itself, as stated after Eq. 7.9.) It is obvious, as we expected, that the second term, 0.27, falls far short of significance as a chi-square with one degree of freedom. Thus, we compute only the first discriminant function, that associated with the larger eigenvalue 6.4171 of $\mathbf{S}_w^{-1}\mathbf{S}_c$.

We substitute $\lambda_1 = 6.4171$ in the expression $\mathbf{S}_w^{-1}\mathbf{S}_c - \lambda\mathbf{I}$, and find the adjoint of the resulting matrix. The outcome is:

$$\mathbf{Adj}(\mathbf{S}_w^{-1}\mathbf{S}_c - 6.4171\mathbf{I}) = \begin{bmatrix} .2029 & .2149 \\ -6.2444 & -6.6149 \end{bmatrix}$$

Normalizing either column of this matrix to unity by dividing the elements by the square root of the sum of the squares of the two elements, and further multiplying each element by -1 (so that the larger element will be positive), we find the normalized eigenvector associated with λ_1 to be

$$\mathbf{v}'_{c(1)} = [-.0325, .9995]$$

Thus, the single significant discriminant function for the column effect is:

$$Y_c = -.0325X_1 + .9995X_2.$$

For all practical purposes, the accuracy score alone suffices to differentiate among the three conditions of distribution of practice.

For computing the discriminant functions associated with the interaction effect, the matrix to be substituted for $\mathbf{S}_e^{-1}\mathbf{S}_h$ in Eq. 7.12 is

$$\mathbf{S}_w^{-1}\mathbf{S}_{rc} = \begin{bmatrix} .09907 & .05621 \\ .83324 & .47298 \end{bmatrix}$$

The characteristic equation is

$$\lambda^2 - .57205\lambda + .00002 = 0$$

Since the constant term is equal to zero within rounding error, we may take

$$\lambda_1 = .5721$$

to be the single nonvanishing eigenvalue.

The associated eigenvector, normalized to unity, is found to be

$$\mathbf{v}'_{rc} = [.1181, .9930]$$

Again, the discriminant function is practically equal to X_2 (the accuracy score) itself.

7.5 OTHER DESIGNS

In the foregoing we discussed the procedures of MANOVA in its simplest application: two-factor designs with equal cell frequencies. We now give a brief guide to making the requisite extensions of the methods of ANOVA to multivariate designs of other types. Simply, the basic principle is always this: Compute the SSCP matrix corresponding to each sum-of-squares in ANOVA; select the appropriate error SSCP matrix (following the same rule as in ANOVA) for testing each null hypothesis, and calculate Λ_h from Eq. 7.7 or 7.8.

For two-factor designs with cell frequencies unequal but proportional to the marginal frequencies, the modifications to the procedures described above are quite minor. Only slight changes in the computational formulas 7.3a to 7.6a need to be introduced, and the subsequent steps remain exactly as they were in the equal-frequencies case. (In fact, the design with equal cell frequencies is simply a special case of the more general proportional-frequencies design, also known as the *orthogonal design.*)

Denoting the frequency, that is, the number of experimental units, in the (r, c)-cell by n_{rc}, we further define:

$$n_{r.} = \sum_{c=1}^{C} n_{rc} \qquad \text{the } r\text{th row frequency}$$

$$n_{.c} = \sum_{r=1}^{R} n_{rc} \qquad \text{the } c\text{th column frequency}$$

and

$$N = \sum_{r=1}^{R} n_{r.} = \sum_{c=1}^{C} n_{.c} \qquad \text{the total sample size}$$

(Porportionality of cell frequencies to the marginals is said to hold when $n_{rc} = n_{r.}n_{.c}/N$ for all cells.) Then, the modifications consist simply in replacing, in the computational formulas, all instances of *n occurring alone* by n_{rc}, all instances of nC by $n_{r.}$, those of nR by $n_{.c}$, and nRC by N. (It is understood, of course, that any subscripted n is included within the scope of summation with regard to the symbol(s) in the subscript.) Thus, the formulas for the general elements of $\mathbf{S}_{rc}$ and $\mathbf{S}_w$ become

$$(\mathbf{S}_{rc})_{\alpha\beta} = \sum_{r=1}^{R}\sum_{c=1}^{C} (T_{\alpha rc.}T_{\beta rc.}/n_{rc}) - \sum_{r=1}^{R} (T_{\alpha r..}T_{\beta r..}/n_{r.}) - \sum_{c=1}^{C} (T_{\alpha .c.}T_{\beta .c.}/n_{.c}) + T_{\alpha ...}T_{\beta ...}/N \tag{7.13}$$

and

$$(\mathbf{S}_w)_{\alpha\beta} = \sum_{r=1}^{R}\sum_{c=1}^{C}\left[\sum_{i=1}^{n} X_{\alpha rci}X_{\beta rci} - T_{\alpha rc.}T_{\beta rc.}/n_{rc}\right] \tag{7.14}$$

respectively. The modified formulas for $(\mathbf{S}_r)_{\alpha\beta}$ and $(\mathbf{S}_c)_{\alpha\beta}$ should be obvious from that for $(\mathbf{S}_{rc})_{\alpha\beta}$ because the latter contains all the "ingredients" for the first two.

When the cell frequencies are disproportionate, the problem becomes more involved; in fact, the significance tests are then only approximate. The two most commonly used methods are the method of least squares and that of unweighted means. We outline the latter method here. The least-squares approach perhaps leads to significance tests that are more nearly exact than does the unweighted-means method described below, but it is computationally far more complex. A discussion of the least-squares approach in the univariate case is given by Winer (1962), and a computer program for the multi-variate case is described by Bock (1963).

For carrying out the unweighted-means analysis in MANOVA, the first step is to compute the vectors of cell means for all the dependent variables. These cell-mean vectors are then used as though they were observation vectors based on one experimental unit per cell for computing all the SSCP matrices except $\mathbf{S}_w$. That is, Eqs. 7.3a to 7.5a, with n set equal to 1, are used for computing $\mathbf{S}_r$, $\mathbf{S}_c$, and $\mathbf{S}_{rc}$. (Be sure to note that the various totals in these formulas are now obtained by the appropriate summing of *cell means*, not the original scores.)

The within-cells SSCP matrix $\mathbf{S}_w$ is computed in accordance with Eq. 7.14 using the original scores $X_{\alpha rci}$. But this matrix is then "scaled down" by dividing it by the harmonic mean of the cell frequencies, thus making it commensurate with the other SSCP matrices that were computed by treating the cell means as though they were single observations. Division by the harmonic mean of the cell frequencies is, of course, the same as multiplication by the *reciprocal* of the harmonic mean, that is, the average of the reciprocal of the cell frequencies. The multiplicative factor is

$$1/\tilde{n} = \left[\sum_{r=1}^{R} \sum_{c=1}^{C} (1/n_{rc})\right] \Big/ RC \tag{7.15}$$

Thus, the actual within-cells SSCP matrix $\mathbf{S}_w$ is replaced by a "scaled-down" matrix,

$$\tilde{\mathbf{S}}_w = (1/\tilde{n})\mathbf{S}_w \tag{7.16}$$

The analysis from this point on proceeds in exactly the same way as for the equal-frequencies design, except that the n.d.f. for $\mathbf{S}_w$ is $N - RC$ instead of the $RC(n - 1)$ shown in Table 7.1. It should be noted, however, that the analysis is now only approximate; hence the outcome must be interpreted with caution, especially if it is one of borderline significance.

Other designs commonly used in educational and psychological research include three-factor designs, randomized-blocks design, repeated-measures design, and nested-factors (or hierarchical) design. We cannot describe the multivariate extensions of all these analyses without making this chapter inordinately long. The reader should have little difficulty in arriving at the correct procedures if he is familiar with the corresponding univariate ANOVA methods and remembers the general principle stated at the outset of this section: to each sum-of-squares in ANOVA there corresponds an SSCP matrix in MANOVA. Thus, in a randomized blocks design, the highest-order interaction SSCP matrix replaces the within-cells SSCP matrix of the corresponding randomized groups design. (This is because a q-factor randomized-blocks design with m blocks is formally equivalent with a $(q + 1)$-factor randomized-groups design.) In a nested-factors design, different error SSCP matrices will have to be used for testing the effects involving nested and crossed factors. A general computer program capable of handling all the commonly used designs was developed by Bock (1965) and put into operational form by Clyde, Cramer, and Sherin (1966).

7.6 OTHER TEST CRITERIA

In this and the preceding chapter we have used the Λ-ratio as our only test criterion in carrying out significance tests. Although this is the oldest and most widely used criterion, it is by no means the only one available. The several alternative criteria that have been proposed are all functions of $\mathbf{S}_e^{-1}\mathbf{S}_h$, just as Λ is. Among them, the two best known ones are the following, where $\lambda_1, \lambda_2, \ldots, \lambda_p$ are the eigenvalues of $\mathbf{S}_e^{-1}\mathbf{S}_h$ in descending order of magnitude: (a) Hotelling's (1951) trace criterion

$$\tau = \sum_{i=1}^{p} (\mathbf{S}_e^{-1}\mathbf{S}_h)_{ii} = \sum_{i=1}^{p} \lambda_i \tag{7.17}$$

and (b) Roy's (1957, Ch. 6) largest root criterion

$$\theta = \lambda_1/(1 + \lambda_1) \tag{7.18}$$

which is the largest eigenvalue of $(\mathbf{S}_e + \mathbf{S}_h)^{-1}\mathbf{S}_h$.

It is seen from Eq. 7.12 that λ_i is also the value of the discriminant criterion for the ith discriminant function associated with the effect being tested. Thus, large values of the λ_i indicate significant effects, and it follows, since both τ and θ are increasing functions of the λ_i, that values of τ and θ that are greater than specified centile points, say $\tau_{1-\alpha}$ and $\theta_{1-\alpha}$, respectively, lead to rejection of the null hypothesis. This is in contrast to the Λ-criterion we have been using. In that case, values smaller than a specified lower centile point Λ_α, for example, indicate a significant effect, because

$$\Lambda = 1/(1 + \lambda_1)(1 + \lambda_2)\cdots(1 + \lambda_p)$$

is a decreasing function of the λ_i.

Pillai (1960) has derived the distribution followed by τ (which he denotes as $U^{(s)}$) under the null hypothesis, and has tabled selected centile points of the distribution, or rather the family of distributions for various combinations of values of the parameters m, n, and s, defined as follows:

$$m = (p - \nu_h - 1)/2 \qquad n = (\nu_e - p - 1)/2$$

and

$$s = \min(\nu_h, p)$$

Similarly, selected centile points for the null distributions of θ are given by Heck (1960) and by Pillai (1960, 1966), also using m, n, and s as parameters.

It should be clear that, when $\nu_h = 1$, both τ and θ reduce to simple functions of Λ, for $\mathbf{S}_e^{-1}\mathbf{S}_h$ has only one nonzero eigenvalue λ_1 in this case, and

$$\tau = \lambda_1 \qquad \theta = \lambda_1/(1 + \lambda_1) \qquad \text{and} \qquad \Lambda = 1/(1 + \lambda_1)$$

From the last of these relations, it follows that

$$\lambda_1 = \frac{1 - \Lambda}{\Lambda}$$

But from Table 4.2, by replacing N by $\nu_e + \nu_h + 1 = \nu_e + 2$, it is seen that

$$\frac{1 - \Lambda}{\Lambda} \frac{\nu_e - p + 1}{p}$$

follows the F-distribution with p and $\nu_e - p + 1$ degrees of freedom. Consequently,

$$\tau \frac{\nu_e - p + 1}{p} \qquad \text{and} \qquad \frac{\theta}{1 - \theta} \frac{\nu_e - p + 1}{p}$$

may be referred to the same F-distribution, and the need for special tables disappears in the special case when $\nu_h = 1$.

When $\nu_h > 1$, however, not only are the tables given by Heck and Pillai, referred to above, necessary for using the trace and largest-root criteria, but the significance tests employing these criteria and that involving the Λ-ratio may all lead to different conclusions. The question naturally arises, which conclusion is the "best"? As might be expected, there is no unique answer to this question, since it all depends on what sort of alternative to the null hypothesis happens to be true.

A Monte Carlo study was carried out by Schatzoff (1966b) to explore this question. Using an index called the expected significance level, he compared the relative sensitivities of six test statistics, including the three discussed here, over a wide variety of population structures. Wilks' Λ-criterion and Hotelling's trace criterion were found to be about equally sensitive for a wide spectrum of alternatives, but Roy's largest-root criterion was poor except when the population structure was such that most of the difference was concentrated in a single dimension. This conclusion is not surprising, and further investigation seems to be needed for determining the conditions under which Λ is superior to τ and vice versa.

Studies illustrating the use of the three test criteria are described by Bock (1966), Bock and Haggard (1968), and L. V. Jones (1966).

7.7 SUGGESTED FURTHER READINGS

In this chapter we have taken what may be called the classical approach to MANOVA. This was done in the belief that most readers would be more familiar with ANOVA treated in this mode, developed by R. A. Fisher, than with an approach that is gradually becoming popular these days as the availability of high-speed computers increases. This new approach treats MANOVA (as well as univariate ANOVA) in the framework of what is called the *general linear hypothesis*, which is essentially a formal multiple linear regression analysis. (Strictly speaking, this approach is not new, for it was treated in Fisher's early writings, but many of its refinements are recent, and its practical use had to await the advent of computers.) Readers who wish to become acquainted with this approach are referred to textbooks by Bock (1971), Mendenhall (1968), Morrison (1967), and Seber (1966).

EXERCISES

1. An experiment was carried out to explore possible differences in grammatical usage among men and women with varying degrees of formal education. Thus, the independent variables were education and sex. Three levels were used for education: (1) Did not complete high school; (2) Graduated from high school; and (3) Attended college for two or more years.

   The dependent variables were measures of relative frequencies of use of: (1) Personal possessive pronouns; (2) Nouns; and (3) Quantifiers in tape-recorded stories told by *S*s in response to a set of 20 cartoons stripped of their captions. (More specifically, the scores were 20 Arcsine $\sqrt{p}$, where p is the observed relative frequency for each grammatical category, and the multiplier 20 was used to get numbers of convenient orders of magnitude.)

   Shown below are the six cell-mean vectors for the three dependent variables, and (in italics) the cell frequencies. The within-cells SSCP matrix $\mathbf{S}_w$, based on the original scores (not shown) is also given. (Adapted from Jones, 1966.)

Carry out a complete MANOVA by the method of unweighted means for these data, including the construction of discriminant functions for those effects that are significant at the 1% level.

Sex / Education	Male	Female
1	[2.09, 4.53, 3.56] *8*	[1.38, 4.17, 3.38] *10*
2	[2.12, 5.35, 3.59] *11*	[1.47, 4.89, 3.12] *8*
3	[2.13, 5.94, 3.51] *9*	[1.74, 5.37, 3.27] *8*

$$\mathbf{S}_w = \begin{bmatrix} 6.9312 & 2.9520 & -.5040 \\ 2.9520 & 19.0512 & 1.5792 \\ -.5040 & 1.5792 & 7.3008 \end{bmatrix}$$

2. An experiment was conducted for comparing three different approaches to teaching arithmetic and language skills to fifth-grade pupils:

(a) "Traditional" [T]—Lecture, discussion, and drill

(b) "Programed" [P]—Entire course taught by programed instruction

(c) "Eclectic" [E]—A combination of the first two approaches, wherein the teacher does the major classroom exposition, and exercises and self-tests are handled through programed material.

Since it is well known that boys and girls of this age differ in their relative performances in these two subjects, sex was used as a second factor.

Specifically, 60 boys and 60 girls were selected at random from the fifth grade of a large school, and were randomly assigned (20 pupils of each sex per group) to the three treatment groups.

At the end of five months of instruction, all 120 pupils were given a standardized achievement test in arithmetic (X_1), and a standardized achievement test in the language skills (X_2).

The total scores for each of the six subgroups on the two tests (in the order [X_1, X_2]) were as given in the cells of the table below, which also shows the marginal and the grand totals:

	T	*P*	*E*	
Boys	[514, 453]	[524, 473]	[536, 490]	[1574, 1416]
Girls	[458, 507]	[473, 518]	[494, 571]	[1425, 1596]
	[972, 960]	[997, 991]	[1030, 1061]	[2999, 3012]

The within-cells SSCP matrix for these data was found to be:

$$\mathbf{S}_w = \begin{bmatrix} 1795.39 & 1230.56 \\ 1230.56 & 1753.50 \end{bmatrix}$$

(a) Carry out a MANOVA to test the significance of the two main effects and the interaction effect on the vector criterion variable, $[X_1, X_2]$. Use the 1% level of significance for all decisions.

(b) Construct the linear discriminant function(s) corresponding to the effect(s) found significant in (a) above.

Chapter Eight

Applications to Classification Problems

In one sense, the problem of classification is as old as science itself. Whether it be minerals, plants, or anthropological specimens, their classification into species, classes, and other taxonomic groups has always been a method for introducing order into an unstructured field. However, the systematic application of statistical (especially multivariate) techniques to the problem is a relatively recent development.

Probably the earliest such application was that by Tildesley (1921), who used Karl Pearson's "coefficient of racial likeness" for classifying prehistoric skeletal remains into racial groups on the basis of several anthropometric measurements. Discriminant analysis has been used for dealing with classification problems in a variety of fields, including politics (classifying U.S. senators into "conservative," "progressive," and other groups on the basis of their voting records); anthropology (classifying Egyptian skulls into one of four series); and educational guidance ("classifying" college students into one of several curricular groups on the basis of a set of test scores). [For references, see Tatsuoka and Tiedeman (1954).]

All of the examples cited above presume the existence of well-defined groups, and deal with deciding how to classify as yet "unlabeled" individuals into one or another of these groups. This is the sense in which we shall use the term *classification problem.* The prior task of discovering or establishing the system of groups, which may, for distinction, be called *taxonomic problems*, will not be treated here. Statistical approaches to the taxonomic problem include Karl Pearson's (1894) "dissection of frequency curves" into normal components, Stephenson's (1953) inverted or *Q*-model factor analysis, McQuitty's (1955) pattern analysis, and Cattell and Coulter's (1966) taxonome method.

8.1 CLASSIFICATION AND THE CONCEPT OF RESEMBLANCE

The classification problems, as delineated above, amounts to seeking an answer to the question: Which of these several groups does this individual "resemble" the most, in terms of a specified set of measurable characteristics? That is, we have at hand a sample from each of K well-defined populations (such as clinical diagnostic categories, occupational groups, or curricular groups), with measures for each individual on p variables that are known or deemed to be important in differentiating among the several populations or groups. Subsequently, we encounter an individual whose group membership is unknown, but for whom we have measures on these same p variables, and we wish to classify him as a member of one or another of these K groups—the one with which he shows greatest "resemblance" in terms of these p measures.

The crux of the matter obviously lies in how we define "resemblance" in this context. Various measures of *profile* (or *pattern*) *similarity* and of distance (that is, *dissimilarity*) have been proposed in the literature [Mahalanobis (1936), Cattell (1949), Du Mas (1949), Cronbach and Gleser (1952)]. We here choose the familiar χ^2 statistic, introduced in Chapter Four (Eq. 4.7), to serve as a measure of dissimilarity. This is a reasonable choice, since the larger the χ^2 value of an individual with reference to a given group, the farther away (in the generalized-distance sense) is the point $[X_{1i}, X_{2i}, \ldots, X_{pi}]$ representing his set of scores from the centroid $[\bar{X}_{1k}, \bar{X}_{2k}, \ldots, \bar{X}_{pk}]$ of that group. Thus, he may be said to be the more deviant from the "average member" of that group, the larger his χ^2 value. Conversely, an individual with a small χ^2 value with reference to a group is "closer" to the average member of that group, and may hence be said to resemble that group. Furthermore, if the reference group is adequately describable by means of a multivariate normal distribution of the p variables, then knowledge of an individual's χ^2 value allows us to estimate the percentage of individuals in the group who are "closer to" or "farther from" the group centroid than is that individual. This is because, as shown in Chapter Four, the χ^2 value determines the particular centile ellipsoid on which a given point lies.

Thus, a simple classification scheme, which may be called the *minimum chi-square rule*, would be as follows:

Compute the χ^2 value of the unclassified individual with respect to each of the K groups, and assign him to that group with respect to which his χ^2 value is the smallest.

This rule has the property of minimizing the probability of misclassifications when the K populations have multivariate normal distributions with equal disperson matrices. If this common dispersion matrix $\mathbf{\Sigma}$ is known, it would, of course, be used in computing each of the K χ^2 values

$$\chi_{ik}^{\ 2} = \mathbf{x}'_{ik}\mathbf{\Sigma}^{-1}\mathbf{x}_{ik}$$

for individual i, where $\mathbf{x}_{ik}$ is the vector of his p scores in deviation form—deviations from the kth population centroid $\boldsymbol{\mu}_k' = [\mu_{1k}, \mu_{2k}, \ldots, \mu_{pk}]$ if this is known; deviations from the kth sample centroid $\overline{X}_k' = [\overline{X}_{1k}, \overline{X}_{2k}, \ldots, \overline{X}_{pk}]$ if $\boldsymbol{\mu}_k$ is unknown.

Since, in practice, both $\boldsymbol{\Sigma}$ and the $\boldsymbol{\mu}_k$ are usually unknown, sample estimates have to be used for both the dispersion matrix and the centroids. In this case, $\boldsymbol{\Sigma}$ is replaced by its within-groups estimate $\mathbf{S}_w/(N - K) = \mathbf{D}_w$, where $\mathbf{S}_w$ is the within-groups SSCP matrix, and $N = n_1 + n_2 + \cdots + n_k$. Thus, the formula for $\chi_{ik}{}^2$ that is of greatest practical use becomes

$$\begin{aligned}\chi_{ik}{}^2 &= [X_{1i} - \overline{X}_{1k}, X_{2i} - \overline{X}_{2k}, \ldots, X_{pi} - \overline{X}_{pk}]\mathbf{D}_w^{-1} \\ &\quad \times [X_{1i} - \overline{X}_{1k}, \ldots, X_{pi} - \overline{X}_{pk}]' \\ &= \mathbf{x}_{ik}'\mathbf{D}_w^{-1}\mathbf{x}_{ik} \qquad (8.1)\end{aligned}$$

EXAMPLE 8.1. In the numerical examples throughout this chapter, we utilize the data first introduced in Chapter Four (p. 90) comprising the results of administering a three-scale Activity Preference Questionnaire (APQ) to samples from three job categories of an airline company. We suppose that the company's personnel officer is faced with the task of assigning several job applicants to one of these categories on the basis of their APQ scores.

The sample sizes and the group means on the three scales were as follows:

		Means On		
k	n_k	X_1	X_2	X_3
1. Passenger agents	85	12.5882	24.2235	9.0235
2. Mechanics	93	18.5376	21.1348	10.1398
3. Operations control men	66	15.5758	15.4545	13.2424

The within-groups dispersion matrix, computed by dividing each element of the within-groups SSCP matrix $\mathbf{W}$ shown on p. 90 by $\sum n_k - 3 = 241$, is as follows:

$$\mathbf{D}_w = \begin{bmatrix} 16.4640 & 1.4590 & .3180 \\ 1.4590 & 18.2832 & .9769 \\ .3180 & .9769 & 11.1341 \end{bmatrix}$$

Suppose that four job applicants (already accepted for employment, but as yet unassigned to particular job categories) had the following scores on the three scales of the APQ:

i	X_{1i}	X_{2i}	X_{3i}
1	15	24	6
2	15	16	11
3	16	23	14
4	19	20	15

The first step for computing the χ^2 value of each individual with respect to each group is to express his three scores as deviations from the means of the group. The results are:

i	$\mathbf{x}_{i1}$			$\mathbf{x}_{i2}$		
1	2.4118	−.2235	−3.0235	−3.5376	2.8602	−4.1398
2	2.4118	−8.2235	1.9765	−3.5376	−5.1398	.8602
3	3.4118	−1.2235	4.9765	−2.5376	1.8602	3.8602
4	6.4118	−4.2235	5.9765	.4624	−1.1398	4.8602

i	$\mathbf{x}_{i3}$		
1	−.5758	8.5455	−7.2424
2	−.5758	.5455	−2.2424
3	.4242	7.5455	.7576
4	3.4242	4.5455	1.7576

Next, we need the inverse $\mathbf{D}_w^{-1}$ of the within-groups dispersion matrix, which is computed to be:

$$\mathbf{D}^{-1} = \begin{bmatrix} 6.1191 & -.4812 & -.1325 \\ -.4812 & 5.5331 & -.4717 \\ -.1325 & -.4717 & 9.0266 \end{bmatrix} \times 10^{-2}$$

We may now carry out the computations for the χ^2's in accordance with Eq. 8.1. Thus, the χ^2 value for individual 1 with respect to Group 1 (passenger agents) is obtained as

$$\chi_{11}^2 = [2.4118 \quad -.2235 \quad -3.0235] \times \begin{bmatrix} .061191 & -.004812 & -.001325 \\ -.004182 & .055331 & -.004717 \\ -.001325 & -.004717 & .090266 \end{bmatrix} \times \begin{bmatrix} 2.4118 \\ -.2235 \\ -3.0235 \end{bmatrix}$$

$$= [.1527 \quad -.0097 \quad -.2751] \begin{bmatrix} 2.4118 \\ -.2235 \\ -3.0235 \end{bmatrix}$$

$$= 1.2022$$

Similarly, this individual's χ^2 values with reference to Group 2 (mechanics) and Group 3 (operations control men) are

$$\chi_{12}^2 = [-3.5376 \quad 2.8602 \quad -4.1398] \times \begin{bmatrix} .061191 & -.004812 & -.001325 \\ -.004182 & .055331 & -.004717 \\ -.001325 & -.004717 & .090266 \end{bmatrix} \times \begin{bmatrix} -3.5376 \\ 2.8602 \\ -4.1398 \end{bmatrix}$$

$$= 2.9355$$

and

$$\chi_{13}^2 = [-.5758 \quad 8.5455 \quad -7.2424] \times \begin{bmatrix} .061191 & -.004812 & -.001325 \\ -.004818 & .055331 & -.004717 \\ -.001325 & -.004717 & .090266 \end{bmatrix} \times \begin{bmatrix} -.5758 \\ 8.5455 \\ -7.2424 \end{bmatrix}$$

$$= 1.8712$$

respectively.

Since his smallest χ^2 value is that with reference to Group 1, our decision according to the minimum chi-square rule would be to assign individual 1 to the passenger agents group.

On computing the three χ^2 values for the remaining three individuals, we may summarize the results and decisions for all four individuals in a table, as follows (where the smallest χ^2 value for each individual is shown in italics):

Individual Number (i)	χ_{i1}^2	χ_{i2}^2	χ_{i3}^2	Decision—Assign to
1	*1.2022*	2.9355	1.8712	Group 1
2	4.7815	2.1690	*.5017*	Group 3
3	3.0836	*1.9339*	3.1276	Group 2
4	7.1235	2.2687	*1.8982*	Group 3

When the K population dispersion matrices are not (or cannot be assumed to be) equal, the separate matrices or their respective sample estimates are used in place of $\mathbf{\Sigma}$ or $\mathbf{D}_w$ in computing χ^2. Thus, the formula for the chi-square statistic now becomes

$$\chi_{ik}^2 = [X_{1i} - \bar{X}_{1k}, X_{2i} - \bar{X}_{2k}, \ldots, X_{pi} - \bar{X}_{pk}]\mathbf{D}_k^{-1} \times [X_{1i} - \bar{X}_{1k}, \ldots, X_{pi} - \bar{X}_{pk}]' \tag{8.2}$$

where $\mathbf{D}_k = \mathbf{S}_k/(n_k - 1)$ is the dispersion matrix of the kth sample. At the same time, the classification rule is modified to be based on minimizing, not χ^2 itself, but an adjusted quantity χ'^2 defined as follows:

$$\chi_{ik}'^2 = \chi_{ik}^2 + \ln|\mathbf{D}_k| \tag{8.3}$$

which is proportional to the natural logarithm of the multivariate normal density function $N(\bar{\mathbf{X}}_k, \mathbf{D}_k)$ evaluated at the point $\mathbf{X}_i' = [X_{1i}, X_{2i}, \ldots, X_{pi}]$. That is, for each individual to be classified, we compute the quantity $\chi_{ik}'^2$, defined by Eqs. 8.2 and 8.3, for each of the K groups, and assign him to that group for which his χ'^2 value is the smallest.[1]

[1]The maximum *centour score* rule proposed by Tiedeman in Tiedeman et al. (1953) (also described in Rulon et al., 1967, p. 167) is intermediate between the minimum χ^2 and minimum χ'^2 rules stated above. An individual's centour score with respect to Group k is found by computing χ_{ik}^2 from Eq. 8.2, determining its centile rank as a chi-square variate with p d.f.'s, and subtracting this value from 100. Thus, the larger an individual's centour score with respect to a given group, the closer is his score point to the centroid of that group.

EXAMPLE 8.2. Continuing with the example given in connection with the minimum chi-square rule, we now discard the assumption that the dispersion matrices in the three populations are equal, and use a separate dispersion matrix $\mathbf{D}_k$ for each group. These are computed as $\mathbf{D}_k = \mathbf{S}_k/(n_k - 1)$, where $\mathbf{S}_k$ is the SSCP matrix for the kth sample. The dispersion matrices and their respective inverses are as follows:

$$\mathbf{D}_1 = \begin{bmatrix} 20.2451 & 4.6170 & -2.4069 \\ 4.6170 & 18.7947 & 2.5066 \\ -2.4069 & 2.5066 & 9.8804 \end{bmatrix}$$

$$\mathbf{D}_1^{-1} = \begin{bmatrix} .055084 & -.015858 & .017442 \\ -.015858 & .059635 & -.018992 \\ .017442 & -.018992 & .110277 \end{bmatrix}$$

$$\mathbf{D}_2 = \begin{bmatrix} 12.7078 & -1.5760 & 2.5436 \\ -1.5760 & 20.7085 & .1759 \\ 2.5436 & .1759 & 10.5129 \end{bmatrix}$$

$$\mathbf{D}_2^{-1} = \begin{bmatrix} .083571 & .006533 & -.020329 \\ .006533 & .048806 & -.002397 \\ -.020329 & -.002397 & .100080 \end{bmatrix}$$

$$\mathbf{D}_3 = \begin{bmatrix} 16.8942 & 1.6727 & .6890 \\ 1.6727 & 14.1902 & .1343 \\ .6890 & .1343 & 13.6326 \end{bmatrix}$$

$$\mathbf{D}_3^{-1} = \begin{bmatrix} .060011 & -.007046 & -.002963 \\ -.007046 & .071305 & -.000346 \\ -.002963 & -.000346 & .073506 \end{bmatrix}$$

The new χ^2 values for individual 1, computed in accordance with Eq. 8.2, are:

$$\chi_{11}{}^2 = [2.4118 \quad -.2235 \quad -3.0235] \times \begin{bmatrix} .055084 & -.015858 & .017442 \\ -.015858 & .059635 & -.018992 \\ .017442 & -.018992 & .110277 \end{bmatrix} \times \begin{bmatrix} 2.4118 \\ -.2235 \\ -3.0235 \end{bmatrix}$$

$$= 1.0686$$

$$\chi_{12}^2 = [-3.5376 \quad 2.8602 \quad -4.1398] \times \begin{bmatrix} .083571 & .006533 & -.020329 \\ .006533 & .048806 & -.002397 \\ -.020329 & -.002397 & .100080 \end{bmatrix} \times \begin{bmatrix} -3.5376 \\ 2.8602 \\ -4.1398 \end{bmatrix} = 2.4896$$

$$\chi_{13}^2 = [-.5758 \quad 8.5455 \quad -7.2424] \times \begin{bmatrix} .060011 & -.007046 & -.002963 \\ -.007046 & .071305 & -.000346 \\ -.002963 & -.000346 & .073506 \end{bmatrix} \times \begin{bmatrix} -.5758 \\ 8.5455 \\ -7.2424 \end{bmatrix} = 9.1699$$

Next, following Eq. 8.3, χ'^2 values are obtained by adding to each χ^2 value the natural logarithm of the determinant of the corresponding dispersion matrix. We find

$$|\mathbf{D}_1| = 3257.0 \qquad |\mathbf{D}_2| = 2605.0 \qquad |\mathbf{D}_3| = 3223.4$$

and

$$\ln|\mathbf{D}_1| = 8.0886 \qquad \ln|\mathbf{D}_2| = 7.8652 \qquad \ln|\mathbf{D}_3| = 8.0782$$

Thus, for individual 1, we obtain

$$\chi'^2_{11} = 1.0686 + 8.0886 = 9.1572$$
$$\chi'^2_{12} = 2.4896 + 7.8652 = 10.3548$$
$$\chi'^2_{13} = 9.1699 + 8.0782 = 17.2481$$

Similar computations for the remaining three individuals yield results and decisions as shown in the following table.

Individual Number (i)	χ'^2_{i1}	χ'^2_{i2}	χ'^2_{i3}	Decision—Assign to
1	*9.1572*	10.3548	17.2481	Group 1
2	14.2854	10.6568	*8.4865*	Group 3
3	12.5058	*10.3660*	12.1401	Group 2
4	18.5103	10.2387	*10.2219*	Group 3

We see that the decisions based on minimum χ'^2 are identical to those based on minimum χ^2, made earlier, so far as these four individuals are concerned.

8.2 TAKING PRIOR PROBABILITIES INTO CONSIDERATION

The preceding classification rules, based on minimum chi-square and the adjusted quantity of Eq. 8.3, do not take into consideration the prior probabilities of group membership—that is, the probability of drawing at random a member of each group from a mixed population of all K groups. Otherwise stated, these procedures assume that the relative frequencies of all groups are equal. Hence, the optimal property of minimum probability of misclassifications holds only when the prior probabilities are in fact all equal. When this is not the case, a further modification on the χ'^2 of Eq. 8.3 becomes necessary in order to take into account the different prior probabilities. (The term "prior" signifies that these are probabilities of group membership before we know an individual's scores on the p predictor variables.)

Let p_k denote the probability that an individual selected at random from a mixed population comprising all K groups is a member of the kth group. Then the appropriate modification of the χ'^2 statistic is given by a constant times the natural logarithm of the multivariate normal density function for group k, multiplied by p_k. That is,

$$\chi''^2_{ik} = \chi'^2_{ik} - 2 \ln p_k \tag{8.4}$$

where χ'^2_{ik} is as defined in Eq. 8.3. Again, the decision rule is to assign the individual to that group for which his χ''^2 value is the smallest.

Although the definition of χ''^2 is straightforward, the estimation of the p_k's often poses a serious problem. It is seldom the case that the K samples at hand are proportional in size to their respective populations. (In fact, the relative sizes of the populations are sometimes undefined, all K of them being theoretically infinite.)

In applications to personnel classification, however, we often have at least a rough idea of what the relative sizes of the various occupational or curricular groups have been in the past, and these may be taken as estimates of p_k. Such a procedure has sometimes been criticized for its tendency to "perpetuate the *status quo*" by keeping traditionally large groups large, and small ones small. This criticism would be justified to the extent that the "traditional" sizes of various vocational and curricular groups have been arbitrarily determined by policy or expediency considerations. No doubt such considerations are a factor in determining the relative sizes of occupational and curricular groups. But there are also certain "natural" limits set by the needs of a society. More physicians are needed, for instance, than biology professors; the number of physicians needed is, in turn, exceeded by the number of electrical engineers employed by our technological society.

Thus, except at times when the structure of society is undergoing a drastic change, there is some justification in using the traditional relative sizes of the various groups as estimates of the prior probabilities of group membership. But what if society *is* undergoing a rapid transition, as from an agrarian to an industrial society? Then surely the relative sizes of, for example, the agricultural expert population and the mechanical engineer population over the past decade will be meaningless in estimating their relative sizes five years hence.

In the above type of situation, the vocational or educational guidance counselor will have to resort to reasonable forecasts of social demands for various occupational categories in the near future. His estimates of prior probabilities of group membership would then be based on such forecasts instead of on the relative sizes of the groups in the recent past.

Let us now return to Eq. 8.4, defining the χ''^2 statistic, to see what the effect of different p_k values is on this statistic, and hence on the classification procedure. Since the term involving p_k is $-2 \ln p_k$, and p_k is a positive number less than 1, the following conclusions may be drawn: (1) the additive component due to p_k is always positive; and (2) the larger the value of p_k, the smaller this additive component. Consequently, if for some two groups j and k, p_k is larger than p_j, then an individual for whom $\chi'^2_{ik} = \chi'^2_{ij}$ will have a *smaller* χ''^2_{ik} than χ''^2_{ij}. Thus, if this individual's χ'^2 value had been smaller for these two groups than for any other group (and hence his assignment to group j or k on the basis of χ'^2 values would have been a matter of toss-up), the effect of the prior probabilities is to break the tie in favor of the group with the larger prior probability, because $\chi''^2_{ik} < \chi''^2_{ij}$.

Although we considered the case when $\chi'^2_{ik} = \chi'^2_{ij}$ above, it is clear that even if $\chi'^2_{ik} > \chi'^2_{ij}$, we may have $\chi''^2_{ik} < \chi''^2_{ij}$ provided p_k is sufficiently larger than p_j. That is, a decision based on χ'^2 values (measuring dissimilarity) may be reversed in favor of a group with a large prior probability of membership (that is, a large group) when χ''^2 values are used as the basis of classification.

We thus see that the role played by prior probabilities is, as it were, to temper our decisions based on resemblance alone with considerations of relative group sizes. Where we might tend to oversupply small groups and undersupply large ones by using resemblance as the sole basis for classification we introduce a corrective effect by taking prior probabilities of group membership into account.

EXAMPLE 8.3. Let us now modify the classification based on χ'^2 values, given in the previous example, by introducing prior-probability considerations. As seen from Eq. 8.4, we have merely to add -2 times the natural logarithm of p_k to each χ'^2_{ik} value in order to get χ''^2_{ik}. For simplicity we use the relative sample sizes $n_k/(n_1 + n_2 + n_3)$ as our estimates of p_k—since this is a fictitious example and we have no other means for estimating the prior probabilities. The values are,

$$p_1 = 85/244 = .34836 \qquad p_2 = 93/244 = .38115$$
$$p_3 = 66/244 = .27049$$

and -2 times their respective natural logarithms give

$$-2 \ln p_k = 2.1090 \qquad 1.9291 \qquad 2.6150$$

for $k = 1, 2, 3$, respectively. We add the appropriate one of these three numbers to each of the three χ'^2_{ik} values already computed for each individual i. Thus, for individual 1,

$$\chi''^2_{11} = 9.1572 + 2.1090 = 11.2662$$
$$\chi''^2_{12} = 10.3548 + 1.9291 = 12.2839$$
$$\chi''^2_{13} = 17.2481 + 2.6150 = 19.8631$$

We similarly obtain the χ''^2 values for individuals 2, 3, and 4. The results and decisions for all four individuals are as shown in the table below.

Individual Number (i)	χ''^2_{i1}	χ''^2_{i2}	χ''^2_{i3}	Decision—Assign to
1	*11.2662*	12.2839	19.8631	Group 1
2	16.3944	12.5859	*11.1015*	Group 3
3	14.6148	*12.2951*	14.7551	Group 2
4	20.6193	*12.1678*	12.8369	Group 2

We note that the decisions are unchanged from those based on the minimum χ'^2 rule, made earlier, for the first three individuals, but individual 4, who was classified as a member of Group 3 previously, is now assigned to Group 2. The larger prior probability of membership in Group 2 (0.38, as against 0.27 for Group 3) has outweighed this individual's greater resemblance (by a narrow margin, to be sure) to the average member of Group 3 than to that of Group 2, as measured by the χ'^2 criterion.

8.3 PROBABILITY OF GROUP MEMBERSHIP

The quantities χ^2 and χ'^2 utilized in Section 8.1 as measures of dissimilarity are closely related to a certain kind of probability, specified below, provided the variables $X_1, X_2, \ldots, X_p$ follow a multivariate normal distribution in each of the K groups. Considering χ'^2 (the more generally applicable of the two statistics), we see from Eq. 8.3 that

$$e^{-\chi_{ik}'^2/2} = |\mathbf{D}_k|^{-1/2} e^{-\chi_{ik}^2/2}$$

and hence

$$(2\pi)^{-p/2} e^{-\chi_{ik}'^2/2} = (2\pi)^{-p/2} |\mathbf{D}_k|^{-1/2} e^{-\chi_{ik}^2/2}$$

which is simply the multivariate normal density function evaluated at the point corresponding to the observed score combination $\mathbf{X}_i$. Therefore, by definition, the quantity

$$(2\pi)^{-p/2} e^{-\chi_{ik}'^2/2} \, dX_1, dX_2, \ldots, dX_p$$

expresses the probability that a randomly drawn member of group k will have a score combination between

$$(X_{1i}, X_{2i}, \ldots, X_{pi})$$

and

$$(X_{1i} + dX_1, X_{2i} + dX_2, \ldots, X_{pi} + dX_p) \tag{8.5}$$

Let us denote this probability by $p(\mathbf{X}_i \mid H_k)$, where H_k stands for the statement: "Individual i is a member of Group k."

The probability alluded to at the outset of this section, and its relationship to χ'^2, may now be stated as follows:

Given that individual i is a randomly selected member of group k, the probability that his score combination lies within the limits displayed in (8.5) is equal to

$$p(\mathbf{X}_i \mid H_k) = (2\pi)^{-p/2} e^{-\chi_{ik}'^2/2} dX_1, dX_2, \ldots, dX_p \tag{8.6}$$

where $\chi_{ik}'^2$ is as defined in Eq. 8.3.

On the other hand, the quantity χ''^2 introduced in the preceding section, taking prior probabilities of group membership into consideration, cannot be related to a conditional probability of the type shown in Eq. 8.6. For the prior probability does not (and cannot) play any role once we confine our attention to a particular group k, as we do in the conditional probability $p(\mathbf{X}_i \mid H_k)$. To take the prior probabilities of group membership into account and still produce a relevant probability after the score combination $\mathbf{X}_i$ has been observed, we have to consider a type of probability which is, as it were, the *inverse* of that displayed in Eq. 8.6: the probability that individual i is a member of group k, given that his score combination is $\mathbf{X}_i$ (or, more precisely, that it lies between the limits stated in 8.5). It will presently be seen that this type of probability, which we denote by $p(H_k|\mathbf{X}_i)$, is functionally related to χ''^2.

In order for $p(H_k \mid \mathbf{X}_i)$ to be definable, we must make one further assumption that was not needed in considering probabilities of the type $p(\mathbf{X}_i \mid H_k)$. This is the requirement that the as yet unclassified individual i must definitely be a member of one or another of the K groups under consideration. That is, the eventuality that he belongs to *none* of these K groups is prohibited[2].

Granted the assumption that the set of statements $H_1, H_2, \ldots, H_k$ exhausts all the possibilities with regard to the group membership of individual i, we may compute his $p(H_k \mid \mathbf{X}_i)$ by means of Bayes' theorem on "inverse probability," or *posterior* probability, as it is more commonly known today. [See, for example, Hays (1963).] The formula is:

$$p(H_k \mid \mathbf{X}_i) = \frac{p_k \cdot p(\mathbf{X}_i \mid H_k)}{\sum_{j=1}^{K} p_j \cdot p(\mathbf{X}_i \mid H_j)} \qquad k = 1, 2, \ldots, K \qquad (8.7)$$

where p_k is the prior probability of membership in Group k, and $p(\mathbf{X}_i \mid H_k)$ is as defined in Eq. 8.6. Substituting from this equation and cancelling the common factors $(2\pi)^{-p/2}$ and $dX_1, dX_2, \ldots, dX_p$ from numerator and denominator, we may write $p(H_k \mid \mathbf{X}_i)$ explicitly in terms of χ'^2_{ik}, as follows:

$$p(H_k \mid \mathbf{X}_i) = \frac{p_k \exp(-\chi'^2_{ik}/2)}{\sum_{j=1}^{K} p_j \exp(-\chi'^2_{ij}/2)}$$

or, further expressing χ'^2 in terms of χ^2 from Eq. 8.3,

$$p(H_k \mid \mathbf{X}_i) = \frac{p_k|\mathbf{D}_k|^{-1/2} \exp(-\chi_{ik}{}^2/2)}{\sum_{j=1}^{K} p_j|\mathbf{D}_j|^{-1/2} \exp(-\chi_{ij}{}^2/2)} \qquad k = 1, 2, \ldots, K \qquad (8.8)$$

[2]In practice, we would be stipulating this requirement even when using probabilities of the type Eq. 8.6 if we make a forced classification of every individual into one or another of the K groups on the basis of the minimum χ^2 (or minimum χ'^2) rule. But there is no logical necessity to do so. We may decide that an individual belongs to none of the K groups if his χ^2 (or χ'^2) values for all of them are fairly large—or, equivalently, if his $p(\mathbf{X}_i \mid H_k)$ is quite small for all k.

Alternatively, we may eliminate χ'^2 in the previous equation in favor of χ''^2 by use of Eq. 8.4, in which case we get

$$p(H_k \mid \mathbf{X}_i) = \frac{\exp(-\chi''^2_{ik}/2)}{\sum_{j=1}^{K} \exp(-\chi''^2_{ij}/2)} \qquad k = 1, 2, \ldots, K \tag{8.8a}$$

We thus see that, whereas $p(\mathbf{X}_i \mid H_k)$ is expressible only in terms of χ^2 or χ'^2, $p(H_k \mid \mathbf{X}_i)$ is related to χ''^2 as well as the two earlier statistics. This is the posterior probability, after observing individual i's score combination $(X_{1i}, X_{2i}, \ldots, X_{pi})$, that he is a member of Group k. For short, we shall refer to this simply as the probability of group membership, omitting the qualifier "posterior" as understood.

We reiterate the distinction between the probability defined in Eq. 8.6 and that in Eq. 8.8, for a clear understanding of this distinction becomes important in the sequel. The former, that is, $p(\mathbf{X}_i \mid H_k)$, represents the proportion of individuals, among members of Group k, who have score combinations in the vicinity of $(X_{1i}, X_{2i}, \ldots, X_{pi})$. The latter, $p(H_k \mid \mathbf{X}_i)$, may be interpreted as the proportion of individuals, among those with the score combination $(X_{1i}, X_{2i}, \ldots, X_{pi})$ or thereabouts, who are members of Group k; here it is assumed that our universal set consists of a mixture of Groups 1, 2, ..., K, and nothing else.

In using the probability of group membership for classification purposes, the decision rule is of course to assign each individual to that group for which his $p(H_k \mid \mathbf{X}_i)$ value is the *largest*. This is a reversal from the rules using χ^2, χ'^2, and χ''^2, in which we sought to minimize the values of these statistics. It should be noted, however, that the numerator of expression 8.8a for $p(H_k \mid \mathbf{X}_i)$ is $\exp(-\chi''^2_{ik}/2)$, which is a monotonically decreasing function of χ''^2_{ik}, and that the denominator does not change with k for any one individual i. Hence, the classification based on maximum $p(H_k \mid \mathbf{X}_i)$ is actually identical with that based on minimum χ''^2. The reader may well wonder why we should bother to compute $p(H_k \mid \mathbf{X}_i)$ at all, if identical decisions are reachable by using χ''^2_{ik} which need to be computed first before getting $p(H_k \mid \mathbf{X}_i)$. The reason will become clear in Section 8.5.

EXAMPLE 8.4. Since the χ''^2 values have already been computed for our four individuals, it is but a short step to getting their $p(H_k \mid \mathbf{X}_i)$ values, which we here abbreviate as $p_{k;i}$ to simplify the notation. Tables of the exponential function usually give e^{-x} as well as e^x as functions of x. If such a table is available, we have only to divide each χ''^2_{ik} by 2 and enter the table with $\chi''^2_{ik}/2$ as argument, making sure to interpolate between tabled values because e^{-x} is quite sensitive to small variations in the argument for small and moderate values of x (= $\chi''^2_{ik}/2$) such as we shall often encounter. If an exponential-function table is not available, we may compute $\exp(-\chi''^2_{ik}/2)$, the basic ingredients of $p_{k;i}$, with the aid of a table of common logarithms, utilizing the following sequence of relations:

$$\begin{aligned}\log[\exp(-\chi''^2_{ik}/2)] &= (-\chi''^2_{ik}/2)\log e \\ &= (-\chi''^2_{ik}/2)(.43429) \\ &= (-.21715)\chi''^2_{ik} = L_{ik} \qquad \text{(say)}\end{aligned}$$

Therefore,

$$\exp(-\chi''^2_{ik}/2) = \text{antilog } L_{ik}$$

That is, we need only to multiply each χ''^2_{ik} value by the constant $-.21715$, and find the number whose common logarithm is equal to this product.

The values of $\exp(-\chi''^2_{ik}/2)$ and their sum (over $k = 1, 2, 3$) for each of the four individuals are as follows:

i	$\exp(-\chi''^2_{i1}/2)$	$\exp(-\chi''^2_{i2}/2)$	$\exp(-\chi''^2_{i3}/2)$	$\sum_k \exp(-\chi''^2_{ik}/2)$
1	.0035777	.0021508	.0000486	.005777
2	.0002755	.0018493	.0038845	.0060093
3	.0006706	.0021388	.0006251	.0034345
4	.0000333	.0022794	.0016312	.0039439

Then, in accordance with Eq. 8.8a, each person's $p_{k;i}$ values are obtained by dividing the first three entries in his row by their sum listed as the rightmost entry. Thus, $p_{1;1} = .0035777/.0057771 = .6193$, and so on. The $p_{k;i}$ values thus computed for the four individuals, and the resulting classificatory decisions, are shown in the table below, where the largest probability value in each row is in italics.

Individual Number (i)	$p_{1;i}$	$p_{2;i}$	$p_{3;i}$	Decision—Assign to
1	*.6193*	.3723	.0084	Group 1
2	.0458	.3077	*.6464*	Group 3
3	.1953	*.6227*	.1820	Group 2
4	.0084	*.5780*	.4136	Group 2

There is, of course, no difference between these decisions and those made earlier on the basis of the minimum χ''^2 rule, for the two decision rules are, as was already mentioned, mathematically equivalent.

8.4 REDUCTION OF DIMENSIONALITY BY DISCRIMINANT ANALYSIS

In the foregoing classification procedures, the relevant statistic—χ^2, χ'^2, χ''^2, or $p(H_k \mid \mathbf{X}_i)$, as the case may be—was computed in terms of a set of p predictor variables. Often the number of predictors is quite large (10 or more, for example) and the computation of these statistics becomes very time consuming, even with the aid of electronic computers, when the number of individuals to be classified is large—as it usually is. It is therefore desirable to reduce the number of variables to be considered—or, in geometric language, to reduce the dimensionality of the space in which we operate.

As discussed extensively by Rulon et al. (1967, Chs. 7–9), and as one might intuitively expect, the technique of multigroup discriminant analysis is the most appropriate means for achieving this reduction of dimensionality (insofar as reduction by linear transformations, that is, rotation of axes, is concerned). In fact, it can be shown (Tatsuoka, 1956) that the results of classification based on the minimum chi-square rule (that is, when we assume equal dispersion matrices in the K populations) are identical regardless of whether we use the original p variables or the $K - 1$ discriminant functions in computing the χ^2 statistic. Thus, if we have 10 predictors and 5 groups, it suffices to compute the χ^2 statistic of Eq. 8.1 from the 4 discriminant-function scores for each person to be classified, instead of from his scores on the original 10 variables. This represents a considerable saving of computational time (even allowing for the fact that the discriminant-function coefficients have first to be computed) if the number of individuals to be classified is large.

When the K population dispersion matrices are not assumed to be equal—that is, when classification is based on χ'^2, χ''^2, or $p(H_k \mid \mathbf{X}_i)$—the identity of classificatory results in the original p-dimensional space and the reduced $(K - 1)$-dimensional space no longer holds. However, experience shows that the classifications in the two spaces yield closely similar results so long as the dispersion matrices are not drastically different. [See Lohnes (1961) and Rulon et al. (1967, p. 317) for empirical evidence of this "robustness" property of discriminant analysis.] Thus, it seems justifiable to use the discriminant functions even when classification is to be based on those statistics that take the separate group dispersions into account.

Apart from computational expediency, however, there is another argument in favor of using discriminant functions rather than the original predictors for computing the decision statistic. This argument, in fact, calls for using only the statistically significant discriminant functions, thus resulting (generally) in an even more drastic reduction of dimensionality than from p to $K - 1$. The idea is that, in using the original p variables (or all $K - 1$ of the discriminant functions), we are doubtless capitalizing to some extent, on chance differences that happen to differentiate the K samples at hand but do not reflect real differences in the corresponding populations. It seems reasonable, then, that by confining our attention to only those discriminant functions that are statistically significant, we would decrease our reliance on apparent differences due to sampling error. If this standpoint is taken, we would insist on using only the significant discriminant functions even when equal population dispersions are assumed. We would do so with the realization that we shall no longer obtain classificatory results identical with those using the original variables, as we could, in this case, get by using all $K - 1$ discriminant functions in computing the χ^2 statistic if our only purpose were that of computational expediency.

The use of discriminant functions, whether we choose to retain all $K - 1$ or only the statistically significant ones, does not entail any essential changes in the classification procedures discussed in the preceding sections. We have merely to replace, in Eqs. 8.1 to 8.8a, the original set of variables $(X_1, X_2, \ldots, X_p)$ and their dispersion matrices by the discriminant functions $Y_1, Y_2, \ldots, Y_r$ and the dispersion matrices for these variables. No new formulas need to be given.

EXAMPLE 8.5. The discriminant functions appropriate to the numerical example we have been carrying in this chapter were already computed in Chapter Six (see p. 166). The normalized coefficients for the two functions (both of which were statistically significant at the .001 level) were there found to be given by the columns of the matrix

$$\mathbf{V} = \begin{bmatrix} .3524 & .9145 \\ -.7331 & .1960 \\ .5818 & -.3540 \end{bmatrix}$$

The two discriminant-function scores Y_1 and Y_2 for each individual are computed in accordance with Eq. 6.1, or, in matrix notation,

$$\mathbf{Y}_i' = \mathbf{X}_i'\mathbf{V}$$

This equation is, of course, used also for computing the group means on the two discriminant functions. The results (given only to two decimal places in Chapter Six) are here shown to four decimals to prevent excessive rounding error in the subsequent series of calculations:

Group 1		Group 2		Group 3	
$\bar{Y}_1$	$\bar{Y}_2$	$\bar{Y}_1$	$\bar{Y}_2$	$\bar{Y}_1$	$\bar{Y}_2$
−8.0723	13.0654	−3.0656	17.5065	1.8636	12.5853

The four individuals' discriminant-function scores, expressed as deviations from the respective group means, are as follows:

i	$\mathbf{y}_{i1}$		$\mathbf{y}_{i2}$		$\mathbf{y}_{i3}$	
1	−.7453	3.2321	−5.7520	−1.2090	−10.6812	3.7122
2	8.0285	−.1059	3.0218	−4.5470	−1.9074	.3742
3	4.9946	1.1186	−.0121	−3.3225	−4.9413	1.5987
4	8.8329	2.9201	3.8262	−1.5210	−1.1030	3.4002

Next, the group dispersion matrices are each computed from an equation typified by 6.2, that is,

$$\mathbf{D}_k(y) = \mathbf{V}'\mathbf{D}_k\mathbf{V}$$

where $\mathbf{D}_k$ is the dispersion matrix of the original predictors for group k. The dispersion matrices, their determinants, and their inverses are as shown below.

With the foregoing data and the prior probabilities of group membership (p_k) given earlier, we can compute successively, χ^2, χ'^2, χ''^2, and $p(H_k \mid \mathbf{Y}_i)$ for each individual with respect to each group. Since the calculations are exactly the same as those shown in the corresponding

k	$\mathbf{D}_k(y)$	$\lvert\mathbf{D}_k(y)\rvert$	$\mathbf{D}_k(y)^{-1}$
1	$\begin{bmatrix} 10.4488 & -1.0316 \\ -1.0316 & 21.7570 \end{bmatrix}$	226.27	$\begin{bmatrix} .096155 & .004559 \\ .004559 & .046178 \end{bmatrix}$
2	$\begin{bmatrix} 17.9734 & 1.0041 \\ 1.0041 & 10.5044 \end{bmatrix}$	187.79	$\begin{bmatrix} .055936 & -.005347 \\ -.005347 & .095709 \end{bmatrix}$
3	$\begin{bmatrix} 13.6425 & -.0772 \\ -.0772 & 16.5172 \end{bmatrix}$	225.33	$\begin{bmatrix} .073302 & .000343 \\ .000343 & .060545 \end{bmatrix}$

numerical examples using the original predictor scores, we illustrate the details only for the first individual with respect to the first group.

$$\chi_{11}{}^2 = [-.7453 \quad 3.2321] \begin{bmatrix} .096155 & .004559 \\ .004559 & .046178 \end{bmatrix} \begin{bmatrix} -.7453 \\ 3.2321 \end{bmatrix}$$

$$= .5140 \qquad \text{(Eq. 8.2)}$$

$$\chi'^2_{11} = \chi_{11}{}^2 + \ln \lvert\mathbf{D}_k(y)\rvert = .5140 + 5.4217 = 5.9357 \qquad \text{(Eq. 8.3)}$$

$$\chi''^2_{11} = \chi'^2_{11} - 2 \ln p_k = 5.9357 + 2.1090 = 8.0447 \qquad \text{(Eq. 8.4)}$$

$$p(H_1 \mid \mathbf{Y}_1) = \frac{\exp(-\chi''^2_{11}/2)}{\sum_{j=1}^{3} \exp(-\chi''^2_{1j}/2)} = \frac{.017910}{.028764}$$

$$= .6227 \qquad \text{(Eq. 8.8a)}$$

The results for the four individuals are as shown below.

Individual Number (i)	χ^2_{i1}	χ^2_{i2}	χ^2_{i3}	Decision—Assign to
1	.5140	1.9164	9.1703	—
2	6.1906	2.6362	.2747	—
3	2.5076	1.0560	1.9393	—
4	8.1306	1.1027	.7867	—

Individual Number (i)	χ'^2_{i1}	χ'^2_{i2}	χ'^2_{i3}	Decision—Assign to
1	*5.9357*	7.1517	14.5879	Group 1
2	11.6123	7.8715	*5.6923*	Group 3
3	7.9293	*6.2913*	7.3569	Group 2
4	13.5523	6.3380	*6.2043*	Group 3
	χ''^2_{i1}	χ''^2_{i2}	χ''^2_{i3}	
1	*8.0447*	9.0808	17.2029	Group 1
2	13.7213	9.8006	*8.3073*	Group 3
3	10.0383	*8.2204*	9.9719	Group 2
4	15.6613	*8.2671*	8.8193	Group 2
	$p_{1;i}$	$p_{2;i}$	$p_{3;i}$	
1	*.6227*	.3710	.0064	Group 1
2	.0433	.3076	*.6490*	Group 3
3	.2215	*.5496*	.2289	Group 2
4	.0139	*.5607*	.4254	Group 2

No decisions are made on the basis of the χ^2_{ik} values, since these are not terminal statistics, but are intermediate to obtaining the χ'^2_{ik} values, having been computed in accordance with Eq. 8.2. As may be ascertained by comparing with previously given results, the decisions based on χ'^2, χ''^2, and $p(H_k \mid \mathbf{Y}_i)$ are respectively identical with those made of the basis of the corresponding quantities computed from the original variables. In particular, the last-mentioned criterion—probability of group membership—not only yields identical decisions but also shows fairly similar numerical values to the $p(H_k \mid \mathbf{X}_i)$ obtained before. This is another instance of verification of the empirical generalization stated earlier, that classificatory results in the original predictor space and the reduced discriminant space tend to be closely similar even when the separate group dispersion matrices are used, provided they are not drastically different.

8.5 JOINT PROBABILITY OF GROUP MEMBERSHIP AND SUCCESS

We now turn to the situation in which considerations of success or productivity in a group (occupational or educational) are of concern besides sheer "belongingness" in the group. This is not a pure classification problem as was defined at the outset of this chapter and was subsequently broadened to take prior probabilities of group membership into account besides resemblance. But it is a natural extension of the classification problem as far as applications to vocational and educational guidance are concerned. Just as it is unreasonable to counsel an individual to enter some occupation on the sole basis of his "resemblance" to current members of that group without regard to societal demands (which lead to different group sizes), it would likewise be remiss to ignore information on the possible degree of success he might enjoy in various occupations, if such information is available.

The customary method for seeking to predict a person's possible degree of success in various occupations has been to use a series of multiple regression equations of some measure of success on a set of predictor variables, one equation having been constructed for each occupational group. The counselee is then advised to consider entering that group for which his predicted degree of success is the highest.[3] But this approach obviously suffers from the fallacy of using regression equations that may not be valid for a given individual. It must be remembered that each regression equation has been constructed on a sample from a particular occupational (or curricular) group, and hence cannot be validly used for anyone who is not a member of that group. In other words, we cannot take at face value the predicted degrees of success estimated from the several regression equations—one for each group—for an individual of unknown group membership. In fact, only one of these equations should, strictly speaking, yield a valid prediction for any one individual—namely, the equation for the particular group of which he is indeed a member.[4]

A possible approach to overcoming the above difficulty lies in utilizing the joint probability of membership and success in a group (Tatsuoka, 1956). In this method, we relinquish the idea of predicting the possible *degree* of

[3]This, of course, is an oversimplified description of the process of vocational or educational guidance. In practice, other considerations besides possible degree of success are no doubt taken into account by the counselor. But it seems that such weighting by other factors has been done largely on an intuitive basis, so that the adequacy of the outcome depends heavily on the expertness of the particular counselor.

[4]Even if we grant that the several occupations or curricula being considered by an individual would often be sufficiently similar to render the predictions from all of the relevant regression equations at least approximately valid, there are other difficulties with using these predictions as the sole bases for counseling. For a discussion of this issue, see Rulon et al. (1967, pp. 323–336).

success as such, and seek instead to estimate an individual's chances of "succeeding" at all, in the sense of exceeding some specified score on the criterion variable. The reason for adopting this more modest goal will presently become clear.

We have already seen, in Section 8.3, how we may compute the probability $p(H_k \mid \mathbf{X}_i)$ that an individual with a predictor-score combination $\mathbf{X}_i$ is a member of Group k. To obtain the joint probability of membership and success, we now need to compute the probability that a member of Group k who has this predictor-score combination $\mathbf{X}_i$ will be "successful"—that is, that he will have a criterion score exceeding an acceptable cut-off point.

Let U_k be a criterion variable appropriate for measuring success in Group k. This may be a measure of productivity, a supervisor rating of efficiency on the job, annual income of a self-employed person, grade-point average in college, or whatever else may be reasonable and available for assessing the degree of success of a member of the kth vocational or curricular group. (We need not have the same measure for all K groups.) Suppose that we have constructed, in the usual manner, a multiple regression equation of U_k on $X_1, X_2, \ldots, X_p$ (the same predictor variables as before) for each group. These equations may be symbolized by

$$\tilde{U}_k = a_k + b_{1k}X_1 + b_{2k}X_2 + \cdots + b_{pk}X_p \qquad (k = 1, 2, \ldots, K)$$

In the course of constructing these equations, we will have computed, among other things, the sum-of-squares $\sum u_k^2$ of the criterion variable (in deviation scores) for the kth sample, and the coefficient of multiple correlation R_{uk} between U_k and $X_1, X_2, \ldots, X_p$. From these, we compute the standard errors of estimate

$$s_{u.x(k)} = \left[\frac{\sum u_k^2}{n_k - 2}(1 - R_{uk}^2)\right]^{1/2} \qquad k = 1, 2, \ldots, K \tag{8.9}$$

Next, for each group, we must decide on a suitable cut-off point U_k^* such that a member of Group k will be called "successful" if and only if his criterion score exceeds this value. Then, with the assumption that the predictors and criterion jointly follow a $(p + 1)$-variate normal distribution in each group, we can estimate the probability that a member of Group k who has a predictor-score combination $\mathbf{X}_i = (X_{1i}, X_{2i}, \ldots, X_{pi})$ will be "successful." The formula is as follows:

$$p(U_{ki} > U_k^* \mid H_k \;\&\; \mathbf{X}_i) = (2\pi)^{-1/2} \int_{z_{ki}^*}^{\infty} \exp(-z^2/2)\, dz \tag{8.10}$$

with

$$z_{ki}^* = \frac{U_k^* - \tilde{U}_k(\mathbf{X}_i)}{s_{u.x(k)}} \tag{8.11}$$

where $\tilde{U}_k(\mathbf{X}_i)$ is the value of U_k predicted from the regression equation for the predictor-score combination $\mathbf{X}_i$. The presence of an integral sign in Eq. 8.10 need not discourage readers who are unfamiliar with the calculus. It is merely a convenient way of indicating that we are to use a table of normal-curve areas to find the area to the right of the point $z = z_{ki}^*$ on the abscissa. So the only actual calculations needed are those involved in determining z_{ki}^* for the given predictor-score combination.

It should be noted that the probability given by Eq. 8.10 is conditioned just as much on the premise that the individual to whom it applies is a member of group k as it is on the fact that his predictor-score combination is $\mathbf{X}_i$. We are not really using the kth group's regression equation for a particular individual i whose group membership is unknown. Rather, we are estimating the proportion, *among Group k members who have a score combination in the vicinity of* $\mathbf{X}_i$, of those who may be expected to have a criterion score exceeding U_k^*.

Now that we have seen how, given a particular predictor-score combination, we may compute for each k $(= 1, 2, \ldots, K)$ the probability of membership in Group k, and the probability of "success" in that group *given* membership therein, it remains only to apply the general multiplication theorem of probability theory in order to obtain the desired probability of the joint event: membership *and* success in a group, given a particular predictor-score combination.

The general form of the multiplication theorem is:

$$p(A \,\&\, B \mid C) = p(A \mid C)p(B \mid A \,\&\, C)$$

In the context of our problem, we let A stand for the event of membership in Group k (that is, the occurrence of H_k), B for "success" in that group (that is, $U_{ki} > U_k^*$), and C for the observation of a particular predictor-score combination $\mathbf{X}_i$. Thus, the desired joint probability may be computed as

$$p(H_k \,\&\, U_{ki} > U_k^* \mid \mathbf{X}_i) = p(H_k \mid \mathbf{X}_i)p(U_{ki} > U_k^* \mid H_k \,\&\, \mathbf{X}_i) \tag{8.12}$$

where the two factors on the right are given by Eq. 8.8 and 8.10, respectively.

Once these probabilities have been computed for each individual to be counseled, the counselor would advise each person to consider entering that group for which his probability of membership and success is the largest. An evaluation of the soundness of this procedure by cross-validation methods admittedly poses certain problems, which it would take us too far afield to discuss here. Suggestions for coping with these problems are discussed in Tatsuoka (1956) and Rulon et al. (1967, Ch. 10).

EXAMPLE 8.6.[5] To illustrate the calculations for obtaining the joint probability of membership and success, we now suppose that scores on a suitable measure U_k of success were available in addition to the APQ scores for all members of each of the three normative samples. We would then be able to construct a multiple regression equation of U_k on X_1, X_2, and X_3 for each group. Suppose that the equations were as follows:

$$\tilde{U}_1 = 1.0603 - .0799X_1 + .1691X_2 + .1443X_3$$
$$\tilde{U}_2 = -.7587 + .2506X_1 + .1232X_2 - .0781X_3$$
$$\tilde{U}_3 = 2.3426 + .1643X_1 - .1440X_2 + .1942X_3$$

The standard errors of estimate associated with these equations, calculated in accordance with Eq. 8.9, were as follows:

$$s_{u.x(1)} = 1.6849 \qquad s_{u.x(2)} = 1.5557$$

and

$$s_{u.x(3)} = 1.5968$$

The predicted success score for a member of each group having a predictor-score combination equal to that of each of our four individuals is then computed from the relevant regression equation. The results are as follows:

i	$\mathbf{X}_i$	$\tilde{U}_1$	$\tilde{U}_2$	$\tilde{U}_3$
1	[15, 24, 6]	4.7860	5.4885	2.5163
2	[15, 16, 11]	4.1547	4.1124	4.6393
3	[16, 23, 14]	5.6914	4.9911	4.3782
4	[19, 20, 15]	5.0887	5.2952	5.4973

(Note that these are not predictions for our four individuals themselves, but for members of the relevant groups with the stated predictor-score combinations, which are equal to those of our four individuals.) The next step is to

[5]The author is indebted to Mr. James M. Kraatz of the University of Illinois Computer-Based Education Research Laboratory for writing a computer program and executing the calculations for this example, including the generation of suitable criterion scores. Details of the calculation of the regression equations and the standard errors of estimate may be found in Rulon et al. (1967, Ch. 10), where this example originally appeared.

decide on the cut-off point U_k^* for "success" in each group. Although the values may, in general, differ from group to group, we here take $U_k^* = 4.5$ for all three groups. We can now determine the z_{ki}^* values in accordance with Eq. 8.11. Thus, for $i = 1$,

$$z_{11}^* = [4.5 - \tilde{U}_1(\mathbf{X}_1)]/1.6849 = (4.5 - 4.7860)/1.6849 = -.1697$$

$$z_{21}^* = [4.5 - \tilde{U}_2(\mathbf{X}_1)]/1.5557 = (4.5 - 5.4885)/1.5557 = -.6354$$

$$z_{31}^* = [4.5 - \tilde{U}_3(\mathbf{X}_1)]/1.5968 = (4.5 - 2.5163)/1.5968 = 1.2423$$

Collecting the results for all four predictor-score combinations $\mathbf{X}_i$ in a single table, we have the following:

i	z_{1i}^*	z_{2i}^*	z_{3i}^*
1	−.1697	−.6354	1.2423
2	.2049	.2491	−.0872
3	−.7071	−.3157	.0763
4	−.3494	−.5112	−.6246

Next, we determine $p(U_{ki} > U_k^* \mid H_k \;\&\; \mathbf{X}_i)$ in accordance with Eq. 8.10 by entering a table of normal-curve areas with each z_{ki}^* value and finding the area to the right of that abscissa point. The results are:

	$p(U_{ki} > U_k^* \mid H_k \;\&\; \mathbf{X}_i)$		
i \ k	1	2	3
1	.5674	.7374	.1071
2	.4188	.4017	.5348
3	.7602	.6239	.4696
4	.6366	.6954	.7339

The final step, following Eq. 8.12, is to multiply each of these probabilities of success (given membership in Group k and the predictor-score combination) by the corresponding probability of group membership, $p(H_k \mid \mathbf{X}_i)$, obtained previously from Eq. 8.8a. The resulting products are the joint probabilities of membership and success in the various groups, given predictor-score combination $\mathbf{X}_i$. For this example, the values of the joint probability, and the "classification" decisions based thereon, are as follows:

$p(H_k \mathbin{\&} U_{ki} > U_k^* \mid \mathbf{X}_i)$

i \ k	1	2	3	Decision—Assign to
1	*.3514*	.2745	.0009	Group 1
2	.0192	.1236	*.3457*	Group 3
3	.1485	*.3885*	.0855	Group 2
4	.0053	*.4019*	.3035	Group 2

We see that the decisions for these four individuals would be exactly the same as those based on the probability of group membership alone. This, of course, need not be true in general.

EXERCISE

Using each of the methods discussed in this chapter, arrive at classification decision for members of the same job-applicant group as that in the numerical examples, whose APQ scores are as follows:

	$\mathbf{X}_i$
Individual # 5:	[17, 19, 15]
Individual # 6:	[16, 20, 11]

Appendix A

Determinants

A.1 DEFINITION AND BASIC COMPUTATIONAL RULES

A determinant of order n is a *number* associated with a square matrix of order n, computed as a function of the latter's elements (which are also called the elements of the determinant) in accordance with the rules given below.

Given an $n \times n$ matrix,

$$\mathbf{A} = \begin{bmatrix} a_{11} & a_{12} & \cdots & a_{1n} \\ a_{21} & a_{22} & \cdots & a_{2n} \\ \vdots & \vdots & & \vdots \\ a_{n1} & a_{n2} & \cdots & a_{nn} \end{bmatrix}$$

Then the determinant of $\mathbf{A}$, symbolized by $|\mathbf{A}|$ or det($\mathbf{A}$), and written in full as

$$\begin{vmatrix} a_{11} & a_{12} & \cdots & a_{1n} \\ a_{21} & a_{22} & \cdots & a_{2n} \\ \vdots & \vdots & & \vdots \\ a_{n1} & a_{n2} & \cdots & a_{nn} \end{vmatrix}$$

is computed as follows:

Step 1. Form all possible products of n factors such that each factor is an element of $\mathbf{A}$, no two of which are from the same row or same column of $\mathbf{A}$. There are $n!$ such products possible. For example, when $n = 3$ the required products are the following six ($= 3!$) quantities:

$$a_{11}a_{22}a_{33} \quad a_{12}a_{23}a_{31} \quad a_{13}a_{21}a_{32} \quad a_{13}a_{22}a_{31} \quad a_{11}a_{23}a_{32} \quad a_{12}a_{21}a_{33}$$

(The requirement that no two factors of a given product be elements of the same row or same column of $\mathbf{A}$ is reflected in the fact that each of the numerals 1, 2, and 3 occurs once and only once as a row subscript, and also once and only once as a column subscript, among the elements that form each product.)

Step 2. Within each product, arrange the factors so that the row subscripts are in natural order $(1, 2, \ldots, n)$. (Each of the six products in the above example with $n = 3$ has already been written in this manner.) Then examine the order in which the column subscripts appear. Specifically, note how many times a larger number precedes a smaller one in the sequence of column subscripts in each product. This number is called the *number of inversions* (hereafter abbreviated as NI) associated with each product. Thus, for the six products for $n = 3$ in the order listed above, the NI's are

$$0, 2, 2, 3, 1 \quad \text{and} \quad 1$$

respectively. (Note that NI is the number of *instances* of larger numbers preceding smaller ones, and *not* the number of larger *numbers* that precede smaller ones. Thus, the NI for $a_{13}a_{21}a_{32}$ is 2, because *3* precedes a smaller number in two instances —preceding *1* and preceding *2*—even though *3* is the only number that precedes a smaller one.)

Step 3. Upon determining the NI's for all $n!$ products, multiply each product with an odd-numbered NI by -1 (that is, change its sign); leave as is all the products with even-numbered NI's (counting 0 as even). Thus, among the six products listed above for $n = 3$, the first three (with NI $= 0, 2, 2$, respectively) will be unaltered, while the last three (with NI $= 3, 1, 1$) will have their signs reversed. (For any n, exactly one-half of the $n!$ products will have their signs reversed).

Step 4. Form the algebraic sum of the products as partially modified in sign in Step 3. This sum is the value of $|\mathbf{A}|$. Thus, for $n = 3$. we have:

$$|\mathbf{A}| = \begin{vmatrix} a_{11} & a_{12} & a_{13} \\ a_{21} & a_{22} & a_{23} \\ a_{31} & a_{32} & a_{33} \end{vmatrix} = \begin{matrix} a_{11}a_{22}a_{33} + a_{12}a_{23}a_{31} + a_{13}a_{21}a_{32} \\ - a_{13}a_{22}a_{31} - a_{11}a_{23}a_{32} - a_{12}a_{21}a_{33} \end{matrix} \tag{A.1}$$

EXAMPLE A.1. Find the value of the following determinant:

$$\begin{vmatrix} 3 & 1 & -2 \\ 2 & 0 & 3 \\ -4 & 2 & 5 \end{vmatrix}$$

Substituting in Eq. A.1, we find

$$\begin{vmatrix} 3 & 1 & -2 \\ 2 & 0 & 3 \\ -4 & 2 & 5 \end{vmatrix} = \begin{matrix} (3)(0)(5) + (1)(3)(-4) + (-2)(2)(2) \\ - (-2)(0)(-4) - (3)(3)(2) - (1)(2)(5) \end{matrix}$$

$$= 0 - 12 - 8 - 0 - 18 - 10 = -48$$

A.2 SPECIAL ROUTINE FOR THIRD-ORDER DETERMINANTS: SARRUS' RULE

For a third-order determinant, the application of Steps 1–4 above may be mechanized as follows:

1. To the right of the determinant, copy the first two columns over again:

$$\begin{vmatrix} a_{11} & a_{12} & a_{13} \\ a_{21} & a_{22} & a_{23} \\ a_{31} & a_{32} & a_{33} \end{vmatrix} \begin{matrix} a_{11} & a_{12} \\ a_{21} & a_{22} \\ a_{31} & a_{32} \end{matrix}$$

2. In the extended three-row, five-column configuration constructed above, connect, by a solid line, each of the three first-row elements of $|\mathbf{A}|$ with the two numbers located "southeast" of it. Similarly, connect by a dotted line, each of the three third-row elements of $|\mathbf{A}|$ with the two numbers located "northeast" of it. Thus:

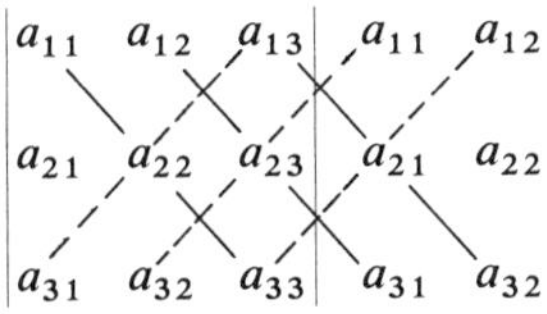

3. Form the product of each triplet of elements connected by a solid line, and prefix the product with a plus sign. Form the product of each triplet of elements connected by a dotted line, and prefix the product with a minus sign:

$$\begin{aligned} &+(a_{11}a_{22}a_{33}) + (a_{12}a_{23}a_{31}) + (a_{13}a_{21}a_{32}) \\ &-(a_{13}a_{22}a_{31}) - (a_{11}a_{23}a_{32}) - (a_{12}a_{21}a_{33}) \end{aligned}$$

The resulting algebraic sum gives the value of $|\mathbf{A}|$.

EXERCISES

Find the value of each of the following determinants. In the case of third-order determinants, evaluate both by explicit use of the definitional rules (Steps 1–4, above), and by Sarrus' rule.

(*a*) $\begin{vmatrix} 3 & 5 \\ 2 & 4 \end{vmatrix}$ (*b*) $\begin{vmatrix} 2 & -3 \\ 1 & 4 \end{vmatrix}$ (*c*) $\begin{vmatrix} 1 & 2 & 3 \\ 3 & 1 & 2 \\ 2 & 3 & 1 \end{vmatrix}$ (*d*) $\begin{vmatrix} 1 & 0 & 0 \\ 3 & 1 & 2 \\ 2 & 3 & 1 \end{vmatrix}$

(*e*) $\begin{vmatrix} 0 & 2 & 0 \\ 3 & 1 & 2 \\ 2 & 3 & 1 \end{vmatrix}$ (*f*) $\begin{vmatrix} 0 & 0 & 3 \\ 3 & 1 & 2 \\ 2 & 3 & 1 \end{vmatrix}$

(*g*) $\begin{vmatrix} 2 & -1 & 3 \\ 3 & 4 & -2 \\ 5 & 3 & 1 \end{vmatrix}$ (*h*) $\begin{vmatrix} 1 & 2 & 0 & -1 \\ 0 & 3 & 1 & 2 \\ -2 & 1 & 4 & 0 \\ 3 & 0 & -1 & 1 \end{vmatrix}$

[Answers: (*a*) 2, (*b*) 11, (*c*) 18, (*d*) −5, (*e*) 2, (*f*) 21, (*g*) 0, (*h*) 72.]

A.3 EXPANSION OF A DETERMINANT IN TERMS OF COFACTORS

Evaluating a determinant of order higher than three is exceedingly tedious, as the reader will have found in doing Exercise (*h*) above. Fortunately, the task can be somewhat simplified by using a derived rule: expanding in terms of cofactors. This rule can be developed, for $n = 3$, by algebraic manipulation of the expression on the right-hand side of Eq. A.1. The extensions to cases when $n > 3$ will be evident.

First rearrange the terms of the expression in question by placing next to each other those products that involve the same first-row element of $|\mathbf{A}|$, thus:

$$|\mathbf{A}| = a_{11}a_{22}a_{33} - a_{11}a_{23}a_{32} + a_{12}a_{23}a_{31} - a_{12}a_{21}a_{33} + a_{13}a_{21}a_{32} - a_{13}a_{22}a_{31}$$

Next, from each adjacent pair of terms we factor out the common first-row element:

$$|\mathbf{A}| = a_{11}(a_{22}a_{33} - a_{23}a_{32}) + a_{12}(a_{23}a_{31} - a_{21}a_{33}) + a_{13}(a_{21}a_{32} - a_{22}a_{31})$$

We then rewrite the expressions in parentheses in the form of determinants, thus:

$$|\mathbf{A}| = a_{11}\begin{vmatrix} a_{22} & a_{23} \\ a_{32} & a_{33} \end{vmatrix} + a_{12}\begin{vmatrix} a_{23} & a_{21} \\ a_{33} & a_{31} \end{vmatrix} + a_{13}\begin{vmatrix} a_{21} & a_{22} \\ a_{31} & a_{32} \end{vmatrix} \tag{A.2}$$

From the definition of the products constituting the terms in evaluating a determinant (Step 1 on p. 243), it is evident that none of the second-order determinants in (A.2) contains elements from the row or column of which the associated multiplier (a_{11}, a_{12}, and a_{13}, respectively) is an element. Thus, the first determinant (whose multiplier is a_{11}) comprises those elements of $|\mathbf{A}|$ that remain after deleting the first row and first column of $|\mathbf{A}|$; the second determinant (with multiplier a_{12}) contains elements that remain on deleting the first row and second column of $|\mathbf{A}|$; and the third (with multiplier a_{13}), those elements that remain after the first row and third column have been deleted.

Next, we pay attention not only to the elements occurring in the three second-order determinants in (A.2), but to the configurations in which they occur. We find that in the first and third determinants, the configuration is just as it would be if we were actually to cross out the appropriate row and column of $|\mathbf{A}|$ and to collect the remaining elements as they stood. In the second determinant, however, the columns are *reversed* from the order in which they occur in $|\mathbf{A}|$. But this anomaly may easily be removed by noting that

$$\begin{vmatrix} a_{23} & a_{21} \\ a_{33} & a_{31} \end{vmatrix} = -\begin{vmatrix} a_{21} & a_{23} \\ a_{31} & a_{33} \end{vmatrix}$$

Making this substitution, the expansion (A.2) of $|\mathbf{A}|$ becomes:

$$|\mathbf{A}| = a_{11}\begin{vmatrix} a_{22} & a_{23} \\ a_{32} & a_{33} \end{vmatrix} - a_{12}\begin{vmatrix} a_{21} & a_{23} \\ a_{31} & a_{33} \end{vmatrix} + a_{13}\begin{vmatrix} a_{21} & a_{22} \\ a_{31} & a_{32} \end{vmatrix} \tag{A.3}$$

The determinant of order $n - 1$ obtained by crossing out the ith row and jth column of an nth order determinant $|\mathbf{A}|$ is known as the *minor determinant* associated with the element a_{ij} of $|\mathbf{A}|$. We will denote this as M_{ij}. Then, the expansion (A.3) of $|\mathbf{A}|$ may be written as

$$|\mathbf{A}| = a_{11}M_{11} - a_{12}M_{12} + a_{13}M_{13} \tag{A.4}$$

It can be shown, by appropriate regroupings of the right-hand side of Eq. A.1, that expansions of $|\mathbf{A}|$ may be made in terms of the minors associated with elements in any row or any column. Thus, for example,

$$|\mathbf{A}| = -a_{21}M_{21} + a_{22}M_{22} - a_{23}M_{23}$$

(where the minors are those associated with the second-row elements) and

$$|\mathbf{A}| = a_{13}M_{13} - a_{23}M_{23} + a_{33}M_{33}$$

(using the third-column minors) are equally qualified with (A.4) as expansions of $|\mathbf{A}|$; they all yield the same numerical value. The rule for attaching a plus or minus sign to each term in any expansion is as follows:

If the sum of the row and column subscripts of the elements appearing in a term is an even number, attach a plus sign; if this sum is an odd number, attach a minus sign.

Symbolically, the sign attached to the term $a_{ij}M_{ij}$ in any expansion is equal to that of $(-1)^{i+j}$. Consequently, we may incorporate $(-1)^{i+j}$ as part of each term, and precede every such augmented term with a plus sign. Thus,

$$\begin{aligned} |\mathbf{A}| &= (-1)^2 a_{11}M_{11} + (-1)^3 a_{12}M_{12} + (-1)^4 a_{13}M_{13} \\ &= (-1)^3 a_{21}M_{21} + (-1)^4 a_{22}M_{22} + (-1)^5 a_{23}M_{23} \\ &= (-1)^4 a_{13}M_{13} + (-1)^5 a_{23}M_{23} + (-1)^6 a_{33}M_{33} \end{aligned}$$

Alternatively, we may let the minors themselves "absorb" the sign of $(-1)^{i+j}$ and denote such "signed minors" by a symbol different from M_{ij}. This, in fact, is how the *cofactors* A_{ij} (introduced on p. 23) are defined. That is,

$$A_{ij} = (-1)^{i+j}M_{ij}$$

is called the cofactor (in $\mathbf{A}$ or $|\mathbf{A}|$) associated with the element a_{ij}. In words, the cofactor of a_{ij} is determined as follows:

Cross out the ith row and jth column of $|\mathbf{A}|$, and write the remaining elements, in their original configuration, as a determinant of one smaller order than $|\mathbf{A}|$. If $i + j$ is even, this determinant itself is A_{ij}; if $i + j$ is odd, then this determinant multiplied by -1 is A_{ij}.

Thus, for example, for the fourth-order determinant in Exercise (*h*),

$$A_{31} = \begin{vmatrix} 2 & 0 & -1 \\ 3 & 1 & 2 \\ 0 & -1 & 1 \end{vmatrix} \quad \text{while} \quad A_{32} = (-1)\begin{vmatrix} 1 & 0 & -1 \\ 0 & 1 & 2 \\ 3 & -1 & 1 \end{vmatrix}$$

EXERCISE

Determine A_{33} and A_{34} for the above example, and evaluate the fourth-order determinant by expansion along the third row.

To summarize, the method of expanding a determinant in terms of the cofactors of its elements along some row or some column, converts the task of evaluating a determinant of order n into one of evaluating n determinants of order $n - 1$, and forming a linear combination of these n determinants. The general rule may be stated as:

$$|\mathbf{A}| = \sum_{j=1}^{n} a_{ij}A_{ij} \qquad \text{(expansion along } i\text{th row)}$$

$$= \sum_{i=1}^{n} a_{ij}A_{ij} \qquad \text{(expansion along } j\text{th column)} \qquad \text{(A.5)}$$

It will be noted that the method described above does not, in general, offer much actual computational saving from the definitional rules except when $n = 4$ (in which case Sarrus' rule for third-order determinants may be used to facilitate computation of the four cofactors needed for an expansion). What it does do for any n, however, is to systematize the computations so that the task of explicitly writing out all $n!$ products required in Step 1 of the definitional rules can be averted.

It should also be evident that a genuine computational saving is effected if the determinant to be evaluated happens to contain one or more zero elements in any row or column. We would then expand it along that row or column which has the largest number of zeros. (In Exercise (*h*), each row and column contained one zero; so it is immaterial which row or column we choose to expand along. Using the third row, as suggested above, it was really unnecessary to determine A_{34} in order to evaluate $|\mathbf{A}|$; A_{34} does not contribute to the expansion because it is multiplied by $a_{34} = 0$.)

A.4 SELECTED PROPERTIES OF DETERMINANTS

As we pointed out above, expansion of a determinant in terms of cofactors represents a real computational saving if there are zero elements in some row or column. Actually, it is always possible to transform a determinant into an equivalent one having at most one nonzero element in some row or column. (*All* elements can be made to vanish, in some row or column, if and only if the value of the determinant is zero.) We shall not describe such transformations in detail, since they are cumbersome to apply to determinants that are likely to be encountered in real problems in multivariate statistical analysis because the elements of such determinants usually are numbers involving many digits. However, some of the properties of determinants on which the transformations are based are important to know for other than computational purposes. We state these properties below without complete proofs, indicating only how they may be proved. In some of the examples given, the determinant of Exercise (*c*), p. 245, is referred to as $|\mathbf{C}|$.

> *Property 1.* If two rows (or two columns) of a determinant are interchanged, the absolute value of the determinant remains unchanged, but the sign is reversed.

(Basis for proof: Any such interchange will reverse the oddness or evenness of the NI associated with each product in the definition of a determinant.)

EXAMPLE A.2. The determinant obtained by interchanging the second and third rows of $|\mathbf{C}|$ is as follows:

$$\begin{vmatrix} 1 & 2 & 3 \\ 2 & 3 & 1 \\ 3 & 1 & 2 \end{vmatrix}$$

Verify that the value of this determinant is -18. (See p. 245 that $|\mathbf{C}| = 18$.)

Property 2. If two rows (or two columns) of a determinant are identical, the value of the determinant is zero.

(Outline of proof: If we interchange the two identical rows (or columns), no change is made in the determinant. But, according to Property 1, the sign should be reversed. The only number for which $x = -x$ is $x = 0$.)

EXAMPLE A.3. Verify that

$$\begin{vmatrix} 1 & 2 & 3 \\ 3 & 1 & 2 \\ 1 & 2 & 3 \end{vmatrix} = 0$$

Property 3. If every element in any row (or any column) of a determinant is multiplied by a constant, the value of the determinant is multiplied by that same constant.

(Basis for proof: Expand the determinant along the row (or column) whose elements are multiplied by the constant.)

EXAMPLE A.4. If we multiply each element of the second column of $|\mathbf{C}|$ by 2, we obtain the following determinant:

$$\begin{vmatrix} 1 & 4 & 3 \\ 3 & 2 & 2 \\ 2 & 6 & 1 \end{vmatrix}$$

Verify that the value of this determinant is $(2)(18) = 36$.

Property 4. If the elements of one row (or column) of a determinant are proportional to those of another row (or column, respectively), the value of the determinant is zero.

(Follows directly from combined use of Properties 2 and 3.)

EXAMPLE A.5. The third column of the following determinant is proportional to its first column:

$$\begin{vmatrix} 1 & 2 & 2 \\ 3 & 1 & 6 \\ 2 & 3 & 4 \end{vmatrix}$$

Verify that the value of this determinant is zero.

Property 5. If the elements of one row (or column) of a determinant are multiplied by a constant, and the results are added to or subtracted from the corresponding elements of another row (or column, respectively), the value of the determinant is unchanged.

Outline of proof: Let

$$|\mathbf{A}| = \begin{vmatrix} a_{11} & a_{12} & a_{13} \\ a_{21} & a_{22} & a_{23} \\ a_{31} & a_{32} & a_{33} \end{vmatrix}$$

Then

$$\begin{vmatrix} a_{11} & a_{12} & a_{13} \\ a_{21} & a_{22} & a_{23} \\ a_{31} + ka_{11} & a_{32} + ka_{12} & a_{33} + ka_{13} \end{vmatrix}$$

$$= (a_{31} + ka_{11})A_{31} + (a_{32} + ka_{12})A_{32} + (a_{33} + ka_{13})A_{33}$$

(by expansion along third row)

$$= a_{31}A_{31} + a_{32}A_{32} + a_{33}A_{33} + (ka_{11}A_{31} + ka_{12}A_{32} + ka_{13}A_{33})$$

$$= |\mathbf{A}| + \begin{vmatrix} a_{11} & a_{12} & a_{13} \\ a_{21} & a_{22} & a_{23} \\ ka_{11} & ka_{12} & ka_{13} \end{vmatrix}$$

(by using Rule A.5 in reverse)

$$= |\mathbf{A}| + 0$$

(by Property 4)

EXAMPLE A.6. Multiplying each element of the second column of $|\mathbf{C}|$ by 2 and subtracting from the corresponding element of the third column, we obtain:

$$\begin{vmatrix} 1 & 2 & -1 \\ 3 & 1 & 0 \\ 2 & 3 & -5 \end{vmatrix}$$

Verify that this determinant has the value 18 (= $|\mathbf{C}|$). Further, multiplying the first row by 5 and subtracting from the third row of this new determinant, we get:

$$\begin{vmatrix} 1 & 2 & -1 \\ 3 & 1 & 0 \\ -3 & -7 & 0 \end{vmatrix}$$

Verify that this determinant also has the value 18.

It is by repeated use of Property 5 that we can successively transform a given determinant to get an equivalent determinant with $n - 1$ zeros in some row or some column, as seen in the above example.

Property 6. If a linear combination is formed of the cofactors of elements along a given row (or column), using the corresponding elements of a *different* row (or column) as coefficients, then the value of this linear combination is 0.

(Outline of proof: The expression in parentheses in the third last line of the proof of Property 5 exemplifies such a linear combination, if we take $k = 1$.)

EXAMPLE A.7. For

$$|\mathbf{C}| = \begin{vmatrix} 1 & 2 & 3 \\ 3 & 1 & 2 \\ 2 & 3 & 1 \end{vmatrix}$$

we have

$$C_{12} = -\begin{vmatrix} 3 & 2 \\ 2 & 1 \end{vmatrix} = 1 \qquad C_{22} = \begin{vmatrix} 1 & 3 \\ 2 & 1 \end{vmatrix} = -5$$

and

$$C_{32} = -\begin{vmatrix} 1 & 3 \\ 3 & 2 \end{vmatrix} = 7$$

as the second column cofactors.

Linearly combining these cofactors with the first-column elements as coefficients, we obtain:

$$(1)(1) + (3)(-5) + (2)(7) = 0$$

Or, using the third-column elements as coefficients, we have:

$$(3)(1) + (2)(-5) + (1)(7) = 0$$

Property 6 forms the basis for the method for finding the inverse of a matrix, described on pp. 23–25. It can readily be verified that the (i, j)-element of the product $\mathbf{A}\,\mathbf{adj}(\mathbf{A})$ (where $\mathbf{adj}(\mathbf{A})$ is the adjoint of $\mathbf{A}$) is a linear combination of the cofactors of the jth row of $|\mathbf{A}|$, using the elements of the ith row as combining weights (coefficients). Hence, the off-diagonal elements (where $i \neq j$) all vanish by virtue of Property 6, while each diagonal element ($j = i$) has the value $|\mathbf{A}|$ because it represents an expansion of $|\mathbf{A}|$ along its ith row.

Property 6 also plays a prominent role in the determination of eigenvectors of a matrix $\mathbf{A}$, described in Chapter Five. If the elements of a vector $\mathbf{v}$ are proportional to any column of the adjoint of $\mathbf{A} - \mu\mathbf{I}$ (where μ is an eigenvalue of $\mathbf{A}$, that is, a root of the characteristic equation, $|\mathbf{A} - \mu\mathbf{I}| = 0$), then it follows that each element of the vector $(\mathbf{A} - \mu\mathbf{I})\mathbf{v}$ is equal to zero. All but one of these equalities hinges on Property 6, and the exceptional one is true because the determinant $|\mathbf{A} - \mu\mathbf{I}|$ has the value 0 by definition of μ.

Property 7. The determinant of the product of two square matrices is equal to the product of their respective determinants. That is,

$$|\mathbf{AB}| = |\mathbf{A}||\mathbf{B}| \tag{A.6}$$

provided $\mathbf{A}$ and $\mathbf{B}$ are square matrices of the same order.

(The proof is lengthy but not difficult. We first form the matrix product in accordance with Eq. 2.5. In evaluating the determinant of this product, we utilize the decomposition rule, as in the proof of Property 5, in combination with Property 4.)

EXAMPLE A.8. Let

$$\mathbf{A} = \begin{bmatrix} 1 & 2 & 3 \\ 3 & 1 & 2 \\ 2 & 3 & 1 \end{bmatrix} \quad \text{and} \quad \mathbf{B} = \begin{bmatrix} 1 & 0 & 0 \\ 3 & 2 & -1 \\ 5 & 2 & 4 \end{bmatrix}$$

Then

$$\mathbf{AB} = \begin{bmatrix} 22 & 10 & 10 \\ 16 & 6 & 7 \\ 16 & 8 & 1 \end{bmatrix} \quad \text{and} \quad |\mathbf{AB}| = 180$$

while

$$|\mathbf{A}| = 18 \quad \text{and} \quad |\mathbf{B}| = 10$$

Therefore

$$|\mathbf{A}||\mathbf{B}| = (18)(10) = 180 = |\mathbf{AB}|$$

Appendix B

Pivotal Condensation Method of Matrix Inversion

Most of the practical matrix inversion routines consist in systematic arrangements of sequences of elementary row operations. By an "elementary row operation" we mean that a given row of a matrix is multiplied by some constant, and the result is subtracted from each of the other rows. (The constant is ordinarily such that the subtraction will make one of each minuend row's elements equal to zero after the operation. Thus, a different multiplier is used for each minuend row.) Bearing this general principle in mind will make it easier to follow the description of the pivotal condensation method, given below with reference to a specific numerical example.

The $p \times p$ matrix **A** whose inverse is sought is written in the upper left-hand $p \times p$ sector of a worksheet such as that in Table B.1 ($p = 4$ in this example). The worksheet should be large enough to accommodate $2p + (p - 1)(p + 2)/2$ rows and $2p + 1$ columns of figures with one or two more significant digits than the elements of **A**. (More significant digits than needed or warranted in the final result must be carried in the intermediate calculations to allow for cumulative rounding errors.)

To the right of **A**, we write the identity matrix of order p (that is, the same order as **A**). We now have $2p$ columns of p rows each. In each row of the last column, we enter the sum of the $2p$ numbers in that row (that is, the elements of that row of **A** *plus* 1 *plus* $p - 1$ zeros). This column is introduced for the purpose of having a summational check on our calculations, step by step. Henceforth, the numbers in this column are treated in exactly the same way as the numbers to their left in each computational step. Then, as each new row is generated, we check to see if the sum of the numbers in columns (1)–($2p$) of that row is equal (within rounding error) to the column-($2p + 1$) entry; if not, some arithmetical error has occurred, and we must correct it before proceeding.

1. Each of the $2p + 1$ numbers in row (0.1) is divided by the left-most number (called the first "pivotal element"), 6.4570, which is underscored in the worksheet. The resulting quotients are written in the corresponding columns of row (1.0), which is called the first "pivotal row."

TABLE B.1 Worksheet for Computing the Inverse of a Matrix.

	(1)	(2)	(3)	(4)	(5)	(6)	(7)	(8)	(9)
(0.1)	6.4570	2.8248	2.6712	2.5704	1	0	0	0	15.5234
(0.2)	3.4912	12.1606	4.1503	6.1378	0	1	0	0	26.9399
(0.3)	2.4961	4.8126	4.9151	3.2634	0	0	1	0	16.4872
(0.4)	3.7817	5.6427	3.8276	8.1125	0	0	0	1	22.3645
(1.0)	1	.437479	.413691	.398080	.154871	0	0	0	2.404120
(1.1)	0	10.633273	2.706022	4.748023	−.540686	1	0	0	18.546633
(1.2)	0	3.720609	3.882486	2.269753	−.386574	0	1	0	10.486274
(1.3)	0	3.988286	2.263145	5.607081	−.585676	0	0	1	13.272836
(2.0)	0	1	.254486	.446525	−.050849	.094044	0	0	1.744207
(2.1)	0	0	2.935643	.608408	−.197385	−.349901	1	0	3.996762
(2.2)	0	0	1.248182	4.826212	−.382876	−.375074	0	1	6.316440
(3.0)	0	0	1	.207249	−.067237	−.119191	.340641	0	1.361462
(3.1)	0	0	0	4.567528	−.298952	−.226302	−.425182	1	4.617092
(4. 0)	0	0	0	1	−.065452	−.049546	−.093088	.218937	1.010851
(4. −1)	0	0	1	0	−.053672	−.108923	.359933	−.045374	1.151964
(4. −2)	0	1	0	0	−.007964	.143887	−.050032	−.086214	.999678
(4. −3)	1	0	0	0	.206614	.001836	−.089957	−.030667	1.087827

$$|\mathbf{A}| = (6.4570)(10.6333)(2.9356)(4.5675) = 920.624$$

2. The numbers in row (1.0) are multiplied by the column-(1) number in row (0.2) (3.4912 here), and the products are subtracted from the corresponding numbers in row (0.2) to yield the elements of row (1.1); similarly, row (1.0) is multiplied by the first element of row (0.3) (2.4961 here), and the result subtracted from row (0.3) to yield row (1.2); and so on through row $(0.p)$.

We now have a set of $p - 1$ rows, (1.1), (1.2), . . . , $(1.p - 1)$, the first element of each of which is 0 by design. [We say that we have "swept out" the first column of **A**, using row (1.0) as pivot.] Steps 1 and 2 are now repeated on these rows, exclusive of column (1). That is, each element of row (1.1) is divided by the second pivotal element 10.633273 to yield the second pivotal row (2.0). This is then used to sweep out column (2) of rows (1.2), (1.3), . . . , $(1.p - 1)$, and the results are recorded as rows (2.1), (2.2), . . . , $(2.p - 2)$, respectively.

The cycle is repeated until we come to row $(p-1.1)$, in which the first $p - 1$ elements are all 0, as in row (3.1) here. The next step still follows rule (1); that is, we divide each element of row (3.1) by the fourth pivotal element (4.567528 here) to get row (4.0), or $(p.0)$ in general. The numbers in columns (5)–(8) [or $(p + 1)-(2p)$ in general] in this row form the *last* row of $\mathbf{A}^{-1}$. The computation of rows $(4.-1)$, $(4.-2)$, and so on, is somewhat different from the preceding steps. This phase is known as the "backward solution."

To generate row $(4.-1)$, we eliminate the column-(4) [or column-(p) in general] element of row (3.0) by subtracting .207249 times row (4.0) from row (3.0). Symbolically,

$$(4.-1) = (3.0) - (3.0)_4 \times (4.0)$$

where $(3.0)_4$ denotes the fourth element of row (3.0). [In the p-dimensional case, this generalizes to: $(p.-1) = (p-1.0) - (p-1.0)_p \times (p.0)$.]

Row $(4.-2)$ is obtained by subtracting .446525 times row (4.0) *and* .254486 times row $(4.-1)$ from row (2.0); that is,

$$(4.-2) = (2.0) - (2.0)_4 \times (4.0) - (2.0)_3 \times (4.-1)$$

which generalizes to

$$(p.-2) = (p-2.0) - (p-2.0)_p \times (p.0) - (p-2.0)_{p-1} \times (p.-1)$$

Thus, the backward solution involves subtracting multiples of successively more and more rows, $(p.0)$, $(p.-1)$, $(p.-2)$, and so forth, from earlier and earlier pivotal rows. The elements in columns $(p + 1)$ through $(2p)$ of the successively generated rows $(p.0)$, $(p.-1)$, $(p.-2)$, . . . , $(p.-(p - 1))$ form rows p, $p - 1$, $p - 2$, . . . , 1, respectively, of $\mathbf{A}^{-1}$. That is to say, the successive rows of $\mathbf{A}^{-1}$ appear in *reverse order* in the lower right-hand $p \times p$ sector of the worksheet. Thus, for this example we have, on rounding the numbers to five decimal places (that is, a maximum of five significant digits),

$$\mathbf{A}^{-1} = \begin{bmatrix} .20661 & .00184 & -.08996 & -.03067 \\ -.00796 & .14389 & -.05003 & -.08621 \\ -.05367 & -.10892 & .35993 & -.04537 \\ -.06545 & -.04955 & -.09309 & .21894 \end{bmatrix}$$

Even though we have carried a step-by-step summational check of our calculations, it is always wise to perform a final, positive check by multiplying the result by the original matrix **A** to see that we obtain the identity matrix. This is left for the reader to carry out.

B.1 OTHER USES OF PIVOTAL CONDENSATION

The method described above may be used for two other purposes besides computing the inverse of a matrix.

1. Solving a system of linear equations $\mathbf{Ax} = \mathbf{c}$, where **A** is a square, nonsingular matrix of coefficients, **x** is the column vector of unknowns, and **c** is a column vector of known constants. For this purpose, we write the vector **c** in column $(p + 1)$, rows (0.1)–$(0.p)$, of our worksheet instead of writing the identity matrix in columns $(p + 1)$ through $(2p)$. Thus, the worksheet will now have only $p + 2$ columns, the last being the summational check column as before. All computations are done exactly as described above, and the column-$(p + 1)$ entries in the final rows $(p.0)$, $(p.-1)$, $(p.-2)$, . . . , $(p.-(p - 1))$ give the values of $x_p, x_{p-1}, x_{p-2}, \ldots, x_1$, respectively, of the solution.

2. Evaluating a determinant $|\mathbf{A}|$. The value of $|\mathbf{A}|$ is readily obtained as a by-product of the process of inverting the matrix **A**, or of solving the system of linear equations $\mathbf{Ax} = \mathbf{c}$. All we have to do is to find the product of all the pivotal elements as used in the process, namely, the first nonzero element of rows (0.1), (1.1), (2.1), . . . , $(p-1.1)$ in the worksheet. (The computation of $|\mathbf{A}|$ for our numerical example is shown at the foot of Table B.1, where the pivotal elements have been rounded to four decimal places.) Consequently, if our sole purpose is to evaluate a given determinant, we need only enter its elements in the upper left-hand $p \times p$ sector of our worksheet, and the row sums in column $(p + 1)$ for carrying out the summational checks. The computations will now terminate with row $(p-1.1)$, whose only nonzero entries will be the last (that is, pth) pivotal element and the entry in the check column, which should equal the former within rounding error.

B.2 EXCEPTIONAL CASES

For the sake of completeness, we indicate a slight modification of the procedure described in the foregoing, which becomes necessary in some exceptional situations —even though these are hardly likely to arise when we seek the inverses of matrices that are of interest in multivariate analysis. The anomaly in question is the occurrence of a 0 in the diagonal position in some step; that is, when an element which should, according to the foregoing instructions, be used as a pivotal element happens to be 0—so that we cannot divide by it to generate the next pivotal row. This may happen (a) when the matrix is nonsingular, in which case we need the modification described below in order to proceed; or (b) when the matrix is singular, in which case we have to abandon our attempt to find its inverse, for none exists. A simple example of each case should suffice to illustrate how to proceed in the first case and how to recognize an instance of the second case.

1. Suppose we set out to find the inverse of

$$\mathbf{B} = \begin{bmatrix} 2.0 & 2.0 & 2.0 \\ 1.5 & 1.5 & 1.0 \\ 2.0 & 3.0 & 1.0 \end{bmatrix}$$

in accordance with the procedure described above. The worksheet is shown in Table B.2 below.

TABLE B.2 Worksheet for Computing Inverse of B.

	(1)	(2)	(3)	(4)	(5)	(6)	(7)
(0.1)	<u>2.0</u>	2.0	2.0	1	0	0	7.0
(0.2)	1.5	1.5	1.0	0	1	0	5.0
(0.3)	2.0	3.0	1.0	0	0	1	7.0
(1.0)	1	1.0	1.0	.5	0	0	3.5
(1.1)	0	0.0	−.50	−.75	1	0	−.25
(1.2)	0	<u>1.0</u>	−1.0	−1.0	0	1	0.0
(2.0)	0	1	−1.0	−1.0	0	1.0	0.0
(2.1)	0	0	−<u>.50</u>	−.75	1	0	−.25
(3. 0)	0	0	1	1.5	−2.0	0.0	.5
(3.−1)	0	1	0	.5	−2.0	1.0	.5
(3.−2)	1	0	0	−1.5	4.0	−1.0	2.5

In row (1.1), obtained by subtracting 1.5 times row (1.0) from row (0.2), we find that not only the column (1) entry is zero (by design), but the column (2) entry is also zero (by peculiarity). [Throughout Table B.2, elements that are zero by design (that is, will be zero in inverting any matrix) are written as "0" without a decimal point, while those that are zero because of a peculiarity of the particular matrix is written as "0.0"; a similar distinction is made between 1 and 1.0.] Ordinarily, we would divide each element of row (1.1) by its column (2) element in order to get the second pivotal row, (2.0). We cannot do so in this case because the column (2) element is zero.

In such a situation, provided that there is at least one nonzero element among the first p (= 3 here) in that row, we adopt the following modification: Use the next row after the anomalous one(s) for generating the next pivotal row. We shall call this step a "row interchange." Thus, in the present example, we divide row

(1.2) by its column (2) element (which happens to be 1.0 here) in order to obtain row (2.0), the second pivotal row. (The divisor 1.0 is underscored as the second pivotal element.) Then, row (2.1) is obtained by multiplying row (2.0) by 0.0 (the column (2) element of row (1.1)) and subtracting the result from row (1.1); in other words, row (1.1) itself is recorded as row (2.1). We then proceed as usual. Of course, if we encounter another anomalous row in the process of inverting a larger matrix, we perform a row interchange again before proceeding.

The desired inverse emerges, as usual, with the rows in reverse order, in columns $(p + 1)$ through $(2p)$ of rows $(p.0), (p.-1), \ldots, (p.-(p - 1))$. Thus for this example,

$$\mathbf{B}^{-1} = \begin{bmatrix} -1.5 & 4.0 & -1.0 \\ .5 & -2.0 & 1.0 \\ 1.5 & -2.0 & 0.0 \end{bmatrix}$$

The value of the determinant $|\mathbf{B}|$, however, may need to have its sign adjusted depending on how many row interchanges were performed. This value is given by the product of all p pivotal elements if an *even* number of row interchanges were made (just as in the case of no interchange); but if an *odd* number of row interchanges were performed, the sign of this product is reversed to get the value of the determinant. In other words, the determinant is equal to $(-1)^m$ times the product of pivotal elements, where m is the number of row interchanges that were made. In the present example, with $m = 1$, we have

$$|\mathbf{B}| = (-1)[(2.0)(1.0)(-.50)] = 1.0$$

2. The case of a singular matrix. As implied above in the proviso that an anomalous row must nevertheless have at least one nonzero element among its first p before we decide to make a row interchange, a singular matrix of order $p \times p$ is identified by the fact that, at some stage or other, there will emerge a row whose first p elements are *all* equal to zero. In the example shown in Table B.3,

TABLE B.3 Worksheet Showing that a Matrix is Singular.

	(1)	(2)	(3)	(4)	(5)	(6)	(7)
(0.1)	1.0	−4.0	−3.0	1	0	0	−5.0
(0.2)	−2.0	5.0	4.0	0	1	0	8.0
(0.3)	3.0	3.0	1.0	0	0	1	8.0
(1.0)	1	−4.0	−3.0	1.0	0	0	−5.0
(1.1)	0	−3.0	−2.0	2.0	1	0	−2.0
(1.2)	0	15.0	10.0	−3.0	0	1	23.0
(2.0)	0	1	2/3	−2/3	−1/3	0	2/3
(2.1)	0	0	0.0	7.0	5.0	1	13.0

where we set out to find the inverse of the 3 × 3 matrix in the upper left-hand sector, we see that the first three elements of row (2.1) are all zeros. We therefore conclude that the matrix is singular and abandon our attempt to find its inverse. Furthermore, we can assert that the rank of the matrix is equal to the total number of pivotal rows obtained, or 2 in this case.

In this example we were able to determine the rank of the matrix as soon as a row with first p elements zero was found, because that row (2.1) was in fact the last possible one before we would normally start the backward solution. This is not true in general, however. We may encounter a row with p initial zeros at an earlier stage, in which case we must proceed further before we can tell what the rank of the matrix is. (Of course, we already know at that point that the rank is smaller than p, and hence that the matrix is singular.) The further steps to be taken are as follows: Delete the row with p initial zeros from further consideration, and proceed "as far as we can go," using the cycle of steps (1) and (2) described earlier. By "as far as we can go" is meant that we continue until either we get a single pivotal row with no other rows to be operated on by it, or when the sweeping out by some pivotal row results in *all* the other rows at that stage becoming "null rows" (through the pth column). In either case, no further pivotal rows can be constructed, so our computation is necessarily terminated. The rank of the matrix is then equal to the total number of pivotal rows, (1.0), (2.0), and so on, that were constructed in all. Table B.4 illustrates the procedure with reference to a 4 × 4 matrix. (We have condensed the worksheet by retaining only the first four columns. The reader should supply the check column in retracing the calculations for practice.)

TABLE B.4 Determining the Rank of a Matrix.

	(1)	(2)	(3)	(4)
(0.1)	<u>2.0</u>	−1.0	4.0	2.0
(0.2)	1.0	0.0	2.0	3.0
(0.3)	2.0	−1/2	4.0	4.0
(0.4)	3.0	2.0	−1.0	2.0
(1.0)	1	−1/2	2.0	1.0
(1.1)	0	<u>1/2</u>	0.0	2.0
(1.2)	0	1/2	0.0	2.0
(1.3)	0	7/2	—7.0	−1.0
(2.0)	0	1	0.0	4.0
(2.1)	0	0	0.0	0.0
(2.2)	0	0	−<u>7.0</u>	−15.0
(3.0)	0	0	1	15/7

Observe that row (2.1), obtained by operating on row (1.2) with the second pivotal row, (2.0), is a null row. However, we do not stop here and conclude that the rank of the matrix is 2 (which is the number of pivotal rows constructed thus far). We would do so only if row (2.2) were also a null row. But since it is not, we use row (2.2) for generating the next pivotal row, (3.0). Now we have no other row on which to operate with this new pivotal row. Our computation necessarily ends at this point, and we have determined that the rank of our matrix is 3.

Appendix C

Symbolic Differentiation by Vectors or Matrices

At several points in this book we made use of what is known as a *symbolic derivative* with respect to a vector (Dwyer and MacPhail, 1948). The process of symbolic differentiation by a vector involves no more (and no less) than finding the partial derivative of a given function with respect to each element of the vector, and then arranging the results in the form of a vector of the same type. Thus, if $f(\mathbf{x})$ is a function of the elements of a vector $\mathbf{x}$, the symbolic derivative $\frac{\partial}{\partial \mathbf{x}} f(\mathbf{x})$ is, by definition, the column vector whose elements are

$$\frac{\partial f}{\partial x_1}, \frac{\partial f}{\partial x_2}, \frac{\partial f}{\partial x_3}, \ldots, \frac{\partial f}{\partial x_p}$$

Similarly,

$$\frac{\partial}{\partial \mathbf{x}'} f(\mathbf{x})$$

is defined to be the row vector with these same elements.

The type of function of vector elements that occurs most frequently in multivariate analysis is the quadratic form $\mathbf{x}'\mathbf{A}\mathbf{x}$, or composite functions of this function, such as $(\mathbf{x}'\mathbf{A}\mathbf{x})/(\mathbf{x}'\mathbf{B}\mathbf{x})$, $\log(\mathbf{x}'\mathbf{A}\mathbf{x})$, $\exp[-(\mathbf{x}'\mathbf{A}\mathbf{x})/2]$, and so on. It is therefore useful to have a general formula for writing out the symbolic derivative of $\mathbf{x}'\mathbf{A}\mathbf{x}$ with respect to $\mathbf{x}$. We derive the formula for the simplest case, when $\mathbf{x}$ is a two-dimensional vector and $\mathbf{A}$ is a 2×2 matrix. It will readily be seen that the derivation generalizes to any dimensionality.

Since

$$\mathbf{x}'\mathbf{A}\mathbf{x} = [x_1, x_2] \begin{bmatrix} a_{11} & a_{12} \\ a_{21} & a_{22} \end{bmatrix} \begin{bmatrix} x_1 \\ x_2 \end{bmatrix} = a_{11}x_1^2 + (a_{12} + a_{21})x_1x_2 + a_{22}x_2^2$$

it follows that

$$\frac{\partial}{\partial x_1}(\mathbf{x}'\mathbf{A}\mathbf{x}) = 2a_{11}x_1 + (a_{12} + a_{21})x_2$$

and

$$\frac{\partial}{\partial x_2}(\mathbf{x}'\mathbf{A}\mathbf{x}) = (a_{12} + a_{21})x_1 + 2a_{22}x_2$$

On collecting these partial derivatives into a column vector, we obtain

$$\begin{aligned}\frac{\partial}{\partial \mathbf{x}}(\mathbf{x}'\mathbf{A}\mathbf{x}) &= \begin{bmatrix} 2a_{11}x_1 + (a_{12} + a_{21})x_2 \\ (a_{12} + a_{21})x_1 + 2a_{22}x_2 \end{bmatrix} \\ &= \begin{bmatrix} 2a_{11} & a_{12} + a_{21} \\ a_{12} + a_{21} & 2a_{22} \end{bmatrix} \begin{bmatrix} x_1 \\ x_2 \end{bmatrix} \\ &= \left(\begin{bmatrix} a_{11} & a_{12} \\ a_{21} & a_{22} \end{bmatrix} + \begin{bmatrix} a_{11} & a_{21} \\ a_{12} & a_{22} \end{bmatrix} \right) \begin{bmatrix} x_1 \\ x_2 \end{bmatrix} = (\mathbf{A} + \mathbf{A}')\mathbf{x}\end{aligned}$$

An exactly parallel derivation can be made when $\mathbf{x}$ is a vector of any dimensionality, say p, and $\mathbf{A}$ is a $p \times p$ matrix, by noting that

$$\mathbf{x}'\mathbf{A}\mathbf{x} = \sum_i \sum_j a_{ij}x_i x_j = \sum_{i=1}^{p} a_{ii}x_i^2 + \sum_{i<j}\sum (a_{ij} + a_{ji})x_i x_j$$

We have thus shown that

$$\frac{\partial}{\partial \mathbf{x}}(\mathbf{x}'\mathbf{A}\mathbf{x}) = (\mathbf{A} + \mathbf{A}')\mathbf{x} \tag{C.1}$$

Similarly, by arranging the partial derivatives with respect to x_i in the form of a row vector, we get

$$\frac{\partial}{\partial \mathbf{x}'}(\mathbf{x}'\mathbf{A}\mathbf{x}) = \mathbf{x}'(\mathbf{A} + \mathbf{A}') \tag{C.1$'$}$$

When the matrix $\mathbf{A}$ of the quadratic form is symmetrical, so that $\mathbf{A}' = \mathbf{A}$, Eqs. C.1 and C.1$'$ reduce, respectively, to the simpler forms

$$\frac{\partial}{\partial \mathbf{x}}(\mathbf{x}'\mathbf{A}\mathbf{x}) = 2\mathbf{A}\mathbf{x} \tag{C.1s}$$

and

$$\frac{\partial}{\partial \mathbf{x}'}(\mathbf{x}'\mathbf{A}\mathbf{x}) = 2\mathbf{x}'\mathbf{A} \tag{C.1$'$s}$$

A special case of a quadratic form with a symmetric matrix is the squared norm $\mathbf{x}'\mathbf{x}$ of a vector $\mathbf{x}$. We have merely to let $\mathbf{A} = \mathbf{I}$ in Eqs. C.1s and C.1$'$s to obtain

$$\frac{\partial}{\partial \mathbf{x}}(\mathbf{x}'\mathbf{x}) = 2\mathbf{x} \tag{C.2}$$

and

$$\frac{\partial}{\partial \mathbf{x}'}(\mathbf{x}'\mathbf{x}) = 2\mathbf{x}' \tag{C.2$'$}$$

Another related type of expression whose symbolic (partial) derivative is often needed is the *bilinear form* $\mathbf{x}'\mathbf{A}\mathbf{y}$ or $\mathbf{y}'\mathbf{A}\mathbf{x}$. It may readily be verified that

$$\frac{\partial}{\partial \mathbf{x}}(\mathbf{x}'\mathbf{A}\mathbf{y}) = \frac{\partial}{\partial \mathbf{x}}(\mathbf{y}'\mathbf{A}\mathbf{x}) = \mathbf{A}\mathbf{y} \qquad \text{(C.3)}$$

and

$$\frac{\partial}{\partial \mathbf{x}'}(\mathbf{x}'\mathbf{A}\mathbf{y}) = \frac{\partial}{\partial \mathbf{x}'}(\mathbf{y}'\mathbf{A}\mathbf{x}) = \mathbf{y}'\mathbf{A} \qquad \text{(C.3')}$$

Although the symbolic differentiation of a quadratic form by a vector (together with the usual "chain rule" for differentiation of a composite function) is the most that is used, in the way of symbolic derivatives, in this book, we present some additional results for the benefit of those readers who wish to pursue the mathematics of multivariate analysis further. The formulas given below are especially useful in connection with obtaining maximum likelihood estimates of the parameters of multivariate normal distributions.

C.1 SYMBOLIC DERIVATIVE OF A QUADRATIC FORM W.R.T. ITS MATRIX

Consider now the situation in which the vector $\mathbf{x}$ of a quadratic form is fixed, and we wish to find the partial derivatives of $\mathbf{x}'\mathbf{A}\mathbf{x}$ w.r.t. (with respect to) the elements a_{ij} of $\mathbf{A}$. The matrix whose (i, j)-element is $\frac{\partial}{\partial a_{ij}}(\mathbf{x}'\mathbf{A}\mathbf{x})$, is called the symbolic derivative of $\mathbf{x}'\mathbf{A}\mathbf{x}$ w.r.t. $\mathbf{A}$. We illustrate the derivation with the two-dimensional case. For short, let us denote $\mathbf{x}'\mathbf{A}\mathbf{x}$ by Q. (We temporarily assume $\mathbf{A}$ to be nonsymmetric.)

$$Q = a_{11}x_1^2 + (a_{12} + a_{21})x_1x_2 + a_{22}x_2^2$$

Therefore,

$$\frac{\partial Q}{\partial a_{11}} = x_1^2 \qquad \frac{\partial Q}{\partial a_{12}} = x_1x_2 \qquad \frac{\partial Q}{\partial a_{21}} = x_1x_2 \qquad \frac{\partial Q}{\partial a_{22}} = x_2^2$$

Collecting these partial derivatives into a 2×2 matrix, we have

$$\frac{\partial Q}{\partial \mathbf{A}} = \begin{bmatrix} x_1^2 & x_1x_2 \\ x_2x_1 & x_2^2 \end{bmatrix} = \begin{bmatrix} x_1 \\ x_2 \end{bmatrix} [x_1, x_2] = \mathbf{x}\mathbf{x}'$$

It may readily be verified that an exactly parallel derivation holds for the p-dimensional case. We thus have, as a general rule,

$$\frac{\partial}{\partial \mathbf{A}}(\mathbf{x}'\mathbf{A}\mathbf{x}) = \mathbf{x}\mathbf{x}' \qquad \text{provided } \mathbf{A} \text{ is nonsymmetrical} \qquad \text{(C.4)}$$

To see that Eq. C.4 does not hold when $\mathbf{A}$ is symmetrical, consider again the two dimensional case, with $a_{12} = a_{21} = b$ (say). The quadratic form now becomes

$$Q = a_{11}x_1^2 + 2bx_1x_2 + a_{22}x_2^2$$

so that

$$\frac{\partial Q}{\partial a_{11}} = x_1^{\,2} \qquad \frac{\partial Q}{\partial a_{12}} = \frac{\partial Q}{\partial a_{21}} = \frac{\partial Q}{\partial b} = 2x_1x_2 \qquad \frac{\partial Q}{\partial a_{22}} = x_2^{\,2}$$

Hence, collecting these results into a matrix, the symbolic derivative becomes

$$\frac{\partial Q}{\partial \mathbf{A}} = \begin{bmatrix} x_1^{\,2} & 2x_1x_2 \\ 2x_2x_1 & x_2^{\,2} \end{bmatrix} = \begin{bmatrix} x_1 \\ x_2 \end{bmatrix} [x_1, x_2] + \begin{bmatrix} 0 & x_1x_2 \\ x_2x_1 & 0 \end{bmatrix}$$

A somewhat neater form of this expression may be obtained by doubling $\mathbf{xx}'$ and subtracting $x_1^{\,2}$ and $x_2^{\,2}$ from the respective diagonal elements. Generalizing to p dimensions, we may write

$$\frac{\partial}{\partial \mathbf{A}} (\mathbf{x}'\mathbf{A}\mathbf{x}) = 2\mathbf{xx}' - \mathbf{D}(x_i^2) \qquad \text{when } \mathbf{A} \text{ is symmetric} \tag{C.4s}$$

where $\mathbf{D}(x_i^{\,2})$ denotes the diagonal matrix with $x_1^{\,2}, x_2^{\,2}, \ldots, x_p^{\,2}$ as its diagonal elements.

C.2 SYMBOLIC DERIVATIVE OF A DETERMINANT W.R.T. ITS MATRIX

Next, let us consider the partial derivative of the determinant $|\mathbf{A}|$ w.r.t. each of its elements. Expanding the determinant along its ith row, we have, in accordance with Eq. A.5 of Appendix A

$$|\mathbf{A}| = a_{i1}A_{i1} + a_{i2}A_{i2} + \cdots + a_{ij}A_{ij} + \cdots + a_{ip}A_{ip} \tag{C.5}$$

where A_{ij} is the cofactor of a_{ij}. Now, by definition of a cofactor, *none* of the cofactors in the above expansion contains the element a_{ij} (since the ith row and some column of $|\mathbf{A}|$ have been deleted in getting each of these cofactors), provided $\mathbf{A}$ is nonsymmetric. Hence, for any particular j, the only term in the above expansion which contains a_{ij} is $a_{ij}A_{ij}$. Therefore, the partial derivative of $|\mathbf{A}|$ w.r.t. a_{ij} is

$$\frac{\partial}{\partial a_{ij}} |\mathbf{A}| = A_{ij} \tag{C.6}$$

Hence, on collecting these partial derivatives for all (i, j) pairs into a $p \times p$ matrix, we have

$$\frac{\partial |\mathbf{A}|}{\partial \mathbf{A}} = (A_{ij}) = [\mathbf{adj}(\mathbf{A})]' \qquad \text{when } \mathbf{A} \text{ is nonsymmetric} \tag{C.7}$$

where $\mathbf{adj}(\mathbf{A})$ denotes the adjoint of $\mathbf{A}$. (Note that the result is the transpose of $\mathbf{adj}(\mathbf{A})$ rather than $\mathbf{adj}(\mathbf{A})$ itself, because by definition, $[\mathbf{adj}(\mathbf{A})]_{ij} = A_{ji}$.)

When $\mathbf{A}$ is a symmetric matrix, $a_{ij}A_{ij}$ is not the only term in expansion (C.5) that contains the element a_{ij} (or, equivalently, a_{ji}). In fact, *every* cofactor except A_{ii} contains a_{ij} (in the "guise" of a_{ji}). Hence, the partial derivative of $|\mathbf{A}|$ w.r.t. a_{ij} becomes more complicated than when $\mathbf{A}$ is nonsymmetric. Let us examine the outcome for the determinant of a symmetric 3×3 matrix $\mathbf{A}$. To indicate the equality of elements in symmetrical positions, we use b, c, and d for the off-diagonal elements. Expanding $|\mathbf{A}|$ along its first row, we have:

$$\begin{vmatrix} a_{11} & b & c \\ b & a_{22} & d \\ c & d & a_{33} \end{vmatrix} = a_{11}\underbrace{\begin{vmatrix} a_{22} & d \\ d & a_{33} \end{vmatrix}}_{A_{11}} - b\underbrace{\begin{vmatrix} b & d \\ c & a_{33} \end{vmatrix}}_{-A_{12}} + c\underbrace{\begin{vmatrix} b & a_{22} \\ c & d \end{vmatrix}}_{A_{13}}$$

The partial derivative of $|\mathbf{A}|$ w.r.t. $a_{13} = a_{31} = c$ is, therefore

$$\begin{aligned} \frac{\partial}{\partial c}|\mathbf{A}| &= (-b)\frac{\partial}{\partial c}(ba_{33} - dc) + \frac{\partial}{\partial c}[c(bd - a_{22}c)] \\ &= (-b)(-d) + (bd - a_{22}c) + c(-a_{22}) = 2(bd - a_{22}c) = 2A_{13} \end{aligned}$$

This result generalizes to symmetrical determinants of any order, although not quite as obviously as in the case of (C.4s); that is,

$$\frac{\partial}{\partial a_{ij}}|\mathbf{A}| = 2A_{ij} \qquad \text{for } i \neq j \qquad \text{when } \mathbf{A} \text{ is symmetric} \tag{C.6s}$$

It should be clear from the expansion displayed above that, for a diagonal element a_{ii}, its occurrence is only the explicit one in $a_{ii}A_{ii}$ (in the expansion of $|\mathbf{A}|$ along row i), even when $\mathbf{A}$ is symmetric. Hence, the partial derivative of $|\mathbf{A}|$ w.r.t. a diagonal element a_{ii} retains the expression in (C.6) with j replaced by i; that is, the factor 2 does not appear as in (C.6s). Collecting the partial derivatives into a matrix, we have the awkward result,

$$\frac{\partial}{\partial \mathbf{A}}|\mathbf{A}| = \begin{bmatrix} A_{11} & 2A_{12} & \cdots & 2A_{1p} \\ 2A_{21} & A_{22} & \cdots & 2A_{2p} \\ \vdots & & & \vdots \\ 2A_{p1} & 2A_{p2} & \cdots & A_{pp} \end{bmatrix} = 2\,\mathbf{adj}(\mathbf{A}) - \mathbf{D}(A_{ii}) \tag{C.7s}$$

where $\mathbf{D}(A_{ii})$ denotes the diagonal matrix with $A_{11}, A_{22}, \ldots, A_{pp}$ as its diagonal elements, when $\mathbf{A}$ is a symmetric matrix. Note, however, that we need not write $[\mathbf{adj}(\mathbf{A})]'$ here, since $\mathbf{A}' = \mathbf{A}$ implies $[\mathbf{adj}(\mathbf{A})]' = \mathbf{adj}(\mathbf{A})$.

C.3 SYMBOLIC "PSEUDODERIVATIVES"

It was seen in Eqs. C.4s and C.7s that symbolic differentiation with respect to a matrix yields a more complicated result when the matrix is symmetric than than when it is not (C.4. and C.7). This is unfortunate, in view of the fact that most of the matrices that play the role of (symbolic) variable of differentiation in multivariate analysis are symmetrical. However, two observations suggest a way to avoid this complication.

First, the complexity of the symbolic derivative w.r.t. a symmetric matrix consists only in the fact that the diagonal elements retain the same expression as in the case of a nonsymmetric matrix, while the off-diagonal elements are *twice* the corresponding expression for the nonsymmetric case. (This was substantiated above only when the function to be differentiated was a quadratic form or the determinant of the matrix, but the same holds true for any scalar function of a matrix.)

Secondly, the situations in which we seek symbolic derivatives w.r.t. matrices are mostly (if not exclusively) when we will subsequently set the result equal to a null matrix—as when we seek the necessary conditions for maximizing or minimizing the function. To set the symbolic derivative equal to a null matrix is, of course, equivalent to setting the partial derivative w.r.t. each element of the matrix, individually, equal to zero. Consequently, the presence of a factor of 2 in the off-diagonal elements of the symbolic derivative w.r.t. a symmetric matrix (as compared to the corresponding expression in the nonsymmetric case) is really immaterial—as long as we are setting the whole thing equal to a null matrix anyway.

We are thus led to the conclusion that—provided our purpose is to set the result equal to a null matrix—we may, without any distortion, construct a matrix whose diagonal elements are the actual partial derivatives of the function w.r.t. the diagonal elements of the matrix of differentiation $\mathbf{A}$, but whose off-diagonal elements are *one-half* of the partial derivatives w.r.t. the off-diagonal elements a_{ij}. Let us call the matrix thus constructed, the symbolic "pseudoderivative" of $f(\mathbf{A})$ w.r.t. $\mathbf{A}$, and denote it by $\dfrac{\partial^*}{\partial \mathbf{A}} f(\mathbf{A})$ to distinguish it from the true symbolic derivative $\dfrac{\partial}{\partial \mathbf{A}} f(\mathbf{A})$. It then follows that the formulas for the symbolic pseudoderivatives of $\mathbf{x}'\mathbf{A}\mathbf{x}$ and $|\mathbf{A}|$ w.r.t. $\mathbf{A}$ are identical with Eqs. C.4 and C.7, respectively (except that, in the latter, we need not take the transpose of $\mathbf{adj}(\mathbf{A})$). Thus,

$$\frac{\partial^*}{\partial \mathbf{A}} (\mathbf{x}'\mathbf{A}\mathbf{x}) = \mathbf{x}\mathbf{x}' \qquad \text{when } \mathbf{A} \text{ is symmetric} \tag{C.4*s}$$

and

$$\frac{\partial^*}{\partial \mathbf{A}} |\mathbf{A}| = (A_{ij}) = (A_{ji}) = \mathbf{adj}(\mathbf{A}) \qquad \text{when } \mathbf{A} \text{ is symmetric} \tag{C.7*s}$$

EXAMPLE C.1. The problem of finding the maximum-likelihood estimates of the centroid $\boldsymbol{\mu}$ and variance-covariance matrix $\boldsymbol{\Sigma}$ of a p-variate normal population on the basis of a random sample of n observation vectors $\mathbf{x}_i$ $(i = 1, 2, \ldots, n)$ amounts to the following: We are to determine those "values" of $\boldsymbol{\mu}$ and $\boldsymbol{\Sigma}$ (treated as indeterminates) that simultaneously maximize the likelihood of the observed sample—that is, the n-fold joint density function evaluated for the sample at hand. In practice, we maximize the natural logarithm of the likelihood function (which is a monotonically increasing function of the likelihood itself), which is given by

$$\begin{aligned} L(\boldsymbol{\mu}, \boldsymbol{\Sigma}) = (-np/2) \ln 2\pi &+ (n/2) \ln |\mathbf{H}| \\ &- (1/2) \sum_{i=1}^{n} (\mathbf{x}_i - \boldsymbol{\mu})'\mathbf{H}(\mathbf{x}_i - \boldsymbol{\mu}) \end{aligned} \tag{C.8}$$

where $\mathbf{H} = \mathbf{\Sigma}^{-1}$, the inverse of the population variance-covariance matrix. The necessary conditions for maximizing $L(\boldsymbol{\mu}, \mathbf{\Sigma})$ are obtained by setting its symbolic derivatives w.r.t. $\boldsymbol{\mu}$ and $\mathbf{\Sigma}$ (or, equivalently w.r.t. $\boldsymbol{\mu}$ and $\mathbf{H}$), respectively, equal to a null vector and a null matrix.

Since $\boldsymbol{\mu}$ is involved only in the last term of L, which is a sum of n quadratic forms, we obtain, by applying Eq. C.1′s to each summand,

$$\frac{\partial}{\partial \boldsymbol{\mu}'} L(\boldsymbol{\mu}, \mathbf{\Sigma}) = (1/2) \sum_{i=1}^{n} 2(\mathbf{x}_i - \boldsymbol{\mu})'\mathbf{H} = \left[\sum_{i=1}^{n} (\mathbf{x}_i - \boldsymbol{\mu})'\right] \mathbf{H}$$

On setting this expression equal to the p-dimensional null vector $\mathbf{0}'$, we get (since $\mathbf{H}$ is nonsingular),

$$\sum_{i=1}^{n} (\mathbf{x}_i - \boldsymbol{\mu})' = \mathbf{0}'$$

from which we obtain

$$\boldsymbol{\mu} = (1/n) \sum_{i=1}^{n} \mathbf{x}_i = \bar{\mathbf{x}} \qquad \text{(the sample centroid)} \qquad \text{(C.9)}$$

as an estimate of $\boldsymbol{\mu}$.

Next, we find the symbolic pseudoderivative of expression (C.8) w.r.t. $\mathbf{H}$:

$$\begin{aligned}\frac{\partial^*}{\partial \mathbf{H}} L(\boldsymbol{\mu}, \mathbf{\Sigma}) &= (n/2) \frac{1}{|\mathbf{H}|} \frac{\partial^*}{\partial \mathbf{H}} |\mathbf{H}| \\ &\quad - (1/2) \sum_{i=1}^{n} \left[\frac{\partial^*}{\partial \mathbf{H}} (\mathbf{x}_i - \boldsymbol{\mu})'\mathbf{H}(\mathbf{x}_i - \boldsymbol{\mu})\right] \\ &= (n/2) \frac{\mathbf{adj(H)}}{|\mathbf{H}|} - (1/2) \sum_{i=1}^{n} (\mathbf{x}_i - \boldsymbol{\mu})(\mathbf{x}_i - \boldsymbol{\mu})'\end{aligned}$$

where we have used Eqs. C.7*s and C.4*s, respectively, for evaluating the symbolic pseudoderivatives of the determinant and quadratic forms w.r.t. the matrix $\mathbf{H}$. Setting the last expression equal to $\mathbf{0}(p \times p)$ after replacing $\mathbf{adj(H)}/|\mathbf{H}|$ by $\mathbf{H}^{-1}$ to which it is equal by definition, we get

$$n\mathbf{H}^{-1} - \sum_{i=1}^{n} (\mathbf{x}_i - \boldsymbol{\mu})(\mathbf{x}_i - \boldsymbol{\mu})' = \mathbf{0} \qquad \text{(C.10)}$$

But, since we have already found an estimate (C.9) for $\boldsymbol{\mu}$ that is necessary in order to maximize $L(\boldsymbol{\mu}, \mathbf{\Sigma})$, we may substitute this estimate, namely $\bar{\mathbf{x}}$, in place of $\boldsymbol{\mu}$ in Eq. C.10. Noting also that the $\mathbf{H}^{-1}$, which satisfies (C.10), is really the desired estimate $\hat{\boldsymbol{\Sigma}}$ of $\boldsymbol{\Sigma}$ (because $\mathbf{H}^{-1} = (\boldsymbol{\Sigma}^{-1})^{-1} = \boldsymbol{\Sigma}$), we finally get

$$\hat{\boldsymbol{\Sigma}} = (1/n) \sum_{i=1}^{n} (\mathbf{x}_i - \overline{\mathbf{x}})(\mathbf{x}_i - \overline{\mathbf{x}})' = (1/n)\mathbf{S} \qquad \text{(C.11)}$$

where $\mathbf{S}$ is the sample SSCP matrix.

Thus, the estimates C.9 and C.11 for $\boldsymbol{\mu}$ and $\boldsymbol{\Sigma}$ are necessary for maximizing $L(\boldsymbol{\mu}, \boldsymbol{\Sigma})$. It can be shown that they are also sufficient. Therefore, the sample centroid and $\mathbf{S}/n$ are the maximum-likelihood estimates for $\boldsymbol{\mu}$ and $\boldsymbol{\Sigma}$ respectively.

Appendix D

Principal Components (or Factors) by Hotelling's Iterative Procedure for Solving Eigenvalue Problems

D.1 BASIC METHOD

In his 1933 paper on principal component analysis, Hotelling developed an iterative procedure for calculating the eigenvalues and vectors of any symmetric matrix. By an iterative method is meant one in which we start with some initial trial value (here a trial vector), perform a certain operation on it (here, premultiplying it by the given matrix) to obtain a second trial value (vector), repeat the same operation on it to get a third trial value, and so on until two successive trial values agree with each other to a specified degree of accuracy. When this occurs, the iteration is said to have converged on the value (vector) sought.

Let $\mathbf{R}$ be the given $p \times p$ symmetric matrix whose eigenvalues and eigenvectors are sought. The computational steps are as follows:

1. Take any p-dimensional column vector $\mathbf{v}_{(0)}$ as the initial trial vector.

2. Form the product $\mathbf{R}\mathbf{v}_{(0)} = \mathbf{u}_{(1)}$, say.

3. Divide $\mathbf{u}_{(1)}$ by its largest (in absolute value) element, and denote the result by $\mathbf{v}_{(1)}$.

4. Use $\mathbf{v}_{(1)}$ as the second trial vector and repeat steps 2 and 3 to obtain $\mathbf{v}_{(2)}$, which then becomes the third trial vector.

5. The foregoing cycle is repeated (iterated) until two successive trial vectors $\mathbf{v}_{(k)}$ and $\mathbf{v}_{(k+1)}$, for instance, are identical within a specified number of decimal places. (That this convergence will eventually occur was proven by Hotelling.)

6. The number by which $\mathbf{R}\mathbf{v}_{(k)} = \mathbf{u}_{(k+1)}$ was divided in order to get $\mathbf{v}_{(k+1)}$ (that is, the largest element of $\mathbf{u}_{(k+1)}$) is the largest eigenvalue λ_1 of $\mathbf{R}$.

7. Normalize the terminal trial vector $\mathbf{v}_{(k)}$ to unity by dividing it by the square root of the sum of the squares of its elements. The result is $\mathbf{v}_1$, the unit eigenvector of $\mathbf{R}$ corresponding to λ_1.

8. Form the matrix product $\mathbf{v}_1\mathbf{v}_1'$, multiply it by λ_1, and subtract the result

from **R**. The resulting difference matrix is called the first residual matrix, denoted $\mathbf{R}_1$. That is,

$$\mathbf{R}_1 = \mathbf{R} - \lambda_1 \mathbf{v}_1 \mathbf{v}_1' \tag{D.1}$$

9. Carry out steps 1–7 using $\mathbf{R}_1$ in place of **R**. The output in steps 6 and 7 are the largest eigenvalue and associated eigenvector of $\mathbf{R}_1$. These are, respectively, the *second* largest eigenvalue λ_2 and its associated eigenvector $\mathbf{v}_2$ of the original matrix **R**. (This is a consequence of the relation

$$\mathbf{A} = \lambda_1 \mathbf{v}_1 \mathbf{v}_1' + \lambda_2 \mathbf{v}_2 \mathbf{v}_2' + \cdots + \lambda_r \mathbf{v}_r \mathbf{v}_r'$$

proved in Exercise 2, p. 155.)

10. Compute the second residual matrix,

$$\mathbf{R}_2 = \mathbf{R}_1 - \lambda_2 \mathbf{v}_2 \mathbf{v}_2'$$

and again carry out steps 1 through 7 to obtain λ_3 and $\mathbf{v}_3$.

The process is continued until the desired number of eigenvalue-eigenvector pairs of **R** have been found. (See discussion at end of Section 5.6*a*.)

D.2 ACCELERATING THE CONVERGENCE

Although the basic method outlined above will always converge, the number of iterations required may sometimes be forbiddingly large, especially when two successive eigenvalues of **R** happen to be close to each other in numerical value. Hotelling showed in a subsequent paper (1936) that the rate of convergence may be greatly accelerated by using a suitable power of **R** instead of **R** itself. That this will lead to the desired results, and will do so more rapidly, may be seen as follows:

Theorem 9 on page 140 shows that if λ_i and $\mathbf{v}_i$ are the *i*th eigenvalue and vector of any matrix **A**, then $\mathbf{v}_i$ is also the *i*th eigenvector of any power $\mathbf{A}^m$ of **A**, with associated eigenvalue λ_i^m. Hence, if we use $\mathbf{R}^4$ (for example) instead of **R** in the basic method, the eigenvectors we get are immediately the eigenvectors of **R** too, and the fourth roots of the eigenvalues will be the corresponding eigenvalues of **R**. Furthermore, since forming the product $\mathbf{R}^4\mathbf{v}_{(j)}$ is the same as operating on any trial vector $\mathbf{v}_{(j)}$ four times by **R** itself, it follows that one iteration of steps 2 and 3 using $\mathbf{R}^4$ is equivalent to *four* iterations using **R**; hence the acceleration in convergence.

It might seem that the economy thus gained would be offset by the labor of having to raise the successive residual matrices $\mathbf{R}_1$, $\mathbf{R}_2$, and so on to a high power for each successive eigenvalue-eigenvector pair. This is not the case. For, once we have raised the original **R** to a given power, the same power of the subsequent residual matrices can be obtained without any "powering," as shown below.

Squaring both sides of Eq. D.1, we get

$$\begin{aligned}
\mathbf{R}_1^2 &= (\mathbf{R} - \lambda_1 \mathbf{v}_1 \mathbf{v}_1')^2 \\
&= \mathbf{R}^2 - 2\lambda_1 \mathbf{R}(\mathbf{v}_1 \mathbf{v}_1') + \lambda_1^2 (\mathbf{v}_1 \mathbf{v}_1')(\mathbf{v}_1 \mathbf{v}_1') \\
&= \mathbf{R}^2 - 2\lambda_1 (\mathbf{R}\mathbf{v}_1)\mathbf{v}_1' + \lambda_1^2 \mathbf{v}_1 (\mathbf{v}_1' \mathbf{v}_1)\mathbf{v}_1'
\end{aligned}$$

where we have used the distributive and associative laws, respectively, in going from the first to the second, and from the second to the third expressions on the right. But, in the last expression, we may replace $\mathbf{R}\mathbf{v}_1$ by $\lambda_1\mathbf{v}_1$ (since $\mathbf{v}_1$ is the eigenvector of $\mathbf{R}$ associated with λ_1), and $\mathbf{v}_1'\mathbf{v}_1$ by 1 (since $\mathbf{v}_1$ is a unit vector). We then have

$$\begin{aligned}\mathbf{R}_1{}^2 &= \mathbf{R}^2 - 2\lambda_1{}^2(\mathbf{v}_1\mathbf{v}_1') + \lambda_1{}^2(\mathbf{v}_1\mathbf{v}_1') \\ &= \mathbf{R}^2 - \lambda_1{}^2(\mathbf{v}_1\mathbf{v}_1')\end{aligned}$$

Similarly, it may be shown that for any positive integer m,

$$\mathbf{R}_1{}^m = \mathbf{R}^m - \lambda_1{}^m(\mathbf{v}_1\mathbf{v}_1') \tag{D.2}$$

and, more generally, for the jth residual matrix,

$$\mathbf{R}_j{}^m = \mathbf{R}_{j-1}^m - \lambda_j{}^m(\mathbf{v}_j\mathbf{v}_j') \tag{D.3}$$

EXAMPLE D.1. Given the following reduced correlation matrix for three variables (estimated communalities in the diagonal), let us determine as many eigenvalue-eigenvector pairs as are necessary "essentially" to account for the observed matrix, and thereby construct the principal-factor matrix.

$$\mathbf{R} = \begin{bmatrix} .6 & -.4 & .1 \\ -.4 & .9 & .3 \\ .1 & .3 & .3 \end{bmatrix}$$

In order to speed up the convergence of the iterations, we first raise $\mathbf{R}$ to the fourth power by squaring twice:

$$\mathbf{R}^4 = (\mathbf{R}^2)^2 = \begin{bmatrix} .6067 & -.9159 & -.2040 \\ -.9159 & 1.5509 & .4171 \\ -.2040 & .4171 & .1394 \end{bmatrix}$$

As our initial trial vector $\mathbf{v}_{(0)}$ (step 1), we take the 3-dimensional column vector whose elements are equal to the row totals of $\mathbf{R}^4$:

$$\mathbf{v}_{(0)} = \begin{bmatrix} -.5132 \\ 1.0521 \\ .3525 \end{bmatrix}$$

(For those who wonder what basis there could be for this choice of $\mathbf{v}_{(0)}$, it may be pointed out that this is equivalent to having taken [1, 1, 1]′ as a trial vector one iteration before $\mathbf{v}_{(0)}$. Lacking all grounds for a logical choice, this vector with uniform elements should—on the average—be the best bet.)

Next, we form the product $\mathbf{R}^4\mathbf{v}_{(0)}$ in accordance with step 2:

$$\mathbf{R}^4\mathbf{v}_{(0)} = [-1.3469, 2.2488, .5927]' = \mathbf{u}_{(1)}$$

where, in order to save space, we have indicated the column vector $\mathbf{u}_{(1)}$ by exhibiting a row vector flagged with the transposition sign. This notation will be used hereunder.

Since the largest absolute-valued element of $\mathbf{u}_{(1)}$ is 2.2488, we divide each element by this number to obtain $\mathbf{v}_{(1)}$, as prescribed in step 3:

$$[-1.3469, 2.2488, .5927]' \div 2.2488 = [-.5989, 1, .2635]' = \mathbf{v}_{(1)}$$

This becomes the second trial vector with which to carry out steps 2 and 3 as stated in 4. For further economy of space, we display the subsequent iterations without explicitly labelling each $\mathbf{u}_{(i)}$, and indicate step 3 (division by the largest element of $\mathbf{u}_{(i)}$) by an arrow (→) leading to $\mathbf{v}_{(i)}$:

$$\mathbf{R}^4\mathbf{v}_{(1)} = [-1.3330, 2.2094, .5760]' \rightarrow [-.6033, 1, .2607]' = \mathbf{v}_{(2)}$$

$$\mathbf{R}^4\mathbf{v}_{(2)} = [-1.3351, 2.2122, .5765]' \rightarrow [-.6035, 1, .2606]' = \mathbf{v}_{(3)}$$

$$\mathbf{R}^4\mathbf{v}_{(3)} = [-1.3352, 2.2123, .5765]' \rightarrow [-.6035, 1, .2606]' = \mathbf{v}_{(4)}$$

We terminate our iterations at this point, following instruction 5, since $\mathbf{v}_{(3)} = \mathbf{v}_{(4)}$ to four decimal places, which is the limit of accuracy to be expected, because we have only four significant digits in the elements of $\mathbf{R}^4$. Hence, according to statement 6, the largest element 2.2123 of $\mathbf{u}_{(4)}$ is the largest eigenvalue of $\mathbf{R}^4$, and its fourth root $(2.2123)^{1/4} = 1.2196$ is the largest eigenvalue λ_1 of $\mathbf{R}$.

Step 7 consists in normalizing $\mathbf{v}_{(3)}$ to unity by dividing each of its elements by $[(-.6035)^2 + (1)^2 + (.2606)^2]^{1/2} = 1.1967$. The result is

$$\mathbf{v}_1 = [-.5043, .8356, .2178]'$$

the unit-length eigenvector of $\mathbf{R}$ associated with its largest eigenvalue $\lambda_1 = 1.2196$.

We are now ready to form the fourth power of the first residual matrix $\mathbf{R}_1$ in accordance with step 8 as modified by replacing Eq. D.1 by D.2, and thence to determine the second eigenvalue-eigenvector pair as instructed in 9. However, let us first consider the question of whether it is necessary to do so. That is, we want to test whether the first factor alone will adequately account for the observed correlation matrix. Several criteria for this purpose were mentioned in Section 5.6a. We will here use the simplest and most intuitive of these: the proportion of common variance accounted for by the first factor. This is assessed by calculating the ratio

$$\lambda_i/\mathrm{tr}(\mathbf{R}) = 1.2196/(.6 + .9 + .3) = .68$$

Thus, only about 68% of the common variance is explained by the first principal factor, so we should certainly extract at least one more factor.

Following Eq. D.2 with $m = 4$, we compute

$$\begin{aligned}\mathbf{R}_1^{\,4} &= \mathbf{R}^4 - \lambda_1^{\,4}(\mathbf{v}_1\mathbf{v}_1') \\ &= \begin{bmatrix} .6067 & -.9159 & -.2040 \\ -.9159 & 1.5509 & .4171 \\ -.2040 & .4171 & .1394 \end{bmatrix} \\ &\quad -(2.2123)\begin{bmatrix} -.5043 \\ .8356 \\ .2178 \end{bmatrix}[-.5043, .8356, .2178] \\ &= \begin{bmatrix} .4411 & .1636 & .3891 \\ .1636 & .0627 & .1446 \\ .3891 & .1446 & .3454 \end{bmatrix} \times 10^{-1}\end{aligned}$$

Using the vector with elements equal to the row totals of $10\mathbf{R}_1^{\,4}$ as the initial trial vector, that is

$$\mathbf{v}_{(0)} = [.9938, .3709, .8791]'$$

we carry out steps 1–7 as instructed in 9, thus:

$$\begin{aligned}10\mathbf{R}_1^{\,4}\mathbf{v}_{(0)} &= [.8411, .3130, .7440]' \rightarrow [1, .3721, .8845]' = \mathbf{v}_{(1)} \\ 10\mathbf{R}_1^{\,4}\mathbf{v}_{(1)} &= [.8461, .3148, .7484]' \rightarrow [1, .3718, .8845]' = \mathbf{v}_{(2)} \\ 10\mathbf{R}_1^{\,4}\mathbf{v}_{(2)} &= [.8461, .3148, .7484]' \rightarrow [1, .3718, .8845]' = \mathbf{v}_{(3)}\end{aligned}$$

The iteration is terminated at this point since $\mathbf{v}_{(2)} = \mathbf{v}_{(3)}$ to four places. Note, however, that the fourth power of the largest eigenvalue of $\mathbf{R}_1$ is not .8461 itself, but one-tenth of this number, or .08461, since we have used $10\mathbf{R}_1^{\,4}$ in our iterations. Hence, the second largest eigenvalue λ_2 of the original correlation matrix $\mathbf{R}$ is $(.08461)^{1/4} = .5393$, and its associated eigenvector is

$$\begin{aligned}\mathbf{v}_2 = \mathbf{v}_{(2)}\big/\sqrt{\mathbf{v}_{(2)}'\mathbf{v}_{(2)}} &= [1, .3718, .8845]'/1.3858 \\ &= [.7216, .2683, .6382]'\end{aligned}$$

We now examine what proportion of the common variance is accounted for by the first two principal factors. The relevant index is

$$(1.2196 + .5393)/1.8 = .978$$

which shows that about 98% of the common variance has already been accounted for. Hence, we may safely conclude that computation of the third factor is unnecessary.

To get a better idea of what it means to say that the first two factors account for practically all of the common variance, let us construct the factor matrix. This is obtained by rescaling each eigenvector to have a length equal to the square root of the corresponding eigenvalue instead of unity. Thus, the factor matrix **F** is given by

$$\mathbf{F} = \mathbf{V}\mathbf{\Lambda}^{1/2}$$

where **V** is the matrix whose columns are $\mathbf{v}_1$ and $\mathbf{v}_2$ (that is, the principal-components transformation matrix), and $\mathbf{\Lambda}$ is the diagonal matrix with diagonal elements λ_1 and λ_2. For the present example,

$$\mathbf{F} = \begin{bmatrix} -.5043 & .7216 \\ .8356 & .2683 \\ .2178 & .6382 \end{bmatrix} \begin{bmatrix} \sqrt{1.2196} & 0 \\ 0 & \sqrt{.5393} \end{bmatrix} = \begin{bmatrix} -.5569 & .5299 \\ .9228 & .1970 \\ .2405 & .4687 \end{bmatrix}$$

This factor matrix postmultiplied by its transpose should essentially reproduce the correlation matrix **R**. Calculating the product, we find

$$\mathbf{F}\mathbf{F}' = \begin{bmatrix} .5909 & -.4095 & .1144 \\ -.4095 & .8904 & .3143 \\ .1144 & .3143 & .2775 \end{bmatrix}$$

whose difference from **R** is

$$\mathbf{R} - \mathbf{F}\mathbf{F}' = \begin{bmatrix} .0091 & .0095 & -.0144 \\ .0095 & .0096 & -.0143 \\ -.0144 & -.0143 & .0225 \end{bmatrix}$$

(This is, of course, the same as the second residual matrix $\mathbf{R}_2$ computed as $\mathbf{R}_2 = \mathbf{R}_1 - \lambda_2\mathbf{v}_2\mathbf{v}_2'$.) The discrepancies, or residuals, are all seen to be of the order of .01, except in the (3, 3)-element, where the difference is about .02. Surely no one would deem it meaningful to extract a third factor from this residual matrix!

In conclusion it may be well to point out another difference between component analysis and factor analysis as they are customarily done—over and above the matter of ones or communalities in the diagonal, and the absence or presence of subsequent rotations. In principal-components analysis, the end product is the transformation matrix **V**, whose columns are the unit eigenvectors of the correlation or SSCP matrix, and which defines the principal components as linear combinations of the observed variables. In factor analysis, on the other hand, the factor matrix (or, more precisely, the *factor pattern* matrix) $\mathbf{V}\mathbf{\Lambda}^{1/2}$ is displayed as the final outcome, and its elements are described as follows: The entries in each column are called the *factor loadings* of the successive variables on each factor, indicating the "influence" of that factor on the several variables; they are *not* weights to be applied to the variables to produce each factor. Linear combinations in the factor-analysis model go the other way: the *factors* (including a factor unique to each variable besides the common factors) are linearly combined to express each variable, and the elements in a given *row* of the factor matrix are the weights to be applied to the successive common factors in expressing that variable. (The unique-factor term is extra, and its weight is not shown in the common-factor matrix.)

Appendix E

Statistical and Numerical Tables

TABLE E.1 Normal Curve Areas.

z	.00	.01	.02	.03	.04	.05	.06	.07	.08	.09
0.0	.0000	.0040	.0080	.0120	.0160	.0199	.0239	.0279	.0319	.0359
0.1	.0398	.0438	.0478	.0517	.0557	.0596	.0636	.0675	.0714	.0753
0.2	.0793	.0832	.0871	.0910	.0948	.0987	.1026	.1064	.1103	.1141
0.3	.1179	.1217	.1255	.1293	.1331	.1368	.1406	.1443	.1480	.1517
0.4	.1554	.1591	.1628	.1664	.1700	.1736	.1772	.1808	.1844	.1879
0.5	.1915	.1950	.1985	.2019	.2054	.2088	.2123	.2157	.2190	.2224
0.6	.2257	.2291	.2324	.2357	.2389	.2422	.2454	.2486	.2517	.2549
0.7	.2580	.2611	.2642	.2673	.2704	.2734	.2764	.2794	.2823	.2852
0.8	.2881	.2910	.2939	.2967	.2995	.3023	.3051	.3078	.3106	.3133
0.9	.3159	.3186	.3212	.3238	.3264	.3289	.3315	.3340	.3365	.3389
1.0	.3413	.3438	.3461	.3485	.3508	.3531	.3554	.3577	.3599	.3621
1.1	.3643	.3665	.3686	.3708	.3729	.3749	.3770	.3790	.3810	.3830
1.2	.3849	.3869	.3888	.3907	.3925	.3944	.3962	.3980	.3997	.4015
1.3	.4032	.4049	.4066	.4082	.4099	.4115	.4131	.4147	.4162	.4177
1.4	.4192	.4207	.4222	.4236	.4251	.4265	.4279	.4292	.4306	.4319
1.5	.4332	.4345	.4357	.4370	.4382	.4394	.4406	.4418	.4429	.4441
1.6	.4452	.4463	.4474	.4484	.4495	.4505	.4515	.4525	.4535	.4545
1.7	.4554	.4564	.4573	.4582	.4591	.4599	.4608	.4616	.4625	.4633
1.8	.4641	.4649	.4656	.4664	.4671	.4678	.4686	.4693	.4699	.4706
1.9	.4713	.4719	.4726	.4732	.4738	.4744	.4750	.4756	.4761	.4767
2.0	.4772	.4778	.4783	.4788	.4793	.4798	.4803	.4808	.4812	.4817
2.1	.4821	.4826	.4830	.4834	.4838	.4842	.4846	.4850	.4854	.4857
2.2	.4861	.4864	.4868	.4871	.4875	.4878	.4881	.4884	.4887	.4890
2.3	.4893	.4896	.4898	.4901	.4904	.4906	.4909	.4911	.4913	.4916
2.4	.4918	.4920	.4922	.4925	.4927	.4929	.4931	.4932	.4934	.4936
2.5	.4938	.4940	.4941	.4943	.4945	.4946	.4948	.4949	.4951	.4952
2.6	.4953	.4955	.4956	.4957	.4959	.4960	.4961	.4962	.4963	.4964
2.7	.4965	.4966	.4967	.4968	.4969	.4970	.4971	.4972	.4973	.4974
2.8	.4974	.4975	.4976	.4977	.4977	.4978	.4979	.4979	.4980	.4981
2.9	.4981	.4982	.4982	.4983	.4984	.4984	.4985	.4985	.4986	.4986
3.0	.4987	.4987	.4987	.4988	.4988	.4989	.4989	.4989	.4990	.4990

This table is abridged from Table 1 of *Statistical Tables and Formulas*, by A. Hald (New York: John Wiley & Sons, Inc., 1952). Reproduced by permission of A. Hald and the publishers.

n \ 1−α	.75	.90	.95	.975	.99	.995	.9995
1	1.000	3.078	6.314	12.706	31.821	63.657	636.619
2	.816	1.886	2.920	4.303	6.965	9.925	31.598
3	.765	1.638	2.353	3.182	4.541	5.841	12.941
4	.741	1.533	2.132	2.776	3.747	4.604	8.610
5	.727	1.476	2.015	2.571	3.365	4.032	6.859
6	.718	1.440	1.943	2.447	3.143	3.707	5.959
7	.711	1.415	1.895	2.365	2.998	3.499	5.405
8	.706	1.397	1.860	2.306	2.896	3.355	5.041
9	.703	1.383	1.833	2.262	2.821	3.250	4.781
10	.700	1.372	1.812	2.228	2.764	3.169	4.587
11	.697	1.363	1.796	2.201	2.718	3.106	4.437
12	.695	1.356	1.782	2.179	2.681	3.055	4.318
13	.694	1.350	1.771	2.160	2.650	3.012	4.221
14	.692	1.345	1.761	2.145	2.624	2.977	4.140
15	.691	1.341	1.753	2.131	2.602	2.947	4.073
16	.690	1.337	1.746	2.120	2.583	2.921	4.015
17	.689	1.333	1.740	2.110	2.567	2.898	3.965
18	.688	1.330	1.734	2.101	2.552	2.878	3.922
19	.688	1.328	1.729	2.093	2.539	2.861	3.883
20	.687	1.325	1.725	2.086	2.528	2.845	3.850
21	.686	1.323	1.721	2.080	2.518	2.831	3.819
22	.686	1.321	1.717	2.074	2.508	2.819	3.792
23	.685	1.319	1.714	2.069	2.500	2.807	3.767
24	.685	1.318	1.711	2.064	2.492	2.797	3.745
25	.684	1.316	1.708	2.060	2.485	2.787	3.725
26	.684	1.315	1.706	2.056	2.479	2.779	3.707
27	.684	1.314	1.703	2.052	2.473	2.771	3.690
28	.683	1.313	1.701	2.048	2.467	2.763	3.674
29	.683	1.311	1.699	2.045	2.462	2.756	3.659
30	.683	1.310	1.697	2.042	2.457	2.750	3.646
40	.681	1.303	1.684	2.021	2.423	2.704	3.551
60	.679	1.296	1.671	2.000	2.390	2.660	3.460
120	.677	1.289	1.658	1.980	2.358	2.617	3.373
∞	.674	1.282	1.645	1.960	2.326	2.576	3.291

[a]Taken from Table III of R. A. Fisher and F. Yates: "Statistical Tables for Biological, Agricultural, and Medical Research," published by Oliver & Boyd Ltd., Edinburgh, by permission of the authors and publishers.

TABLE E.3 Percentage Points of the Chi-square Distribution[a]

$1-\alpha$ / υ	.005	.010	.025	.050	.100	.250	.500
1	$.0^4393$	$.0^3157$	$.0^3982$	$.0^2393$	.0158	.102	.455
2	.0100	.0201	.0506	.103	.211	.575	1.386
3	.0717	.115	.216	.352	.584	1.213	2.366
4	.207	.297	.484	.711	1.064	1.923	3.357
5	.412	.554	.831	1.145	1.610	2.675	4.351
6	.676	.872	1.237	1.635	2.204	3.455	5.348
7	.989	1.239	1.690	2.167	2.833	4.255	6.346
8	1.344	1.647	2.180	2.733	3.490	5.071	7.344
9	1.735	2.088	2.700	3.325	4.168	5.899	8.343
10	2.156	2.558	3.247	3.940	4.865	6.737	9.342
11	2.603	3.053	3.816	4.575	5.578	7.584	10.341
12	3.074	3.571	4.404	5.226	6.304	8.438	11.340
13	3.565	4.107	5.009	5.892	7.042	9.299	12.340
14	4.075	4.660	5.629	6.571	7.790	10.165	13.339
15	4.601	5.229	6.262	7.261	8.547	11.037	14.339
16	5.142	5.812	6.908	7.962	9.312	11.912	15.339
17	5.697	6.408	7.564	8.672	10.085	12.792	16.338
18	6.265	7.015	8.231	9.390	10.865	13.675	17.338
19	6.844	7.633	8.907	10.117	11.651	14.562	18.338
20	7.434	8.260	9.591	10.851	12.443	15.452	19.337
21	8.034	8.897	10.283	11.591	13.240	16.344	20.337
22	8.643	9.542	10.982	12.338	14.042	17.240	21.337
23	9.260	10.196	11.689	13.091	14.848	18.137	22.337
24	9.886	10.856	12.401	13.848	15.659	19.037	23.337
25	10.520	11.524	13.120	14.611	16.473	19.939	24.337
26	11.160	12.198	13.844	15.379	17.292	20.843	25.337
27	11.808	12.879	14.573	16.151	18.114	21.749	26.336
28	12.461	13.565	15.308	16.928	18.939	22.657	27.336
29	13.121	14.257	16.047	17.708	19.768	23.567	28.336
30	13.787	14.954	16.791	18.493	20.599	24.478	29.336
40	20.707	22.164	24.433	26.509	29.051	33.660	39.335
50	27.991	29.707	32.357	34.764	37.689	42.942	49.335
60	35.535	37.485	40.482	43.188	46.459	52.294	59.335
z	−2.576	−2.326	−1.960	−1.645	−1.282	−.675	.000

For $\nu > 60$, use the approximation $\chi^2 = \frac{1}{2}[z - \sqrt{2\nu - 1}]^2$, where z is the corresponding percent point of the unit normal distribution, shown in last line.

TABLE E.3 (continued) Percentage Points of the Chi-square Distribution

$1-\alpha$ / υ	.750	.900	.950	.975	.990	.995	.999
1	1.323	2.706	3.841	5.024	6.635	7.879	10.828
2	2.773	4.605	5.991	7.378	9.210	10.597	13.816
3	4.108	6.251	7.815	9.348	11.345	12.838	16.266
4	5.385	7.779	9.488	11.143	13.277	14.860	18.467
5	6.626	9.236	11.071	12.833	15.086	16.750	20.515
6	7.841	10.645	12.592	14.449	16.812	18.548	22.458
7	9.037	12.017	14.067	16.013	18.475	20.278	24.322
8	10.219	13.362	15.507	17.535	20.090	21.955	26.125
9	11.389	14.684	16.919	19.023	21.666	23.589	27.877
10	12.549	15.987	18.307	20.483	23.209	25.188	29.588
11	13.701	17.275	19.675	21.920	24.725	26.757	31.264
12	14.845	18.549	21.026	23.337	26.217	28.300	32.909
13	15.984	19.812	22.362	24.736	27.688	29.820	34.528
14	17.117	21.064	23.685	26.119	29.141	31.319	36.123
15	18.245	22.307	24.996	27.488	30.578	32.801	37.697
16	19.369	23.542	26.296	28.845	32.000	34.267	39.252
17	20.489	24.769	27.587	30.191	33.407	35.719	40.790
18	21.605	25.989	28.869	31.526	34.805	37.157	42.312
19	22.718	27.204	30.144	32.852	36.191	38.582	43.820
20	23.828	28.412	31.410	34.170	37.566	39.997	45.315
21	24.935	29.615	32.671	35.479	38.932	41.401	46.797
22	26.039	30.813	33.924	36.981	40.289	42.796	48.268
23	27.141	32.007	35.173	38.076	41.638	44.181	49.728
24	28.241	33.196	36.415	39.364	42.980	45.559	51.179
25	29.339	34.382	37.653	40.647	44.314	46.928	52.618
26	30.435	35.563	38.885	41.923	45.642	48.290	54.052
27	31.528	36.741	40.113	43.195	46.963	49.645	55.476
28	32.621	37.916	41.337	44.461	48.278	50.993	56.892
29	33.711	39.088	42.557	45.722	49.588	52.336	58.301
30	34.800	40.256	43.773	46.979	50.892	53.672	59.703
40	45.616	51.805	55.759	59.342	63.691	66.766	73.402
50	56.334	63.167	67.505	71.420	76.154	79.490	86.661
60	66.982	74.397	79.082	83.298	88.379	91.952	99.607
z	.675	1.282	1.645	1.960	2.326	2.576	3.090

[a]Abridged from Catherine M. Thompson: Tables of percentage points of the incomplete beta function and of the chi-square distribution, *Biometrika*, vol. 32 (1941), pp. 187–191, and is published here with the permission of the editor of *Biometrika*.

TABLE E.4 Table of F for .05 (roman), .01 (*italic*), and .001 (bold face) Levels of Significance[a]

n_2 \ n_1	1	2	3	4	5	6	8	12	24	∞
1	161	200	216	225	230	234	239	244	249	254
	4052	*4999*	*5403*	*5625*	*5724*	*5859*	*5981*	*6106*	*6234*	*6366*
	405284	**500000**	**540379**	**562500**	**576405**	**585937**	**598144**	**610667**	**623497**	**636619**
2	18.51	19.00	19.16	19.25	19.30	19.33	19.37	19.41	19.45	19.50
	98.49	*99.01*	*99.17*	*99.25*	*99.30*	*99.33*	*99.36*	*99.42*	*99.46*	*99.50*
	998.5	**999.0**	**999.2**	**999.2**	**999.3**	**999.3**	**999.4**	**999.4**	**999.5**	**999.5**
3	10.13	9.55	9.28	9.12	9.01	8.94	8.84	8.74	8.64	8.53
	34.12	*30.81*	*29.46*	*28.71*	*28.24*	*27.91*	*27.49*	*27.05*	*26.60*	*26.12*
	167.5	**148.5**	**141.1**	**137.1**	**134.6**	**132.8**	**130.6**	**128.3**	**125.9**	**123.5**
4	7.71	6.94	6.59	6.39	6.26	6.16	6.04	5.91	5.77	5.63
	21.20	*18.00*	*16.69*	*15.98*	*15.52*	*15.21*	*14.80*	*14.37*	*13.93*	*13.46*
	74.14	**61.25**	**56.18**	**53.44**	**51.71**	**50.53**	**49.00**	**47.41**	**45.77**	**44.05**
5	6.61	5.79	5.41	5.19	5.05	4.95	4.82	4.68	4.53	4.36
	16.26	*13.27*	*12.06*	*11.39*	*10.97*	*10.67*	*10.27*	*9.89*	*9.47*	*9.02*
	47.04	**36.61**	**33.20**	**31.09**	**29.75**	**28.84**	**27.64**	**26.42**	**25.14**	**23.78**
6	5.99	5.14	4.76	4.53	4.39	4.28	4.15	4.00	3.84	3.67
	13.74	*10.92*	*9.78*	*9.15*	*8.75*	*8.47*	*8.10*	*7.72*	*7.31*	*6.88*
	35.51	**27.00**	**23.70**	**21.90**	**20.81**	**20.03**	**19.03**	**17.99**	**16.89**	**15.75**

	5.59	4.74	4.35	4.12	3.97	3.87	3.73	3.57	3.41	3.23
7	*12.25*	*9.55*	*8.45*	*7.85*	*7.46*	*7.19*	*6.84*	*6.47*	*6.07*	*5.65*
	29.22	**21.69**	**18.77**	**17.19**	**16.21**	**15.52**	**14.63**	**13.71**	**12.73**	**11.69**
	5.32	4.46	4.07	3.84	3.69	3.58	3.44	3.28	3.12	2.93
8	*11.26*	*8.65*	*7.59*	*7.01*	*6.63*	*6.37*	*6.03*	*5.67*	*5.28*	*4.86*
	25.42	**18.49**	**15.83**	**14.39**	**13.49**	**12.86**	**12.04**	**11.19**	**10.30**	**9.34**
	5.12	4.26	3.86	3.63	3.48	3.37	3.23	3.07	2.90	2.71
9	*10.56*	*8.02*	*6.99*	*6.42*	*6.06*	*5.80*	*5.47*	*5.11*	*4.73*	*4.31*
	22.86	**16.39**	**13.90**	**12.56**	**11.71**	**11.13**	**10.37**	**9.57**	**8.72**	**7.81**
	4.96	4.10	3.71	3.48	3.33	3.22	3.07	2.91	2.74	2.54
10	*10.04*	*7.56*	*6.55*	*5.99*	*5.64*	*5.39*	*5.06*	*4.71*	*4.33*	*3.91*
	21.04	**14.91**	**12.55**	**11.28**	**10.48**	**9.92**	**9.20**	**8.45**	**7.64**	**6.76**
	4.84	3.98	3.59	3.36	3.20	3.09	2.95	2.79	2.61	2.40
11	*9.65*	*7.20*	*6.22*	*5.67*	*5.32*	*5.07*	*4.74*	*4.40*	*4.02*	*3.60*
	19.69	**13.81**	**11.56**	**10.35**	**9.58**	**9.05**	**8.35**	**7.63**	**6.85**	**6.00**
	4.75	3.88	3.49	3.26	3.11	3.00	2.85	2.69	2.50	2.30
12	*9.33*	*6.93*	*5.95*	*5.41*	*5.06*	*4.82*	*4.50*	*4.16*	*3.78*	*3.36*
	18.64	**12.97**	**10.80**	**9.63**	**8.89**	**8.38**	**7.71**	**7.00**	**6.25**	**5.42**

[a]Table E.4 is reprinted, in rearranged form, from Table V of R. A. Fisher and F. Yates: "Statistical Tables for Biological, Agricultural, and Medical Research," published by Oliver & Boyd Ltd., Edinburgh, by permission of the authors and publishers.

TABLE E.4 (continued) **Table of F for .05** (roman), **.01** (*italic*), **and .001 (bold face) Levels of Significance**[a]

n_2 \ n_1	1	2	3	4	5	6	8	12	24	∞
13	4.67 *9.07* **17.81**	3.80 *6.70* **12.31**	3.41 *5.74* **10.21**	3.18 *5.20* **9.07**	3.02 *4.86* **8.35**	2.92 *4.62* **7.86**	2.77 *4.30* **7.21**	2.60 *3.96* **6.52**	2.42 *3.59* **5.78**	2.21 *3.16* **4.97**
14	4.60 *8.86* **17.14**	3.74 *6.51* **11.78**	3.34 *5.56* **9.73**	3.11 *5.03* **8.62**	2.96 *4.69* **7.92**	2.85 *4.46* **7.43**	2.70 *4.14* **6.80**	2.53 *3.80* **6.13**	2.35 *3.43* **5.41**	2.13 *3.00* **4.60**
15	4.54 *8.68* **16.59**	3.68 *6.36* **11.34**	3.29 *5.42* **9.34**	3.06 *4.89* **8.25**	2.90 *4.56* **7.57**	2.79 *4.32* **7.09**	2.64 *4.00* **6.47**	2.48 *3.67* **5.81**	2.29 *3.29* **5.10**	2.07 *2.87* **4.31**
16	4.49 *8.53* **16.12**	3.63 *6.23* **10.97**	3.24 *5.29* **9.00**	3.01 *4.77* **7.94**	2.85 *4.44* **7.27**	2.74 *4.20* **6.81**	2.59 *3.89* **6.19**	2.42 *3.55* **5.55**	2.24 *3.18* **4.85**	2.01 *2.75* **4.06**
17	4.45 *8.40* **15.72**	3.59 *6.11* **10.66**	3.20 *5.18* **8.73**	2.96 *4.67* **7.68**	2.81 *4.34* **7.02**	2.70 *4.10* **6.56**	2.55 *3.79* **5.96**	2.38 *3.45* **5.32**	2.19 *3.08* **4.63**	1.96 *2.65* **3.85**
18	4.41 *8.28* **15.38**	3.55 *6.01* **10.39**	3.16 *5.09* **8.49**	2.93 *4.58* **7.46**	2.77 *4.25* **6.81**	2.66 *4.01* **6.35**	2.51 *3.71* **5.76**	2.34 *3.37* **5.13**	2.15 *3.00* **4.45**	1.92 *2.57* **3.67**

19	4.38 *8.18* **15.08**	3.52 *5.93* **10.16**	3.13 *5.01* **8.28**	2.90 *4.50* **7.26**	2.74 *4.17* **6.61**	2.63 *3.94* **6.18**	2.48 *3.63* **5.59**	2.31 *3.30* **4.97**	2.11 *2.92* **4.29**	1.88 *2.49* **3.52**
20	4.35 *8.10* **14.82**	3.49 *5.85* **9.95**	3.10 *4.94* **8.10**	2.87 *4.43* **7.10**	2.71 *4.10* **6.46**	2.60 *3.87* **6.02**	2.45 *3.56* **5.44**	2.28 *3.23* **4.82**	2.08 *2.86* **4.15**	1.84 *2.42* **3.38**
21	4.32 *8.02* **14.59**	3.47 *5.78* **9.77**	3.07 *4.87* **7.94**	2.84 *4.37* **6.95**	2.68 *4.04* **6.32**	2.57 *3.81* **5.88**	2.42 *3.51* **5.31**	2.25 *3.17* **4.70**	2.05 *2.80* **4.03**	1.81 *2.36* **3.26**
22	4.30 *7.94* **14.38**	3.44 *5.72* **9.61**	3.05 *4.82* **7.80**	2.82 *4.31* **6.81**	2.66 *3.99* **6.19**	2.55 *3.76* **5.76**	2.40 *3.45* **5.19**	2.23 *3.12* **4.58**	2.03 *2.75* **3.92**	1.78 *2.31* **3.15**
23	4.28 *7.88* **14.19**	3.42 *5.66* **9.47**	3.03 *4.76* **7.67**	2.80 *4.26* **6.69**	2.64 *3.94* **6.08**	2.53 *3.71* **5.65**	2.38 *3.41* **5.09**	2.20 *3.07* **4.48**	2.00 *2.70* **3.82**	1.76 *2.26* **3.05**
24	4.26 *7.82* **14.03**	3.40 *5.61* **9.34**	3.01 *4.72* **7.55**	2.78 *4.22* **6.59**	2.62 *3.90* **5.98**	2.51 *3.67* **5.55**	2.36 *3.36* **4.99**	2.18 *3.03* **4.39**	1.98 *2.66* **3.74**	1.73 *2.21* **2.97**
25	4.24 *7.77* **13.88**	3.38 *5.57* **9.22**	2.99 *4.68* **7.45**	2.76 *4.18* **6.49**	2.60 *3.86* **5.88**	2.49 *3.63* **5.46**	2.34 *3.32* **4.91**	2.16 *2.99* **4.31**	1.96 *2.62* **3.66**	1.71 *2.17* **2.89**

[a]Table E.4 is reprinted, in rearranged form, from Table V of R. A. Fisher and F. Yates: "Statistical Tables for Biological, Agricultural, and Medical Research," published by Oliver & Boyd Ltd., Edinburgh, by permission of the authors and publishers.

TABLE E.4 (continued) Table of F for .05 (roman), .01 (*italic*), and .001 (bold face) Levels of Significance[a]

n_2 \ n_1	1	2	3	4	5	6	8	12	24	∞
26	4.22	3.37	2.98	2.74	2.59	2.47	2.32	2.15	1.95	1.69
	7.22	*5.53*	*4.64*	*4.14*	*3.82*	*3.59*	*3.29*	*2.96*	*2.58*	*2.13*
	13.74	**9.12**	**7.36**	**6.41**	**5.80**	**5.38**	**4.83**	**4.24**	**3.59**	**2.82**
27	4.21	3.35	2.96	2.73	2.57	2.46	2.30	2.13	1.93	1.67
	7.68	*5.49*	*4.60*	*4.11*	*3.78*	*3.56*	*3.26*	*2.93*	*2.55*	*2.10*
	13.61	**9.02**	**7.27**	**6.33**	**5.73**	**5.31**	**4.76**	**4.17**	**3.52**	**2.75**
28	4.20	3.34	2.95	2.71	2.56	2.44	2.29	2.12	1.91	1.65
	7.64	*5.45*	*4.57*	*4.07*	*3.75*	*3.53*	*3.23*	*2.90*	*2.52*	*2.06*
	13.50	**8.93**	**7.19**	**6.25**	**5.66**	**5.24**	**4.69**	**4.11**	**3.46**	**2.70**
29	4.18	3.33	2.93	2.70	2.54	2.43	2.28	2.10	1.90	1.64
	7.60	*5.42*	*4.54*	*4.04*	*3.73*	*3.50*	*3.20*	*2.87*	*2.49*	*2.03*
	13.39	**8.85**	**7.12**	**6.19**	**5.59**	**5.18**	**4.64**	**4.05**	**3.41**	**2.64**
30	4.17	3.32	2.92	2.69	2.53	2.42	2.27	2.09	1.89	1.62
	7.56	*5.39*	*4.51*	*4.02*	*3.70*	*3.47*	*3.17*	*2.84*	*2.47*	*2.01*
	13.29	**8.77**	**7.05**	**6.12**	**5.53**	**5.12**	**4.58**	**4.00**	**3.36**	**2.59**
40	4.08	3.23	2.84	2.61	2.45	2.34	2.18	2.00	1.79	1.51
	7.31	*5.18*	*4.31*	*3.83*	*3.51*	*3.29*	*2.99*	*2.66*	*2.29*	*1.80*
	12.61	**8.25**	**6.60**	**5.70**	**5.13**	**4.73**	**4.21**	**3.64**	**3.01**	**2.23**

	4.00	3.15	2.76	2.52	2.37	2.25	2.10	1.92	1.70	1.39
60	*7.08*	*4.98*	*4.13*	*3.65*	*3.34*	*3.12*	*2.82*	*2.50*	*2.12*	*1.60*
	11.97	**7.76**	**6.17**	**5.31**	**4.76**	**4.37**	**3.87**	**3.31**	**2.69**	**1.90**
	3.92	3.07	2.68	2.45	2.29	2.17	2.02	1.83	1.61	1.25
120	*6.85*	*4.79*	*3.95*	*3.48*	*3.17*	*2.96*	*2.66*	*2.34*	*1.95*	*1.38*
	11.38	**7.31**	**5.79**	**4.95**	**4.42**	**4.04**	**3.55**	**3.02**	**2.40**	**1.56**
	3.84	2.99	2.60	2.37	2.21	2.09	1.94	1.75	1.52	1.00
∞	*6.64*	*4.60*	*3.78*	*3.32*	*3.02*	*2.80*	*2.51*	*2.18*	*1.79*	*1.00*
	10.83	**6.91**	**5.42**	**4.62**	**4.10**	**3.74**	**3.27**	**2.74**	**2.13**	**1.00**

[a]Table E.4 is reprinted, in rearranged form, from Table V of R. A. Fisher and F. Yates: "Statistical Tables for Biological, Agricultural, and Medical Research," published by Oliver & Boyd Ltd., Edinburgh, by permission of the authors and publishers.

TABLE E.5 Four-Place Logarithms.

n	0	1	2	3	4	5	6	7	8	9
10	0000	0043	0086	0128	0170	0212	0253	0294	0334	0374
11	0414	0453	0492	0531	0569	0607	0645	0682	0719	0755
12	0792	0828	0864	0899	0934	0969	1004	1038	1072	1106
13	1139	1173	1206	1239	1271	1303	1335	1367	1399	1430
14	1461	1492	1523	1553	1584	1614	1644	1673	1703	1732
15	1761	1790	1818	1847	1875	1903	1931	1959	1987	2014
16	2041	2068	2095	2122	2148	2175	2201	2227	2253	2279
17	2304	2330	2355	2380	2405	2430	2455	2480	2504	2529
18	2553	2577	2601	2625	2648	2672	2695	2718	2742	2765
19	2788	2810	2833	2856	2878	2900	2923	2945	2967	2989
20	3010	3032	3054	3075	3096	3118	3139	3160	3181	3201
21	3222	3243	3263	3284	3304	3324	3345	3365	3385	3404
22	3424	3444	3464	3483	3502	3522	3541	3560	3579	3598
23	3617	3636	3655	3674	3692	3711	3729	3747	3766	3784
24	3802	3820	3838	3856	3874	3892	3909	3927	3945	3962
25	3979	3997	4014	4031	4048	4065	4082	4099	4116	4133
26	4150	4166	4183	4200	4216	4232	4249	4265	4281	4298
27	4314	4330	4346	4362	4378	4393	4409	4425	4440	4456
28	4472	4487	4502	4518	4533	4548	4564	4579	4594	4609
29	4624	4639	4654	4669	4683	4698	4713	4728	4742	4757
30	4771	4786	4800	4814	4829	4843	4857	4871	4886	4900
31	4914	4928	4942	4955	4969	4983	4997	5011	5024	5038
32	5051	5065	5079	5092	5105	5119	5132	5145	5159	5172
33	5185	5198	5211	5224	5237	5250	5263	5276	5289	5302
34	5315	5328	5340	5353	5366	5378	5391	5403	5416	5428
35	5441	5453	5465	5478	5490	5502	5514	5527	5539	5551
36	5563	5575	5587	5599	5611	5623	5635	5647	5658	5670
37	5682	5694	5705	5717	5729	5740	5752	5763	5775	5786
38	5798	5809	5821	5832	5843	5855	5866	5877	5888	5899
39	5911	5922	5933	5944	5955	5966	5977	5988	5999	6010
40	6021	6031	6042	6053	6064	6075	6085	6096	6107	6117
41	6128	6138	6149	6160	6170	6180	6191	6201	6212	6222
42	6232	6243	6253	6263	6274	6284	6294	6304	6314	6325
43	6335	6345	6355	6365	6375	6385	6395	6405	6415	6425
44	6435	6444	6454	6464	6474	6484	6493	6503	6513	6522
45	6532	6542	6551	6561	6571	6580	6590	6599	6609	6618

TABLE E.5 (Continued). Four-Place Logarithms

n	0	1	2	3	4	5	6	7	8	9
46	6628	6637	6646	6656	6665	6675	6684	6693	6702	6712
47	6721	6730	6739	6749	6758	6767	6776	6785	6794	6803
48	6812	6821	6830	6839	6848	6857	6866	6875	6884	6893
49	6902	6911	6920	6928	6937	6946	6955	6964	6972	6981
50	6990	6998	7007	7016	7024	7033	7042	7050	7059	7067
51	7076	7084	7093	7101	7110	7118	7126	7135	7143	7152
52	7160	7168	7177	7185	7193	7202	7210	7218	7226	7235
53	7243	7251	7259	7267	7275	7284	7292	7300	7308	7316
54	7324	7332	7340	7348	7356	7364	7372	7380	7388	7396
55	7404	7412	7419	7427	7435	7443	7451	7459	7466	7474
56	7482	7490	7497	7505	7513	7520	7528	7536	7543	7551
57	7559	7566	7574	7582	7589	7597	7604	7612	7619	7627
58	7634	7642	7649	7657	7664	7672	7679	7686	7694	7701
59	7709	7716	7723	7731	7738	7745	7752	7760	7767	7774
60	7782	7789	7796	7803	7810	7818	7825	7832	7839	7846
61	7853	7860	7868	7875	7882	7889	7896	7903	7910	7917
62	7924	7931	7938	7945	7952	7959	7966	7973	7980	7987
63	7993	8000	8007	8014	8021	8028	8035	8041	8048	8055
64	8062	8069	8075	8082	8089	8096	8102	8109	8116	8122
65	8129	8136	8142	8149	8156	8162	8169	8176	8182	8189
66	8195	8202	8209	8215	8222	8228	8235	8241	8248	8254
67	8261	8267	8274	8280	8287	8293	8299	8306	8312	8319
68	8325	8331	8338	8344	8351	8357	8363	8370	8376	8382
69	8388	8395	8401	8407	8414	8420	8426	8432	8439	8445
70	8451	8457	8463	8470	8476	8482	8488	8494	8500	8506
71	8513	8519	8525	8531	8537	8543	8549	8555	8561	8567
72	8573	8579	8585	8591	8597	8603	8609	8615	8621	8627
73	8633	8639	8645	8651	8657	8663	8669	8675	8681	8686
74	8692	8698	8704	8710	8716	8722	8727	8733	8739	8745
75	8751	8756	8762	8768	8774	8779	8785	8791	8797	8802
76	8808	8814	8820	8825	8831	8837	8842	8848	8854	8859
77	8865	8871	8876	8882	8887	8893	8899	8904	8910	8915
78	8921	8927	8932	8938	8943	8949	8954	8960	8965	8971
79	8976	8982	8987	8993	8998	9004	9009	9015	9020	9025
80	9031	9036	9042	9047	9053	9058	9063	9069	9074	9079

Table E.5 (Continued). Four-Place Logarithms

n	0	1	2	3	4	5	6	7	8	9
81	9085	9090	9096	9101	9106	9112	9117	9122	9128	9133
82	9138	9143	9149	9154	9159	9165	9170	9175	9180	9186
83	9191	9196	9201	9206	9212	9217	9222	9227	9232	9238
84	9243	9248	9253	9258	9263	9269	9274	9279	9284	9289
85	9294	9299	9304	9309	9315	9320	9325	9330	9335	9340
86	9345	9350	9355	9360	9365	9370	9375	9380	9385	9390
87	9395	9400	9405	9410	9415	9420	9425	9430	9435	9440
88	9445	9450	9455	9460	9465	9469	9474	9479	9484	9489
89	9494	9499	9504	9509	9513	9518	9523	9528	9533	9538
90	9542	9547	9552	9557	9562	9566	9571	9576	9581	9586
91	9590	9595	9600	9605	9609	9614	9619	9624	9628	9633
92	9638	9643	9647	9652	9657	9661	9666	9671	9675	9680
93	9685	9689	9694	9699	9703	9708	9713	9717	9722	9727
94	9731	9736	9741	9745	9750	9754	9759	9763	9768	9773
95	9777	9782	9786	9791	9795	9800	9805	9809	9814	9818
96	9823	9827	9832	9836	9841	9845	9850	9854	9859	9863
97	9868	9872	9877	9881	9886	9890	9894	9899	9903	9908
98	9912	9917	9921	9926	9930	9934	9939	9943	9948	9952
99	9956	9961	9965	9969	9974	9978	9983	9987	9991	9996

TABLE E.6 Squares, Square Roots, and Reciprocals.

n	n^2	$\sqrt{n}$	$\sqrt{10n}$	$1/n$	n	n^2	$\sqrt{n}$	$\sqrt{10n}$	$1/n$
1	1	1.000	3.162	1.00000	36	1296	6.000	18.974	.02778
2	4	1.414	4.472	.50000	37	1369	6.083	19.235	.02703
3	9	1.732	5.477	.33333	38	1444	6.164	19.494	.02632
4	16	2.000	6.325	.25000	39	1521	6.245	19.748	.02564
5	25	2.236	7.071	.20000	40	1600	6.325	20.000	.02500
6	36	2.449	7.746	.16667	41	1681	6.403	20.248	.02439
7	49	2.646	8.367	.14286	42	1764	6.481	20.494	.02381
8	64	2.828	8.944	.12500	43	1849	6.557	20.736	.02326
9	81	3.000	9.487	.11111	44	1936	6.633	20.976	.02273
10	100	3.162	10.000	.10000	45	2025	6.708	21.213	.02222
11	121	3.317	10.488	.09091	46	2116	6.782	21.448	.02174
12	144	3.464	10.954	.08333	47	2209	6.856	21.679	.02128
13	169	3.606	11.402	.07692	48	2304	6.928	21.909	.02083
14	196	3.742	11.832	.07143	49	2401	7.000	22.136	.02041
15	225	3.873	12.247	.06667	50	2500	7.071	22.361	.02000
16	256	4.000	12.649	.06250	51	2601	7.141	22.583	.01961
17	289	4.123	13.038	.05882	52	2704	7.211	22.804	.01923
18	324	4.243	13.416	.05556	53	2809	7.280	23.022	.01887
19	361	4.359	13.784	.05263	54	2916	7.348	23.238	.01852
20	400	4.472	14.142	.05000	55	3025	7.416	23.452	.01818
21	441	4.583	14.491	.04762	56	3136	7.483	23.664	.01786
22	484	4.690	14.832	.04545	57	3249	7.550	23.875	.01754
23	529	4.796	15.166	.04348	58	3364	7.616	24.083	.01724
24	576	4.899	15.492	.04167	59	3481	7.681	24.290	.01695
25	625	5.000	15.811	.04000	60	3600	7.746	24.495	.01667
26	676	5.099	16.125	.03846	61	3721	7.810	24.698	.01639
27	729	5.196	16.432	.03704	62	3844	7.874	24.900	.01613
28	784	5.292	16.733	.03571	63	3969	7.937	25.100	.01587
29	841	5.385	17.029	.03448	64	4096	8.000	25.298	.01562
30	900	5.477	17.321	.03333	65	4225	8.062	25.495	.01538
31	961	5.568	17.607	.03226	66	4356	8.124	25.690	.01515
32	1024	5.657	17.889	.03125	67	4489	8.185	25.884	.01493
33	1089	5.745	18.166	.03030	68	4624	8.246	26.077	.01471
34	1156	5.831	18.439	.02941	69	4761	8.307	26.268	.01449
35	1225	5.916	18.708	.02857	70	4900	8.367	26.458	.01429

(*continued*)

TABLE E.6 (continued) Squares, Square Roots, and Reciprocals

n	n^2	$\sqrt{n}$	$\sqrt{10n}$	$1/n$	n	n^2	$\sqrt{n}$	$\sqrt{10n}$	$1/n$
71	5041	8.426	26.646	.01408	86	7396	9.274	29.326	.01163
72	5184	8.485	26.833	.01389	87	7569	9.327	29.496	.01149
73	5329	8.544	27.019	.01370	88	7744	9.381	29.665	.01136
74	5476	8.602	27.203	.01351	89	7921	9.434	29.833	.01124
75	5625	8.660	27.386	.01333	90	8100	9.487	30.000	.01111
76	5776	8.718	27.568	.01316	91	8281	9.539	30.166	.01099
77	5929	8.775	27.749	.01299	92	8464	9.592	30.332	.01087
78	6084	8.832	27.928	.01282	93	8649	9.644	30.496	.01075
79	6241	8.888	28.107	.01266	94	8836	9.695	30.659	.01064
80	6400	8.944	28.284	.01250	95	9025	9.747	30.822	.01053
81	6561	9.000	28.460	.01235	96	9216	9.798	30.984	.01042
82	6724	9.055	28.636	.01220	97	9409	9.849	31.145	.01031
83	6889	9.110	28.810	.01205	98	9604	9.899	31.305	.01020
84	7056	9.165	28.983	.01190	99	9801	9.950	31.464	.01010
85	7225	9.220	29.155	.01176	100	10000	10.000	31.623	.01000

Appendix F

Answers to Exercises

CHAPTER TWO

p. 13

1. (a) $\begin{bmatrix} 5 & 0 & -15/2 \\ 5 & -4 & 1/2 \end{bmatrix}$ (c) $\begin{bmatrix} 6 & 0 & -9 \\ 11/2 & -29/6 & 2/3 \end{bmatrix}$

   (d) $\begin{bmatrix} -1 & -1/2 \\ 0 & 5/6 \\ 3/2 & -1/6 \end{bmatrix}$

   (b) and (e) are meaningless because the two matrices are of different orders.

2. $\mathbf{D} = \begin{bmatrix} 2 & 0 & -3 \\ 1 & -5/3 & 1/3 \end{bmatrix}$ $\mathbf{E} = \begin{bmatrix} -2 & 0 & 3 \\ 4 & 2 & -1 \end{bmatrix}$

3. $\mathbf{D} + \mathbf{E} = \begin{bmatrix} 0 & 0 & 0 \\ 5 & 1/3 & -2/3 \end{bmatrix} = 2\begin{bmatrix} 0 & 0 & 0 \\ 5/2 & 1/6 & -1/3 \end{bmatrix} = 2\mathbf{C}'$

p. 20

1. $\mathbf{A}(\mathbf{B} + \mathbf{C}) = \begin{bmatrix} 3 & 1 \\ 2 & 0 \\ -1 & 2 \end{bmatrix}\begin{bmatrix} 3 & 1 & 0 \\ 0 & 0 & 2 \end{bmatrix} = \begin{bmatrix} 9 & 3 & 2 \\ 6 & 2 & 0 \\ -3 & -1 & 4 \end{bmatrix}$

   $\mathbf{AB} = \begin{bmatrix} 2 & -4 & 4 \\ 2 & -4 & 2 \\ -3 & 6 & 1 \end{bmatrix}$ $\mathbf{AC} = \begin{bmatrix} 7 & 7 & -2 \\ 4 & 6 & -2 \\ 0 & -7 & 3 \end{bmatrix}$

2. $(\mathbf{B} + \mathbf{C})\mathbf{A} = \begin{bmatrix} 3 & 1 & 0 \\ 0 & 0 & 2 \end{bmatrix}\begin{bmatrix} 3 & 1 \\ 2 & 0 \\ -1 & 2 \end{bmatrix} = \begin{bmatrix} 11 & 3 \\ -2 & 4 \end{bmatrix}$

   $\mathbf{BA} = \begin{bmatrix} -2 & 3 \\ 0 & 1 \end{bmatrix}$ $\mathbf{CA} = \begin{bmatrix} 13 & 0 \\ -2 & 3 \end{bmatrix}$

3. $$(\mathbf{AB})\mathbf{C}' = \begin{bmatrix} 2 & -4 & 4 \\ 2 & -4 & 2 \\ -3 & 6 & 1 \end{bmatrix} \begin{bmatrix} 2 & 1 \\ 3 & -2 \\ -1 & 1 \end{bmatrix} = \begin{bmatrix} -12 & 14 \\ -10 & 12 \\ 11 & -14 \end{bmatrix}$$

$$\mathbf{A}(\mathbf{BC}') = \begin{bmatrix} 3 & 1 \\ 2 & 0 \\ -1 & 2 \end{bmatrix} \begin{bmatrix} -5 & 6 \\ 3 & -4 \end{bmatrix} = \begin{bmatrix} -12 & 14 \\ -10 & 12 \\ 11 & -14 \end{bmatrix}$$

Verifies associative law of matrix multiplication (Eq. 2.8).

4. $$[(\mathbf{AB})\mathbf{C}']' = \begin{bmatrix} -12 & -10 & 11 \\ 14 & 12 & -14 \end{bmatrix}$$ (from answer to Exercise 3)

$$(\mathbf{CB}')\mathbf{A}' = \begin{bmatrix} -5 & 3 \\ 6 & -4 \end{bmatrix} \begin{bmatrix} 3 & 2 & -1 \\ 1 & 0 & 2 \end{bmatrix} = \begin{bmatrix} -12 & -10 & 11 \\ 14 & 12 & -14 \end{bmatrix}$$

Verifies Eqs. 2.1 and 2.11.

5. $$[1 \quad 2 \quad -1] \begin{bmatrix} 2 & -4 & 4 \\ 2 & -4 & 2 \\ -3 & 6 & 1 \end{bmatrix} \begin{bmatrix} 1 \\ 2 \\ -1 \end{bmatrix} = [9 \quad -18 \quad 7] \begin{bmatrix} 1 \\ 2 \\ -1 \end{bmatrix} = -34$$

Using Eq. 2.9, $(2)1^2 + (-4)2^2 + (1)(-1)^2 + (-4 + 2)(1)(2) + (4 - 3)(1)(-1) + (2 + 6)(2)(-1) = -34$

6. $$\mathbf{DC} = \begin{bmatrix} 4 & 6 & -2 \\ 3 & -6 & 3 \end{bmatrix}$$

The first diagonal element of **D** multiplies the first row of **C**; the second diagonal element of **D** multiplies the second row of **C**.

7. $$\mathbf{AD} = \begin{bmatrix} 6 & 3 \\ 4 & 0 \\ -2 & 6 \end{bmatrix}$$

The first diagonal element of **D** multiplies the first column of **A**; the second diagonal element of **D** multiplies the second column of **A**.

8. $$\mathbf{EA} = \begin{bmatrix} 15 & 5 \\ 10 & 0 \\ -5 & 10 \end{bmatrix}$$

The common diagonal element of **E** multiplies each row of **A**. Hence every element of **A** gets multiplied by the common diagonal element 5.

p. 25 1. (a) $\begin{bmatrix} 2 & -3/2 \\ -3 & 5/2 \end{bmatrix}$ (b) $\begin{bmatrix} -3 & -2 \\ -2 & -1 \end{bmatrix}$

(c) singular (d) $\begin{bmatrix} 1/3 & 0 \\ 0 & -1/5 \end{bmatrix}$

2. (a) $\frac{1}{2}\begin{bmatrix} 2 & 4 & 3 \\ -3 & 2 & 1 \\ -1 & 3 & 2 \end{bmatrix}$ (b) $\frac{1}{6}\begin{bmatrix} 3 & 1 & 1 \\ 0 & 2 & -4 \\ 3 & -3 & 3 \end{bmatrix}$

   (c) singular

3. The inverse of a diagonal matrix is the diagonal matrix whose diagonal elements are each equal to the reciprocal of the corresponding diagonal element of the original. Proof follows from Eqs. 2.12a and 2.12b (pre- and postmultiplying by a diagonal matrix).

4. $(\mathbf{A}^{-1})(\mathbf{A}^{-1})^{-1} = \mathbf{I}$ (definition of inverse matrix)

   $\mathbf{A}(\mathbf{A}^{-1})(\mathbf{A}^{-1})^{-1} = \mathbf{AI} = \mathbf{A}$ (premultiplying by **A** and definition of identity matrix)

   $(\mathbf{AA}^{-1})(\mathbf{A}^{-1})^{-1} = \mathbf{A}$ (associative law)

   $(\mathbf{A}^{-1})^{-1} = \mathbf{A}$ (definitions of inverse and identity)

   Equation 2.18 follows immediately on multiplying $c\mathbf{A}$ by $(1/c)\mathbf{A}^{-1}$.

5. $\mathbf{A}^{-1} = \frac{1}{4}\begin{bmatrix} 4 & -2 \\ -8 & 5 \end{bmatrix}$ $\therefore$ $\mathbf{v}'\mathbf{A}^{-1}\mathbf{v} = (1/4)[-2 \quad 6]\begin{bmatrix} 4 & -2 \\ -8 & 5 \end{bmatrix}\begin{bmatrix} -2 \\ 6 \end{bmatrix}$

   $= 79$

6. $(\mathbf{AB})(\mathbf{B}^{-1}\mathbf{A}^{-1}) = \mathbf{A}(\mathbf{BB}^{-1})\mathbf{A}^{-1}$ (associative law)

   $= \mathbf{A}(\mathbf{I})\mathbf{A}^{-1}$ (definition of inverse)

   $= \mathbf{AA}^{-1} = \mathbf{I}$ (associative law, definitions of identity and inverse)

CHAPTER THREE

p. 61 $F_\beta < 1$. Therefore accept null hypothesis that regression planes are parallel in the three populations.

$F_\mu = 25.13 > {}_{.99}F_{10}{}^2 = 14.91$. Therefore reject null hypothesis concerning equality of "adjusted" population means; conclude that significant differences among group means persist after covariance adjustments. Adjusted X_0 means in the three groups are: 10.03, 16.55, and 21.81.

CHAPTER FOUR

p. 72 1. Find $|\mathbf{\Sigma}|$ and $\mathbf{\Sigma}^{-1}$, substitute these in Eq. 4.13, and show that the expression factors as indicated in the problem.

2. $(X_1 - 15)^2 + (X_2 - 20)^2 - 1.2(X_1 - 15)(X_2 - 20) = 44.37$

3. $[(5 - 15)^2 + (8 - 20)^2 - 1.2(5 - 15)(8 - 20)]/16 = 6.25$

   which is slightly greater than the 95th centile of the chi-square distribution with 2 d.f. Hence, slightly more than 95% of the population lies within the ellipse. (Interpolation in Table E.3 gives about 95.5%.)

p. 76 $\chi_2^2(.95) = 5.991$. Therefore, by Eq. 4.19, the equation of the desired ellipse is

$$[Y_1 - 18, \quad Y_2 - 15]\begin{bmatrix} 25 & 12 \\ 12 & 16 \end{bmatrix}^{-1}\begin{bmatrix} Y_1 - 18 \\ Y_2 - 15 \end{bmatrix} = 5.991/20$$

Or, in scalar notation,

$$(1.25)(Y_1 - 18)^2 + (1.9531)(Y_2 - 15)^2 - (1.875)(Y_1 - 18)(Y_2 - 15) = 5.991$$

The ellipse may be plotted by computing pairs of Y_2 values, for each of several Y_1 values, from this equation. A better method will be described in Chapter Five.

p. 91 1. $\mathbf{d} = \begin{bmatrix} .7 \\ 2.6 \end{bmatrix}$ $\mathbf{S}_d = \begin{bmatrix} 22.1 & 2.8 \\ 2.8 & 90.4 \end{bmatrix}$ $\mathbf{S}_d^{-1} = \begin{bmatrix} 4.5427 & -.1407 \\ -.1407 & 1.1106 \end{bmatrix} \times 10^{-2}$

Hence

$$\mathbf{T}^2 = (10)(9)[.7 \quad 2.6]\begin{bmatrix} 4.5427 & -.1407 \\ -.1407 & 1.1106 \end{bmatrix}\begin{bmatrix} .7 \\ 2.6 \end{bmatrix} \times 10^{-2} = 8.299$$

The final test statistic is $F = 3.688 < {}_{.95}F_8{}^2 = 4.46$. The null hypothesis cannot be rejected at the 5% level.

2. $$\mathbf{T}^2 = [(20)(30)(48)/50][-.9 \quad .6]\begin{bmatrix} 95.4 & 11.6 \\ 11.6 & 29.2 \end{bmatrix}^{-1}\begin{bmatrix} -.9 \\ .6 \end{bmatrix} = 15.32$$

$$F = [47/(48)(2)](15.32) = 7.50 > {}_{.99}F_{47}{}^2 = 5.11$$

We conclude, at the 1% level, that the population centroids for drug addicts and alcoholics on these two tests are different.

3. The determinants of **W** and **T** are approximately 5903×10^4 and $15{,}152 \times 10^4$. Hence, $\Lambda = 5903/15{,}152 = .3896$. Since $K = 3$ and $p = 2$ in this problem, we may use either the second or the fourth expression in Table 4.2 to obtain $F = 16.86$, which far exceeds the 99th centile of the F_{112}^4 distribution. The null hypothesis of equality of centroids is rejected.

CHAPTER FIVE

p. 105 Substitute $\mathbf{Y}' = \mathbf{X}'\mathbf{V}$ in $\mathbf{Z}' = \mathbf{Y}'\mathbf{U}$ to obtain $\mathbf{Z}' = (\mathbf{X}'\mathbf{V})\mathbf{U} = \mathbf{X}'(\mathbf{V}\mathbf{U})$. It remains only to prove that **VU** is an orthogonal matrix. That is, show that $(\mathbf{VU})(\mathbf{VU})' = \mathbf{I}$ and that $|\mathbf{VU}| = 1$. Use Eqs. 2.11 and A.6 (Appendix A).

p. 114 2. The major axis of the ellipse makes an angle of $-36°52'$ (i.e., about 37° in the clockwise direction) with the positive X_1 (horizontal) axis of the reference system.

3.
$$\text{Var}(\mathbf{Y}) = [.8 \quad -.6]\begin{bmatrix} 23 & -12 \\ -12 & 16 \end{bmatrix}\begin{bmatrix} .8 \\ -.6 \end{bmatrix} = 32.00$$

Axes 10° away from OY in the clockwise and counterclockwise directions make angles of $-46°52'$ and $-26°52'$, respectively, with OX_1. For each of these angles, we look up the cosine and sine in a table of trigonometric functions, and use these as the first and second elements, respectively, of the transformation vector (i.e., in the place .8 and $-.6$ above). We thus find

$$\text{Var}(Y') = [.7298 \quad -.6837]\begin{bmatrix} 23 & -12 \\ -12 & 16 \end{bmatrix}\begin{bmatrix} .7298 \\ -.6837 \end{bmatrix} = 31.70$$

$$\text{Var}(Y'') = [.8921 \quad -.4519]\begin{bmatrix} 23 & -12 \\ -12 & 16 \end{bmatrix}\begin{bmatrix} .8921 \\ -.4519 \end{bmatrix} = 31.25$$

The variance of Y is greater than that of either Y' or Y''.

4. The minor axis makes an angle of $(-36°52' + 90°) = 53°8'$ with the positive X_1 axis.

5. Calling the variable defined by the minor axis Y_2, and those by axes 10° away from it in the clockwise and counterclockwise directions, Y_2' and Y_2'', respectively, we find (in the same way as in Exercise 3): $\text{Var}(Y_2) = 7.00$, $\text{Var}(Y_2') = 7.75$, $\text{Var}(Y_2'') = 7.76$.

p. 119 The characteristic equation is $\lambda^3 - 17\lambda^2 + 86\lambda - 112 = 0$, and the eigenvalues are $\lambda_1 = 8$, $\lambda_2 = 7$, and $\lambda_3 = 2$.

p. 121 $\mathbf{v}_2' = [.8944 \quad -.4472 \quad 0]$ $\qquad$ $\mathbf{v}_3' = [-.1825 \quad -.3651 \quad .9128]$

p. 124 1. The eigenvalues are $\lambda_1 = 173.52$, $\lambda_2 = 58.90$, $\lambda_3 = 16.67$. The orthogonal transformation matrix, whose columns are the corresponding eigenvectors $\mathbf{v}_1$, $\mathbf{v}_2$, and $\mathbf{v}_3$, is:

$$\mathbf{V} = \begin{bmatrix} .8181 & .4091 & -.4042 \\ .4997 & -.1576 & .8517 \\ -.2846 & .8988 & .3334 \end{bmatrix}$$

2. For the first transformed variable, we have

$$\sigma_{y_1}{}^2 = [.8181 \quad .4997 \quad -.2846]\begin{bmatrix} 128.7 & 61.4 & -21.0 \\ 61.4 & 56.9 & -28.3 \\ -21.0 & -28.3 & 63.5 \end{bmatrix}\begin{bmatrix} .8181 \\ .4997 \\ -.2846 \end{bmatrix} = 173.52$$

The other two cases are verified similarly.

p. 135 1. Collecting the three vectors of Example 5.7 (as columns) in a matrix, we have

$$\mathbf{A} = \begin{bmatrix} 1 & 3 & -3 \\ 2 & -1 & 15 \\ -1 & 0 & -6 \end{bmatrix}$$

The characteristic equation $|\mathbf{A} - \lambda \mathbf{I}| = 0$ is found to be

$$-\lambda^3 - 6\lambda^2 + 10\lambda = 0$$

which clearly has one zero root. The rank of **A** is 2, so the given set of vectors is linearly dependent.

2. The characteristic equation of the matrix comprising the vectors in Example 5.8 is $-\lambda^3 + (7/2)\lambda^2 + (11/2)\lambda = 0$. The set is therefore linearly dependent.

3. The set of vectors in Example 5.9 form a diagonal matrix, and hence the characteristic equation is readily found to be

$$(d_1 - \lambda)(d_2 - \lambda)(d_3 - \lambda) = 0$$

whose roots are d_1, d_2, and d_3, none of which is zero. Therefore the given set is linearly independent. This result obviously generalizes to higher dimensions: Any set of vectors which, taken in suitable order, forms a diagonal matrix (with no zero in the diagonal position) is linearly independent.

4. The characteristic equation of the matrix comprising the vectors in Example 5.10 is $-\lambda^3 - 2\lambda^2 + 16\lambda + 28 = 0$ which clearly has no zero root. Therefore, the set is linearly independent.

p. 142 1. $\lambda_1 = 10,\ \lambda_2 = 8,\ \lambda_3 = 3;\qquad \mathbf{v}_1' = [.2672 \quad .8018 \quad .5345],$
$\mathbf{v}_2' = [.9487 \quad -.3162 \quad 0],\qquad \mathbf{v}_3' = [-.1690 \quad -.5070 \quad .8452]$

The verifications in parts (a)–(d) are simple. To verify (e) numerically, assume (for simplicity) that the centroid is (0, 0, 0). Then take any $\mathbf{x}'$, for example [1, 2, 3], and compute $\mathbf{x}'\boldsymbol{\Sigma}\mathbf{x}$. Next, using the transformation matrix with $\mathbf{v}_1, \mathbf{v}_2, \mathbf{v}_3$ as columns, transform **x** into **y** and $\boldsymbol{\Sigma}$ into $\boldsymbol{\Sigma}_y$; compute $\mathbf{y}'\boldsymbol{\Sigma}\mathbf{y}$ and compare with the value of $\mathbf{x}'\boldsymbol{\Sigma}\mathbf{x}$ previously obtained.

2. The principal minor determinants of $\boldsymbol{\Sigma}$ are: 8, 8, 5, $\begin{vmatrix} 8 & 3 \\ 3 & 5 \end{vmatrix} = 31$, $\begin{vmatrix} 8 & 1 \\ 1 & 5 \end{vmatrix} = 39$, $\begin{vmatrix} 8 & 0 \\ 0 & 8 \end{vmatrix} = 64$, and $|\boldsymbol{\Sigma}| = 240$. And $\boldsymbol{\Sigma}$ is symmetric. The required factor matrix **B**, following Eq. 5.36, is

$$\mathbf{B} = \begin{bmatrix} .8481 & 2.5355 & 1.6902 \\ 2.6833 & -.8943 & 0 \\ -.2927 & -.8782 & 1.4640 \end{bmatrix}$$

p. 154 1. The eigenvalues are .8, .7, and .2. The eigenvectors are

$$[.4082 \quad .8165 \quad .4082], \qquad [.8944 \quad -.4472 \quad 0],$$
$$[-.1825 \quad -.3651 \quad .9128]$$

4. The two matrices whose common eigenvalues and respective eigenvectors we need to obtain are:

$$\mathbf{A'A} = \begin{bmatrix} 100 & 0 \\ 0 & 50 \end{bmatrix} \quad \text{and} \quad \mathbf{AA'} = \begin{bmatrix} 73 & 15 & 36 \\ 15 & 25 & -20 \\ 36 & -20 & 52 \end{bmatrix}, \quad \text{with} \quad \begin{matrix} \lambda_1 = 100 \\ \lambda_2 = 50 \end{matrix}$$

Their eigenvectors are the columns of

$$\mathbf{U} = \begin{bmatrix} 1 & 0 \\ 0 & 1 \end{bmatrix} \quad \text{and} \quad \mathbf{V} = \begin{bmatrix} .8 & 3/\sqrt{50} \\ 0 & 5/\sqrt{50} \\ .6 & -4/\sqrt{50} \end{bmatrix}, \quad \text{respectively.}$$

Hence, in accordance with Eq. 5.44, the left general inverse of **A** is

$$\mathbf{X} = \mathbf{U} \begin{bmatrix} \sqrt{\lambda_1} & 0 \\ 0 & \sqrt{\lambda_2} \end{bmatrix}^{-1} \mathbf{V'} = \begin{bmatrix} .08 & .0 & .06 \\ .06 & .10 & -.08 \end{bmatrix}$$

CHAPTER SIX

p. 191 1.

$$\mathbf{W}^{-1}\mathbf{B} = \begin{bmatrix} .13762 & .06940 & -.10411 \\ -.00880 & .02493 & .00739 \\ -.11964 & -.05638 & .09056 \end{bmatrix}$$

The characteristic equation is $\lambda^3 - .25311\lambda^2 + .00672\lambda = 0$, with roots $\lambda_1 = .22297$, $\lambda_2 = .03015$, and $\lambda_3 = 0$. The eigenvectors associated with the two nonzero roots are:

$$\mathbf{v}_1' = [.7529 \quad -.0579 \quad -.6556], \qquad \mathbf{v}_2' = [-.2607 \quad .9071 \quad .3304]$$

Significance test: $V = (296)(2.3026)[\log 1.223 + \log 1.030]$

$$= 59.52 + 8.74 = 68.26$$

$$68.26 > \chi_6^2(.95) = 12.59 \quad \text{and} \quad 8.74 > \chi_2^2(.95) = 5.99$$

Both discriminant functions are significant at the 5% level.

The discriminant-function means for the three groups are:

	Group A	Group B	Group C
Y_1-mean	3.22	3.87	3.08
Y_2-mean	3.76	3.97	4.09

On the first discriminant dimension, Group B (Physical Science) is much higher than the other two groups. Noting that Y_1 has a large positive weight for the S.A.T. Mathematics Test, and a substantial negative weight for the persuasiveness scale, we may interpret this as an abstract and non-"person-oriented" factor.

The second discriminant function has a high positive weight for the S.A.T. Verbal test, and a moderate weight for the persuasiveness scale. It is understandable that the Industrial Management group should average highest on this dimension. We might call this a verbal and "people-oriented" factor.

2. The characteristic equation of $\mathbf{W}^{-1}\mathbf{B}$ is $\lambda^3 - .3575\lambda^2 + .0216\lambda = 0$, with roots $\lambda_1 = .2807$, $\lambda_2 = .0768$, and $\lambda_3 = 0$.
The eigenvectors associated with the two nonzero roots are:

$$\mathbf{v}_1' = [.5439. \quad .8361 \quad -.0717], \qquad \mathbf{v}_2' = [-.3633 \quad .8999 \quad -.2415].$$

Significance test: $V = (96)(2.3026)[\log 1.2807 + \log 1.0768]$

$$= 23.75 + 7.10 = 30.85$$

The overall V (as a chi-square with 6 d.f.'s) is significant at the .001 level, and the residual 7.10 after removing the effect of the first discriminant is (as a chi-square with 2 d.f.'s) significant at the .05 level.

3. The SSCP matrix for the three predictor variables and the two group-membership "criterion" variables Y_1 and Y_2 in the total sample is:

$$\mathbf{S} = \begin{array}{ccc|cc} 271.34 & 40.38 & 55.43 & -31.00 & 7.80 \\ 40.38 & 69.61 & 5.49 & -15.50 & 12.90 \\ 55.43 & 5.49 & 59.80 & -5.00 & -.60 \\ \hline -31.00 & -15.50 & -5.00 & 25.00 & -15.00 \\ 7.80 & 12.90 & -.60 & -15.00 & 21.00 \end{array}$$

The upper left-hand 3×3 submatrix $\mathbf{S}_{pp}$ is the sum of $\mathbf{W}$ and $\mathbf{B}$ given in the question. The lower right-hand 2×2 submatrix $\mathbf{S}_{cc}$ is computed by noting that $\Sigma Y_1 = \Sigma Y_1{}^2 = n_1 = 50$, $\Sigma Y_1 Y_2 = 0$, and $\Sigma Y_2 = \Sigma Y_2{}^2 = n_2 = 30$, with $N = n_1 + n_2 + n_3 = 100$. For computing the upper right-hand 3×2 submatrix $\mathbf{S}_{pc}$, we first have to retrieve the group totals and grand total on X_1, X_2, and X_3 from the group means given. Then note that $\Sigma X_i Y_1 = T_{i1}$ and $\Sigma X_i Y_2 = T_{i2}$ (the totals of X_i in Groups 1 and 2, respectively). The matrix whose eigenvectors yield the discriminant weights is

$$\mathbf{S}_{pp}^{-1}\mathbf{S}_{pc}\mathbf{S}_{cc}^{-1}\mathbf{S}_{cp} = \begin{bmatrix} .156826 & .043267 & .032550 \\ .138548 & .129894 & .009831 \\ -.000329 & -.018279 & .003687 \end{bmatrix}$$

The two nonzero eigenvalues are $\mu_1{}^2 = .2193$ and $\mu_2{}^2 = .0712$. The corresponding eigenvectors are, of course, equal within rounding error to those found in Exercise 2 above.

CHAPTER SEVEN

p. 214 1.

Source	SSCP Matrix	*n.d.f.*	$\lvert S_h + S_e \rvert$ or $\lvert S_e \rvert$	Λ	*F*-ratio
Education	$\begin{bmatrix} .042 & .255 & -.012 \\ .255 & 1.721 & -.116 \\ -.012 & -.116 & .014 \end{bmatrix}$	2	2.357	.5325	5.68
Sex	$\begin{bmatrix} .510 & .405 & .260 \\ .405 & .322 & .206 \\ .260 & .206 & .132 \end{bmatrix}$	1	2.355	.5330	13.44
Interaction	$\begin{bmatrix} .029 & -.017 & .004 \\ -.017 & .011 & .003 \\ .004 & .003 & .023 \end{bmatrix}$	2	1.361	.9220	.63
Error	$\begin{bmatrix} .782 & .333 & -.057 \\ .333 & 2.150 & .178 \\ -.057 & .178 & .824 \end{bmatrix}$	48	1.255	—	—

(The elements of the SSCP matrices have been rounded to three decimal places to save space, but five places were carried in the calculations.) The F for education (with 6 and 92 d.f.'s), and the F for sex (3 and 46 d.f.'s) are both significant at the .001 level. Discriminant functions for the education and sex effects:

$$\mathbf{S}_e^{-1}\mathbf{S}_{\text{educ}} = \begin{bmatrix} -.00111 & -.05408 & .01171 \\ .12207 & .83614 & -.05836 \\ -.04105 & -.32559 & .03030 \end{bmatrix} \quad \begin{matrix} \lambda_1 = .85070 \\ \lambda_2 = .01463 \end{matrix}$$

$$\mathbf{v}_1' = [-.06398 \quad .92867 \quad -.36533]$$

$$\mathbf{S}_e^{-1}\mathbf{S}_{\text{sex}} = \begin{bmatrix} .65291 & .51859 & .33205 \\ .05862 & .04656 & .02981 \\ .34748 & .27600 & .17673 \end{bmatrix} \quad \lambda_1 = .87620$$

$$\mathbf{v}_1' = [\ .88001 \quad .07901 \quad .46834]$$

2.

Source	SSCP Matrix	*n.d.f.*	$\lvert S_h + S_e \rvert$ or $\lvert S_e \rvert$ $(\times 10^{-3})$	Λ	*F*-ratio	*P*
Sex	$\begin{bmatrix} 185.01 & -223.50 \\ -223.50 & 270.00 \end{bmatrix}$	1	2993	.5460	39.93	< .01
Method	$\begin{bmatrix} 42.32 & 74.53 \\ 74.53 & 133.85 \end{bmatrix}$	2	1765	.9257	2.23	> .01

(continued)

Interaction	$\begin{bmatrix} 2.51 & 5.47 \\ 5.47 & 17.55 \end{bmatrix}$	2	1656	.9867	.38	> .01
Within-cells	$\begin{bmatrix} 1795.39 & 1230.56 \\ 1230.56 & 1753.50 \end{bmatrix}$	114	1634			

Only the sex effect is significant at the 1% level. The discriminant function for this effect is computed as follows:

$$\mathbf{S}_w^{-1}\mathbf{S}_{\text{sex}} = \begin{bmatrix} .3669 & -.4432 \\ -.3849 & .4650 \end{bmatrix} \qquad \lambda_1 = .8319$$

$$\mathbf{v}_1 = [\ .6899 \quad -.7238]$$

The discriminant-function means are: $\bar{Y}_{\text{boys}} = 1.09$, $\bar{Y}_{\text{girls}} = -2.73$.

It will be instructive to plot the centroids for the original variables and the discriminant-function means in the same graph.

CHAPTER EIGHT

p. 242

Individual Number (i)	χ_{i1}^2	χ_{i2}^2	χ_{i3}^2	Decision—Assign to
5	6.3714	2.6162	*.9844*	Group 3
6	2.2514	*.5198*	1.6884	Group 2
	$\chi_{i1}^{\prime 2}$	$\chi_{i2}^{\prime 2}$	$\chi_{i3}^{\prime 2}$	
5	17.5633	11.0470	*9.2331*	Group 3
6	11.2337	*8.6721*	9.9175	Group 2
	$\chi_{i1}^{\prime\prime 2}$	$\chi_{i2}^{\prime\prime 2}$	$\chi_{i3}^{\prime\prime 2}$	
5	19.6723	12.9761	*11.8481*	Group 3
6	13.3427	*10.6012*	12.5325	Group 2
		$p(H_k \,\&\, U_{ki} > U_k^* \mid \mathbf{X}_i)$		
5	.0080	.1947	*.3743*	Group 3
6	.0869	*.3613*	.1007	Group 2

References

ANDERSON, T. W. *An introduction to multivariate statistical analysis.* New York: Wiley, 1958.

BARTLETT, M. S. Multivariate analysis. *Journal of the Royal Statistical Society, Series B,* 1947, **9**, 176–197.

BARTLETT, M. S. A further note on tests of significance in factor analysis. *British Journal of Psychology, Statistical Section,* 1951, **4**, 1–2.

BOCK, R. D. Programming univariate and multivariate analysis of variance. *Technometrika,* 1963, **5**, 95–117.

BOCK, R. D. A computer program for univariate and multivariate analysis of variance. In *Proceedings of the IBM scientific computing symposium on statistics.* White Plains, N.Y.: IBM Data Processing Division, 1965.

BOCK, R. D. Contributions of multivariate statistical methods to educational research. In R. B. Cattell (Ed.), *Handbook of multivariate experimental psychology.* Chicago: Rand McNally, 1966.

BOCK, R. D. *Multivariate statistical methods in behavioral research.* New York: McGraw-Hill, 1971.

BOCK, R. D., AND HAGGARD, E. A. The use of multivariate analysis of variance in behavioral research. In D. K. Whitla (Ed.), *Handbook of measurement and assessment in behavioral sciences.* Reading, Mass.: Addison-Wesley, 1968.

BROWN, G. W. Discriminant functions. *Annals of Mathematical Statistics,* 1947, **18**, 514–528.

CARROLL, J. B., AND WILSON, G. F. An interactive-computer program for the Johnson-Neyman technique in the case of two groups, two predictor variables, and one criterion variable. *Educational and Psychological Measurement,* 1970, **30**, 121–132.

CATTELL, R. B. r_p and other coefficients of pattern similarity. *Psychometrika,* 1949, **14**, 279–298.

CATTELL, R. B. The scree test for the number of factors. *Multivariate Behavioral Research,* 1966, **1**, 245–276.

CATTELL, R. B., AND COULTER, M. A. Principles of behavioral taxonomy and the mathematical basis of the taxonome computer program. *The British Journal of Mathematical and Statistical Psychology,* 1966, **19**, 237–269.

CATTELL, R. B., AND EBER, H. W. *The Sixteen Personality Factor Questionnaire.* Champaign, Ill.: Institute for Personality and Ability Testing, 1967–1968.

CLYDE, D. J., CRAMER, E. M., AND SHERIN, R. J. *Multivariate statistical programs.* Coral Gables, Florida: Biometric Laboratory of the University of Miami, 1966.

COOLEY, W. W., AND LOHNES, P. R. *Multivariate Data Analysis.* New York: Wiley, 1971.

CRONBACH, L. J., AND GLESER, G. C. Assessing profile similarity. *Psychological Bulletin,* 1953, **50**, 456–473.

DRAPER, N. R., AND SMITH, H. *Applied regression analysis.* New York: Wiley, 1966.

DU MAS, F. M. The coefficient of profile similarity. *Journal of Clinical Psychology,* 1949, **5**, 121–131.

DWYER, P. S., AND MACPHAIL, M. S. Symbolic matrix derivatives. *Annals of Mathematical Statistics,* 1948, **19**, 517–534.

EDWARDS, A. L. *Experimental design in psychological research* (3rd ed.). New York: Holt, Rinehart, and Winston, 1968.

FERGUSON, G. A. *Statistical analysis in psychology and education* (2nd ed.). New York: McGraw-Hill, 1966.

FISHER, R. A. The use of multiple measurements in taxonomic problems. *Annals of Eugenics,* 1936, **7**, 179–188.

GARRETT, H. E. The discriminant function and its use in psychology. *Psychometrika,* 1943, **8**, 65–79.

GLASS, G. V, AND STANLEY, J. C. *Statistical methods in education and psychology.* Englewood Cliffs, N.J.: Prentice-Hall, 1970.

HARMAN, H. H. *Modern factor analysis* (Rev. ed.). Chicago: University of Chicago Press, 1967.

HAYS, W. L. *Statistics.* New York: Holt, Rinehart, and Winston, 1963.

HECK, D. L. Charts of some upper percentage points of the distribution of the largest characteristic root. *Annals of Mathematical Statistics,* 1960, **31**, 625–642.

HORST, P. *Matrix algebra for social scientists.* New York: Holt, Rinehart, and Winston, 1963.

HOTELLING, H. The generalization of Student's ratio. *Annuals of Mathematical Statistics,* 1931, **2**, 360–378.

HOTELLING, H. Analysis of a complex of statistical variables into principal components. *Journal of Educational Psychology,* 1933, **24**, 417–441, 498–520.

HOTELLING, H. The most predictable criterion. *Journal of Educational Psychology,* 1935, **26**, 139–142.

HOTELLING, H. A generalized *T*-test and measure of multivariate dispersion. *Proceedings of the Second Berkeley Symposium of Mathematical Statistics and Probability,* 1951, **2**, 23–41.

JOHNSON, P. O., AND JACKSON, R. W. B. *Modern statistical methods: Descriptive and inferential.* Chicago: Rand McNally, 1959.

JONES, L. V., Analysis of variance in its multivariate developments. In R. B. Cattell (Ed.), *Handbook of multivariate experimental psychology.* Chicago: Rand McNally, 1966.

KAISER, H. F. The application of electronic computers to factor analysis. *Educational and Psychological Measurement,* 1960, **20**, 141–151.

KELLEY, T. L. *Essential traits of mental life.* Cambridge, Mass.: Harvard University Press, 1935.

LAWLEY, D. N. The estimation of factor loadings by the method of maximum likelihood. *Proceedings of the Royal Society of Edinburgh, Series A,* 1940, **60**, 64–82.

LAWLEY, D. N., AND MAXWELL, A. E. *Factor analysis as a statistical method.* London: Butterworth, 1963.

LOHNES, P. R. Test space and discriminant space classification and related significance tests. *Educational and Psychological Measurement,* 1961, **21**, 559–574.

MAHALANOBIS, P. C. On the generalized distance in statistics. *Proceedings of the National Institute of Science, India*, 1936, **12**, 49–55.

MCKEON, J. J. Canonical analysis: Some relations between canonical correlation, factor analysis, discriminant function analysis, and scaling theory. *Psychometric Monograph*, 1964 (whole of No. 13).

MCNEMAR, Q. *Psychological statistics* (4th ed.). New York: Wiley, 1969.

MCQUITTY, L. L. *A method of pattern analysis for isolating typological and dimensional constructs*. Lackland AFB, Texas: Air Force Personnel and Training Center, Report No. TN-55-62, 1955.

MENDENHALL, W. *Introduction to linear models and the design and analysis of experiments*. Belmont, Calif.: Wadsworth, 1968.

MORRISON, D. F. *Multivariate statistical methods*. New York: McGraw-Hill, 1967.

PEARSON, K. Contributions to the mathematical theory of evolution: I. On the dissection of asymmetrical frequency-curves. *Philosophical Transactions of the Royal Society, Series A*, 1894, **185**, 71–90.

PILLAI, K. C. S. *Statistical tables for tests of multivariate hypotheses*. Manila: Statistical Service Center, University of the Philippines, 1960.

PILLAI, K. C. S. On the distribution of the largest characteristic root in multivariate analysis. *Biometrika*, 1965, **52**, 405–415.

RAO, C. R. *Advanced statistical methods in biometric research*. New York: Wiley, 1952.

RAO, C. R. *Linear Statistical inference and its applications*. New York: Wiley, 1965.

RICHARDSON, M. *College algebra* (3rd ed.). Englewood Cliffs, N.J.: Prentice-Hall, 1966.

ROY, S. N. *Some aspects of multivariate analysis*. New York: Wiley, 1957.

RULON, P. J., TIEDEMAN, D. V., TATSUOKA, M. M., AND LANGMUIR, C. R. *Multivariate statistics for personnel classification*. New York: Wiley, 1967.

SCHATZOFF, M. Exact distributions of Wilks' likelihood ratio criterion and comparisons with competitive tests. Unpublished doctoral dissertation, Harvard University, 1964.

SCHATZOFF, M. Exact distributions of Wilks' likelihood ratio criterion. *Biometrika*, 1966, **53**, 347–358. (a)

SCHATZOFF, M. Sensitivity comparisons among tests of the general linear hypothesis. *Journal of the American Statistical Association*, 1966, **61**, 415–435. (b)

SEARLE, S. R. *Matrix algebra for the biological sciences (including applications in statistics)*. New York: Wiley, 1966.

SEBER, G. A. F. *The linear hypothesis: A general theory*. New York: Hafner, 1966.

STEPHENSON, W. *The study of behavior: Q-technique and its methodology*. Chicago: University of Chicago Press, 1953.

TATSUOKA, M. M. *The relationship between canonical correlation and discriminant analysis*. Cambridge, Mass.: Educational Research Corporation, 1953.

TATSUOKA, M. M. Joint probability of membership in a group and success therein: An index which combines the information from discriminant and regression analyses. Unpublished doctoral dissertation, Harvard University, 1956.

TATSUOKA, M. M. *Discriminant analysis: The study of group differences*. Champaign, Ill.: Institute for Personality and Ability Testing, 1970.

TATSUOKA, M. M., AND TIEDEMAN, D. V. Discriminant analysis. *Review of Educational Research*, 1954, **25**, 402–420.

THURSTONE, L. L. *The vectors of the mind*. Chicago: University of Chicago Press, 1935.

THURSTONE, L. L. *Multiple factor analysis*. Chicago: University of Chicago Press, 1947.

TIEDEMAN, D. V., BRYAN, J. G., AND RULON, P. J. *The utility of the Airman Classification Battery for assignment of airmen to eight air force specialties*. Cambridge, Mass.: Educational Research Corporation, 1953.

TILDESLEY, M. L. A first study of the Burmese skull. *Biometrika*, 1921, **13**, 176–262.

TUCKER, L. R In L. L. Thurstone, *Primary mental abilities. Psychometric Monograph*, 1938 (whole of No. 1).

WALKER, H. M., AND LEV, J. *Statistical inference*. New York: Holt, Rinehart, and Winston, 1953.

WHERRY, R. J. Multiple bi-serial and multiple point bi-serial correlation. *Psychometrika*, 1947, **12**, 189–195.

WILKS, S. S. Certain generalizations in the analysis of variance. *Biometrika*, 1932, **24**, 471–494.

WINER, B. J. *Statistical principles in experimental design*. New York: McGraw-Hill, 1962.

Author Index

Subject Index